Lecture Notes in Physics

Springer
Berlin
Heidelberg
New York
Barcelona
Hong Kong
London
Milan
Paris
Tokyo

Physics and Astronomy

ONLINE LIBRARY

http://www.springer.de/phys/

Editorial Policy

The series *Lecture Notes in Physics* (LNP), founded in 1969, reports new developments in physics research and teaching -- quickly, informally but with a high quality. Manuscripts to be considered for publication are topical volumes consisting of a limited number of contributions, carefully edited and closely related to each other. Each contribution should contain at least partly original and previously unpublished material, be written in a clear, pedagogical style and aimed at a broader readership, especially graduate students and nonspecialist researchers wishing to familiarize themselves with the topic concerned. For this reason, traditional proceedings cannot be considered for this series though volumes to appear in this series are often based on material presented at conferences, workshops and schools (in exceptional cases the original papers and/or those not included in the printed book may be added on an accompanying CD ROM, together with the abstracts of posters and other material suitable for publication, e.g. large tables, colour pictures, program codes, etc.).

Acceptance

A project can only be accepted tentatively for publication, by both the editorial board and the publisher, following thorough examination of the material submitted. The book proposal sent to the publisher should consist at least of a preliminary table of contents outlining the structure of the book together with abstracts of all contributions to be included.
Final acceptance is issued by the series editor in charge, in consultation with the publisher, only after receiving the complete manuscript. Final acceptance, possibly requiring minor corrections, usually follows the tentative acceptance unless the final manuscript differs significantly from expectations (project outline). In particular, the series editors are entitled to reject individual contributions if they do not meet the high quality standards of this series. The final manuscript must be camera-ready, and should include both an informative introduction and a sufficiently detailed subject index.

Contractual Aspects

Publication in LNP is free of charge. There is no formal contract, no royalties are paid, and no bulk orders are required, although special discounts are offered in this case. The volume editors receive jointly 30 free copies for their personal use and are entitled, as are the contributing authors, to purchase Springer books at a reduced rate. The publisher secures the copyright for each volume. As a rule, no reprints of individual contributions can be supplied.

Manuscript Submission

The manuscript in its final and approved version must be submitted in camera-ready form. The corresponding electronic source files are also required for the production process, in particular the online version. Technical assistance in compiling the final manuscript can be provided by the publisher's production editor(s), especially with regard to the publisher's own Latex macro package which has been specially designed for this series.

Online Version/ LNP Homepage

LNP homepage (list of available titles, aims and scope, editorial contacts etc.):
http://www.springer.de/phys/books/lnp/
LNP online (abstracts, full-texts, subscriptions etc.):
http://link.springer.de/series/lnp/

Th. Henning (Ed.)

Astromineralogy

Springer

Editor

Thomas K. Henning
Max-Planck-Institut für Astronomie
Königstuhl 17
69117 Heidelberg, Germany

Cover Picture: (see contribution "The Astromineralogy of Interplanetary Dust Particles"
by J. Bradley in this volume)

Cataloging-in-Publication Data applied for

A catalog record for this book is available from the Library of Congress.

Bibliographic information published by Die Deutsche Bibliothek

Die Deutsche Bibliothek lists this publication in the Deutsche Nationalbibliografie;
detailed bibliographic data is available in the Internet at http://dnb.ddb.de

ISSN 0075-8450
ISBN 3-540-44323-1 Springer-Verlag Berlin Heidelberg New York

Springer-Verlag Berlin Heidelberg New York
a member of BertelsmannSpringer Science+Business Media GmbH

http://www.springer.de

© Springer-Verlag Berlin Heidelberg 2003
Printed in Germany

Typesetting: Camera-ready by the authors/editor
Camera-data conversion by Steingraeber Satztechnik GmbH Heidelberg
Cover design: *design & production*, Heidelberg
Printed on acid-free paper
SPIN: 10890493   54/3141/du - 5 4 3 2 1 0

Preface

The space between stars is not empty, but filled with interstellar matter, mostly composed of atomic, ionized or molecular hydrogen, depending on the temperature and density of the various phases of this medium. About 1% of the total mass of interstellar matter in our galaxy is contained in small solid particles, ranging in size from a few nanometers to many microns and even millimeters in planet-forming disks around young stars. This cosmic dust undergoes a complicated lifecycle, with fresh material produced in the outflows of evolved stars and supernovae, modified by shocks and cosmic rays in the diffuse interstellar medium and provding the surface for the formation of molecular ices in cold and relatively dense molecular clouds.

Although elemental abundances, formation routes, extinction curves, and early infrared and ultraviolet spectroscopy pointed to the presence of silicates, carbides, and graphitic material in space, it was only recently determined that a new field – astromineralogy – emerged from astronomical observations. The unprecedented spectral resolution and wavelength coverage provided by the *Infrared Space Observatory*, together with dedicated experimental work in the field of laboratory astrophysics, provided the unambiguous identification of minerals in space, ranging from olivines and pyroxenes to various carbides.

Cosmic mineralogy already provided earlier results on minerals in space by the in situ study of minerals in our solar system. Meanwhile, we see a bridge between astronomically-identified minerals and stardust minerals found by their isotopic signatures in meteorites and interplanetary dust. This may lead to a wider definition of astromineralogy that includes these investigations.

Early seminars on astromineralogy were held at the Universities of Amsterdam and Heidelberg. They showed the large interest of astronomers, physicists, chemists, and mineralogists in this new field and encouraged the publication of this book. New infrared missions such as SIRTF and NGST will provide more empirical material in this fascinating field.

I thank the authors for their contributions to this book and their willingness to follow my recommendations. In addition, I thank my former colleagues Drs. J. Dorschner, J. Gürtler, C. Jäger, and H. Mutschke for many years of fruitful collaboration in the experimental characterization of cosmic dust analogues, which paved the way for the identification of minerals in space.

Finally, I would like to thank J. Weiprecht. Without his technical help, the production of this book would not have been possible.

Heidelberg and Jena   Thomas Henning
December 2002

Contents

List of Contributors

John Bradley
Institute for Geophysics and Planetary
Physics,
Lawrence Livermore National
Laboratory,
Livermore, CA 94551, USA
jbradley@igpp.ucllnl.org

Johann Dorschner
University of Jena
Astrophysical Institute and University
Observatory
Schillergässschen 2-3
07745 Jena, Germany
dorsch@astro.uni-jena.de

Hans-Peter Gail
University of Heidelberg
Institute for Theoretical Astrophysics
Tiergartenstr. 15
69121 Heidelberg, Germany
gail@ita.uni-heidelberg.de

Martha S. Hanner
Jet Propulsion Laboratory
California Institute of Technology
Pasadena, CA 91109, USA
msh@galah.jpl.nasa.gov

Thomas Henning
Max Planck Institute for Astronomy
Königsstuhl 17
69117 Heidelberg, Germany
henning@mpia-hd.mpg.de

Elmar K. Jessberger
University of Münster
Institute for Planetology
Wilhelm-Klemm-Str. 10
48149 Münster, Germany
ekj@nwz.uni-münster.de

Ingrid Mann
University of Münster
Institute for Planetology
Wilhelm-Klemm-Str. 10
48149 Münster, Germany
imann@uni-münster.de

Frank Molster
ESTEC
Solar System Division
Keplerlaan 1
2201 AZ Nordwijk, Netherlands
fmolster@so.estec.esa.nl

Ulrich Ott
Max Planck Institute for Chemistry
Postfach 3060
55020 Mainz, Germany
ott@mpch-mainz.mpg.de

L.B.F.M. Waters
University of Amsterdam
Astronomical Institute Anton
Pannekoek
Kruislaan 403
1098 SJ Amsterdam, Netherlands
rensw@astro.uva.nl

From Dust Astrophysics Towards Dust Mineralogy – A Historical Review

Johann Dorschner

University of Jena, Astrophysical Institute and University Observatory
Schillergässchen 3, D-07745 Jena, Germany

Abstract. Via meteorite research, mineralogy became the first discipline among the earth sciences that developed a cosmic branch. Cosmic mineralogy is concerned with all solids accessible to mineralogical techniques in the solar system and even beyond. The current link connecting astronomy and mineralogy is the dust in the Galaxy.

Interstellar dust reached astrophysical topicality around 1930. In their first dust models astrophysicists took the assumption for granted that the grains consist of minerals related to those of the solar system. An exception formed the "ice" model prevailing in the decade 1950–1960. It rested on hydrogen-rich volatiles that were assumed to condense in the HI clouds of the diffuse interstellar medium.

Scrutinizing discussions of the weak points of the ice model and new observations in the UV spectral range paved the way for considering refractory materials as grain constituents and stars and planetary systems as the dust suppliers in the Galaxy. In the decade 1960–1970 many refractory dust materials heuristically came into discussion, e.g. graphite and meteoritic silicates. The strong solid-state band detected at 217 nm in the interstellar extinction curves was commonly assigned to graphite grains.

Beginning in 1968, observations of the vibrational bands of the SiO_4 group in circumstellar as well as in interstellar dust provided ample evidence for the ubiquitous occurrence of silicate grains in the Galaxy. Bands of silicates and of some other solids opened the new era of dust diagnostics resting on IR spectroscopy and laboratory simulation experiments. In most cases, the observed spectral features indicated heavily distorted structure of the grain material.

In 1996–1998 the Infrared Space Observatory (ISO) surprisingly discovered stardust bands of crystalline silicates. This opened the chance of unambiguously identifying dust minerals via astronomical spectroscopy in combination with laboratory work (astromineralogy).

In the last two decades, mineralogists have discovered "fossil" dust grains in primitive meteorites, which had been preserved from the parent cloud of the solar system. This way, cosmic mineralogy could directly contribute to astrophysical dust research, and confirm astromineralogical conclusions.

The collaboration between these fields will improve understanding of the solid cosmic matter and influence the mineral definition according to the new galactic requirements.

1 Astrophysics and Cosmic Mineralogy

For centuries, astronomy and mineralogy have been well separated science disciplines that, for a long time, had nothing in common with each other. It is self-evident that in former times mineralogists devoted little interest to the material of other celestial bodies because samples of them were not attainable to practical inspection. Astronomers, who were familiar with studying unattainable phenomena, had focused since antiquity their scientific interests to the mathematical description of the observable movements on the sky. They, at the best, had speculative views on the nature of celestial bodies, which vigorously flourished after the invention of the telescope. However, before the emergence of the astrophysics there was no base for the scientific approach to such questions.

Mineralogists have got much earlier the chance to make scientific statements on extraterrestrial material, because it fell down from the sky as meteorites. However, in the time of Enlightenment leading scientific authorities refused reports on the fall of meteorites as products of superstition and required removal of meteorites from the collections. The turnaround occurred, when the well-witnessed meteorite rain of L'Aigle in 1803 convinced the authorities of Paris academy that the arguments for the cosmic origin of meteorites published at that time in a series of startling publications by the physicist E. F. F. Chladni [1] deserve attention. From that time the way towards the systematic chemical and mineralogical study of meteorites was open, and mineralogy became already in the 19th century the first discipline among the earth sciences that developed a "cosmic branch".

However, only in 1962 this development found its official expression, when the International Mineralogical Association (IMA) formed the working group "Cosmic Mineralogy" [2]. Meanwhile, cosmic mineralogy has considerably extended direct laboratory work to additional extraterrestrial materials (cf. [3]). Further, in-situ soil analyses on Mars and Venus and mass spectrometry of coma dust of comet P1/Halley contributed to cosmic mineralogy. Mineralogical information was also gathered by astronomical spectroscopy of asteroids and comets and recently also on dust in the circumstellar environment of other stars. This last development was summarized by the term "astromineralogy".

Reliable identification of circumstellar dust minerals via spectroscopy became only recently possible. The Infrared Space Observatory (ISO, operating 1995-98) obtained spectra over the whole IR range, which, for the first time, contained bands of crystalline silicates. This "crystalline revolution" (cf. [4]) laid the base for the first identification of circumstellar silicate mineral grains that, with a high degree of certainty, are composed of magnesium-rich minerals of the olivine and pyroxene groups, as comparisons with laboratory-based spectra of such minerals showed. This discovery manifests the coronation of the hitherto astrophysical efforts to investigate cosmic dust mineralogy and puts signs that unambiguous mineral diagnosis of cosmic dust grains via spectroscopic observations in combination with targeted laboratory work is possible.

The history of the endeavours to unravel the nature of dust grains beyond the solar system began more than 70 years ago, as soon as it had become obvious that the selective extinction of starlight must be due to scattering by interstellar dust grains. About 40 years ago a former suspicion won more and more conclusiveness, namely that cool stellar atmospheres and circumstellar envelopes are dust-producing environments. Further, special dust components were detected in the birth clouds of stars, the composition of which was different from the widely spread interstellar and circumstellar dust. In the today dust picture, several different dust populations in the Galaxy are distiguished (for review see [5, 6]).

For some decades the astrophysical methodology to investigate the nature of the grains simply rested on the method of trial and error: a plausible dust model was constructed that had successfully to pass the confrontation with all observed effects attributed to the interaction of dust with star light. The availability of new observations, each time afresh, became the crucial test of a dust model. If it turned out to be compatible with the new requirements, possibly after some modifications, then it survived, otherwise it had to be renounced, and an improved modelling attempt had to be undertaken.

In this overview, we sketch the troublesome way of the astrophysical efforts to understand the nature of the cosmic dust particles. Since already in the earliest dust models, minerals played an important role as grains' constituents, this historical contribution simultaneously is a representation of the endeavours of the astrophysics to gain insight in the interstellar and circumstellar "mineralogy". However, as long as diagnostic spectral features of the grain materials could not be observed, mineral assignments had more or less speculative character without mineralogical authenticity, even if the proposals appeared very plausible. The aforementioned heuristic modelling procedure could not provide unambiguous identification of dust constituents as long as the models were based solely on observations of continuous dust spectra.

This epistemological situation changed in the 1960s with the advent of UV and IR spectroscopy in the astrophysics, which detected characteristic spectral features of the grain materials. On this new observational base, the dust models could lose much of their speculative character. The new spectral "fingerprints" reliably proved the presence of particular chemical bonds of some dust materials. Even if this did not suffice for an exact mineral identification, terms like "interstellar mineralogy" enjoyed increasing popularity among astrophysicists. Further, primitive meteorite and interplanetary dust mineralogy began to play an important role as guide to the solids beyond the solar system.

Correct identification of observed spectral features required targeted laboratory work on analogue materials to be adopted in the dust models. In the 1970s the necessary connection between astrophysics and laboratory work got its institutionalization in the new branch of "laboratory astrophysics". This lead to much more realistic dust modelling. However, as a matter of principle, laboratory material simulation of the heavily disordered or amorphous interstellar dust materials cannot provide ultimate clarity on the identity of the cosmic solids with their laboratory analogues. Nevertheless, experience won by the comparison of spectra of laboratory analogues with observed cosmic dust spectra laid the base of what could be called "interstellar material science".

Great support to the astrophysical efforts to understand the nature of cosmic dust was given by an unexpected discovery in the field of cosmic mineralogy:

From primitive meteorites and probably also from interplanetary dust, components could be isolated that turned out to be "fossil" dust grains from the parent cloud of the solar system. Such presolar solids opened the possibility to undergo intrinsic interstellar material to laboratory analyses. This amazing development will even be surpassed, when in the future interplanetary spacecrafts succeed in collecting dust grains from the local ISM that just pass through the solar system. All of these new possibilities are treated in special chapters of this book.

They pave the way to a close collaboration between astrophysics and cosmic mineralogy with great mutual use. From the analyses of stardust grains in meteorites, astrophysics gets improved knowledge on the processes about circumstellar dust formation, even on nucleosynthesis in evolved stars. On the other hand, astrophysical dust modelling does not become dispensable because the cosmic grains available to laboratory research are (and will be) restricted only to few species of refractory dust solids, not to the main stream dust involved in the manifold evolutionary processes in the Galaxy. Improved astrophysical mineral identification via spectroscopy will give the cosmic mineralogy the chance to widen its scope over the whole Galaxy.

2   Interstellar Dust on the Way to Astrophysical Topicality

2.1   Light Extinction Between the Stars

Interstellar dust research articulated itself as a field of astrophysics around 1930. Interaction of small grains with starlight turned out to be the most plausible explanation of the accumulating observational evidence in favour of selective light extinction in space. Speculations on the occurrence of a general cosmic light extinction have a long history. In 1744 Loys de Chéseaux [7] proposed an absorbing fluid or ether in the space between the stars in order to settle the old problem that in an infinite universe, uniformly filled with stars, the night sky could never be dark. A century later this paradox of dark night sky was named after Olbers, who had proposed a similar solution [8]. However, dust extinction has proved to be unable to solve this cosmological problem.

At the end of the 19th century, light extinction (then commonly addressed as "light absorption") in the interstellar space anew attained astronomical topicality One impact was given by the enigmatic starvoids visible in the Milky Way, the investigation of which flourished by applying celestial photography. The realization that these voids are simply pretended by obscuring dust clouds in front of the starfield, was by no means a matter of course, mainly because W. Herschel's opinion was still dominating. In his report "On the Construction of the Heavens" to the Royal Society in London in 1785 Herschel [9] had described such a starvoid as an *"opening in the heavens"* formed by collection of stars in a nearby cluster, which had left the observed vacancy. For Herschel's often quoted anectodal exclamation on the "hole in the sky", see [10]. In particular E. E. Barnard, one of the pioneers of photographic techniques, usually denoted the starvoid as "holes", sometimes also "black holes" (see, e.g., [11, 12]). He never used the term "dark nebulae". Admittedly, in his large catalogue [13] that contained 182 such objects he more cautiously called them "dark markings" (for details of his curious reluctance to accept the explanation by obscuring clouds, see [14]). It is not clear, whether he did not know or refused to take notice of the view of prominent astronomers, e.g. A. Secchi who already in 1853 had parted with the picture of the holey Milky way and accepted obscuring nebulous matter as the simplest explanation [10, 14].

Another pioneer of Milky Way photography, Max Wolf, also used misleading terms, such as "dunkle Höhlen" (dark caves) or "Sternleeren" (star voids), but already in his early papers, e.g. [15], he was open for the obscuring cloud explanation. In Wolf's famous paper [16] on the statistical method for estimating total absorptions, distances and spatial extensions of such dark clouds, he also presented morphological evidence for a close connection between dark and bright nebulosity. Wolf was convinced that the obscuring medium is dust, however, speculations on the nature of the dust grains were out of his scope.

Extinction of starlight was also suggested by stellar-statistical investigations that resulted in a systematic decrease of the star density in the Galaxy with increasing distance. The state of the knowledge on interstellar extinction reached around 1920 was critically reviewed by Kienle [17], who summarized that claims of observed interstellar light dispersion (dependence of phase velocity on wavelength) had no longer a serious base, that the interstellar light absorption should be smaller than 0.002 mag/pc, and that there are only qualitative indications, but no convincing evidence for the selectivity (wavelength dependence) of this absorption. Kienle concluded that the light absorption should be due to inhomogeneously distributed non-luminous gas masses and dark dust clouds, whereby he stressed that cosmic dust is quite different from terrestrial one because the former should also contain large rocks, apparently in analogy to the meteorites.

Kienle [17] probably did not know Pannekoek's [18] paper, who in 1920 had rejected obscuration due to Rayleigh scattering by gas because this mechanism yielded unrealistically high masses of the dark nebulae. Pannekoek [18] quoted a suggestion by de Sitter that the cloud mass problem vanishes if the extinction is due to dust grains. However, de Sitter's suggestion was "obscured" by the wrong assumption that dust extinction is always neutral. This combination of the right mechanism with the wrong dust model confused in the following time observers that had found clear evidence for selective extinction. They again reverted to Rayleigh scattering.

Still at the end of the 1920s leading astrophysicists had very different opinions on the role of dust. Even capacities as A. S. Eddington underestimated the role of dust. In the review of Eddington's famous book "The Internal Constitution of the Stars" (1926) H. N. Russell [19] critically annotated to the last chapter on diffuse matter in space: *"The fact that...fine dust is enormously the most effective obscuring agent, appears to be more important than any a priori impression. The allied fact that such dust, if smaller than the wavelength will produce selective scattering, following Rayleighs law, but vastly greater per unit mass than in the case of gas, is not mentioned."*

2.2   Interstellar Dust Gets a Physical Face

Pannekoek's [18] hope that scrutinized photometric work would settle the questions on the interstellar extinction could be fulfilled not earlier than about 1929/30. Conclusive measurements of the interstellar extinction coefficient were not only handicaped by the low accuracy of the photographic photometry and the problems of distance determination, but also by the (unknown) distribution of the absorbing medium in the Galaxy. This latter fact explains why the apparently best suited objects for extinction determination, namely the most remote ones, i.e. globular clusters and spiral nebulae, failed to give the expected success because they simply were not observable in low galactic latitudes where the dust was concentrated.

In 1929 Schalén [20] from observations in Cepheus and Cassiopeia determined the mean interstellar extinction to be 0.5 mag/kpc. The values towards Cygnus were much larger and led to an upper limit of 2 mag/kpc. Schalén [20] was the first to speculate on the nature of the dust grains. Based on Hoffmeister's [21] results on the statistics of meteors, which, at that time, were assumed to contain a considerable proportion of interstellar particles, he suggested that the absorber in the dark nebulae is "meteoric dust", the spatial density in the clouds he estimated to be about 10^{-23} g/cm^3.

Schalén's [20] extinction values were confirmed by Trümpler's [23] extensive study of distances, dimensions, and space distribution of open star clusters. These objects turned out to be the best possible test objects for extinction determinations. They are concentrated in low galactic latitudes, where the dust accumulates, and cover all longitudes, and they allow accurate distance determination. Trümpler's [23] investigation proved the occurrence of a general interstellar extinction at low galactic latitudes with a mean photographic coefficient of 0,67 mag/kpc towards all directions. He also found evidence that the brightest cluster stars showed growing discrepancy between colour equivalents and spectral types with increasing distance and, thus, indicated selective extinction that reddened the stars. Trümpler's [23] paper manifests the breakthrough to the final proof of a general selective interstellar extinction "landscape", out of which the dark clouds loom as the "summits"

At the same time, Öhman [22] published a spectrophotometric study of 882 B, A and F type stars. Most of the B stars showed colour equivalents too large to be in agreement with the temperatures corresponding to the spectral type. This disagreement between stellar continuum and lines led some astrophysicists, e.g. Gerasimovič [24] and Unsöld [25], to the premature definition of a "superexcitation" of these atmospheres due to a UV excess radiation. However, Öhman [22] proposed an alternative explanation and considered the low colour temperatures of the observed stellar continuum as a spurious effect caused by the reddening of the star light due to selective interstellar extinction. Figure 1

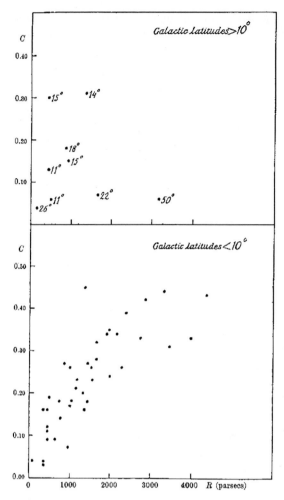

Fig. 1. Öhman's [22] colour-equivalent C as function of stellar distances for early B stars, clear evidence for selective interstellar extinction

shows Öhman's [22] Fig. 25 indicating the tight correlation of the colour equivalents C (precursor of colour excess) for early B stars near the galactic plane with their distances. The extinction coefficient derived from the C-values was in satisfactory agreement with Trümplers results. From his results, Öhman also concluded that the interstellar extinction increases less steeply with decreasing wavelength than Rayleigh scattering would require. This trend was confirmed by later observers and, finally, resulted in a λ^{-1}-law. At the end of that decade, Stebbins et al. [26] could demonstrate the approximative validity of the λ^{-1}-law for the whole optical wavelength range from 380 to 800 nm.

In 1931 Öpik [27] used Öhman's [22] spectrophotometric results for a thorough theoretical analysis of the observed colour-temperature discrepancy. He concluded that the hypothesis of dust extinction is *"the most probable, being at the same time the simplest and well under control of our present physical knowledge"*. The last phrase referred to the light scattering theory developed by Mie [28] in 1908 that had apparently escaped the attention of the astronomers for more than two decades. Öpik learned this theory from papers by Blumer [29] who had painstakingly bothered to compute scattering functions for spherical dielectric particles (refractive indices m=1.25, 1.333, 1,466, 1.5, and ∞) for the characteristic Mie parameter $x=2\pi a/\lambda$ (a being the particle radius) in the range from 0.1 to 10. From Blumer's substantial computational work Öpik [27] recognized the key role of the particle size distribution for the mathematical shape of the wavelength dependence of the extinction coefficient. Since the refractive index of the particle material was apparently of minor influence (dielectric grains!), he concluded that the explanation of the observed reddening law is primarily a question of finding the suitable particle size distribution function. Power laws as derived from meteor observations seemed to solve the problem. Öpik [27] excluded Rayleigh scattering and presented arguments for a λ^{-1}-dependence. Even if he did not explicitly address the chemical nature of interstellar dust, it can be assumed that he thought of "meteoric grains", not only because he preferred a "meteoric" size distribution. In the following years, he became an outstanding advocator of the interstellar origin of the sporadic meteors.

The first who explicitly applied the Mie theory to extinction observations have been Schalén [30] and Schoenberg and Jung [31], probably without knowing Öpik's [27] paper, for they focussed their attention to metal grains. The iron-nickel alloy of the meteorites and the wavelength dependence of the absorption coefficient of small metal grains ($\propto 1/\lambda$) seemed to have triggered this orientation. Schalén [30] calculated absorption coefficients for small iron spheres (diameters in the range 10^{-6} to 10^{-5}cm) at the wavelengths λ=395 and 440 nm and compared the theoretical colour differences with the observed ones derived from objective prism spectra of stars in obscured regions of Auriga, Cygnus, and Cepheus. The result was that the particles in this range could reproduce the observations. The total cloud masses got realistic values and the mass density of the dust clouds resulted in values of the order of 10^{-26} g/cm^3, which was absolutely compatible with Oort's [32] mass limit of interstellar matter near the sun. Similar conclusions were drawn by Schoenberg and Jung [31]. In later papers the calculations were also extended to other metals ([33],[34]).

2.3 Renunciation from Meteor/Meteorite Analogy

As already mentioned, the iron dust model was merely based on analogy considerations from iron meteorites and on the suitable wavelength dependence of the absorption coefficient of metal particles. Further support could be derived from the interstellar meteors. Observations seemed to confirm that more than 60% of the sporadic meteors and fireballs had hyperbolic heliocentric velocities. Even if some astrophysical authorities drastically expressed their doubts on this view (Eddington was alleged to have declared in 1932 *that he would as lief believe in ghosts as in hyperbolic meteors* (cf. [35]), the experts of meteor research could point to confirming observational facts in favour of hyperbolic velocities, e.g. the results of Harvard Meteor Expedition to Arizona in 1931–33. During the whole decade from 1930 to 1940 the "hardliners" of the concept of hyperbolic meteors, Hoffmeister and Öpik, though entangled in violent debates and mutual criticism of their data reduction procedures, agreed with the interstellar origin (for review see [36, 37, 35]).

The criticism of the interstellar meteors was steadily increasing and culminated in 1943/44 with Porter's [38] clear statement based on the analyses of the British meteor data: *"There is no direct evidence for the existence of an excess of hyperbolic velocities. With a few doubtful exceptions, all meteors are members of the solar system... The hyperbolic theory has hindered the progress of meteoric astronomy for many years, and in the opinion of the writer should be abandoned."*. For the review of fallacies leading to the overestimated meteor velocities see Lovell's monograph "Meteor Astronomy" [39]. Because of the rejection of the hypothesis of interstellar meteors analogy arguments connecting meteorites and interstellar solids lost their weight.

The other argument supporting metal grains because of the λ^{-1} dependence of the absorption coefficient is absolutely not cogent. The efficiency factor of extinction (being the sum of absorption and scattering efficiency) of small spheres according to the Mie theory can be expanded as a power series of the characteristic parameter $x=2\pi a/\lambda$ (this was the approach to the problem by Schoenberg and Jung [31, 34]). The first term in this series, which is the decisive term for grains with $x \ll 1$, is $\propto \lambda^{-1}$ for absorbing and $\propto \lambda^{-4}$ for dielectric grains. For somewhat larger grains, terms with higher powers of x also become significant. Apart from this, the coefficients of the terms contain the refractive index of the material, which in the case of metals is strongly wavelength-dependent so that the wavelength dependence of relative small metal particles leaves the simple $1/\lambda$-law. Already Öpik [27] in his early paper had called attention to the fact that dielectric grains can also reproduce the observed interstellar reddening law if combined with a suitable particle size distribution function.

As a matter of fact, after 1938 dielectric grains plaid an increasing role in the discussions, in particular in connection with theoretical studies of reflection nebulae and the diffuse galactic light [40, 41, 42], which led to the realization that the albedo of interstellar grains (the ratio of scattering to extinction cross-section) had to be assumed larger than that of metal grains. However, alone on the basis of extinction and albedo observations in comparison with Mie calculations a final decision between dielectric and metallic grains was not possible (cf. the analysis by Güttler [43]). After the break-down of the interstellar meteor hypothesis, finding plausible mechanisms of grain formation became a new important key for the evaluation of grain models. This "cosmogonic" criterion was applied in dust modelling only after World War II.

2.4   Early Approaches to Dust Formation

In 1935 Lindblad [44] confronted the dust community with the idea of condensation processes in interstellar gas leading to the formation of the meteoric matter in the interstellar clouds. His estimate intended to show that in the age of meteorites (adopted to be about 10^9 years) the condensation of the metal content of the interstellar gas could explain the growth of particles up to the size required. The interstellar particles should continuously grow because the metals in the hot gas ($T \approx 10000$ K) must precipitate on the cool particle surfaces ($T \approx 3$ K). The paper found little resonance with the dust activists. The only immediate response was given by Jung [45] who discussed objections because of charging effects on the grains and found Lindblads timescale for the present interstellar conditions too optimistic.

Nevertheless, it was Lindblad's apparently unexciting note that revolutionized the dust picture. This approach, originally thought to give a physical explanation for the formation of refractory solids in space, a decade later, entirely ruled out the early speculations on mineral dust and focussed the attention to non-mineral volatiles as the significant dust constituents. Before this new dust model dominated the dust astrophysics some other points in connexion with dust condensation have to be briefly reported.

In 1933 Wildt [46] discussed the possibility of condensation of solid grains and their effect as opacity source in the atmospheres of very cool stars. The motivation to this study was the longstanding claim that the variability of Mira stars could be caused by periodical condensation and resublimation of aerosols in their atmospheres (veil theory, see, e.g., Merrill [47]). By the discussion of the vapor pressure curves for selected elements and oxides and the thermal properties of some carbides and nitrides, Wildt drew the conclusion that solid particles of carbon, Al_2O_3, CaO, and several highly refractory carbides (SiC, TiC, ZrC) and nitrides (TiN, ZrN) might form in N stars. He excluded SiO_2 and did also not consider highly refractory silicates.

Wildt's paper was the first approach to the formation of mineral grains in stellar environment. Some of his candidates today are among the dust materials that have been detected in presolar grains with an isotope signature pointing to their stellar origin. However, at Wildt's time it was not imaginable that stellar aerosol particles could be conveyed out of the star and, thus, contribute to the interstellar dust.

Another impact to stellar condensation hypotheses was given by the light curves of R CrB stars. In 1935 Loreta [48] suggested that the irregular deep minima of these variables could be caused by eruptions of "dark matter" transiently obscuring the star. The lacking chemical precision of Loreta's hypothesis was appended by O'Keefe [49], who concluded from the exceptionally high carbon abundance of these stars that the aerosol particles should be graphite crystallites. The sharp drop to the minimum should be caused by the fast graphite formation, whereas the slow recovery to the normal light reflects the dissipation of the ejected soot cloud. Estimates of the expected optical properties and the formation time of the grains showed rough agreement with the observations. However, also O'Keefe's considerations did not include any interstellar significance of the stellar graphite grains.

2.5   The Cold Way to Interstellar Dust

While Wildt's [46] and O'Keefe's [49] stellar mineral grains did not inspire the interstellar dust researchers for more than two decades, Lindblad's [44] condensation idea did it already some years after its publication. During World War II in the Netherlands Lindblad's [44] suggestion has provoked intensive studies of interstellar condensation culminating in a new dust model founded by Oort and van de Hulst [50]. In his Ph.D. thesis, van de Hulst [52] had thoroughly studied the optics of small grains. This laid the basis for some improvements of the original model [50], which appeared in 1949 [51], as well as for van de Hulst's famous monograph "Light Scattering by Small Grains" [53] used for about three decades as a standard book by astrophysicists, meteorologists and other people occupied with any sorts of small droplets or grains.

In contrast to Lindblad's [44] note, the Oort-van de Hulst model [50, 51] aimed at the formation of volatile grains consisting of what was later commonly called "dirty ice". The ice grains should grow in HI regions (according to Strömgren's [54] new classification) by gas accretion. The equilibrium temperature ($T \approx 10$–20 K) of the dust grains in the interstellar radiation field was low enough that gas species like C, N, O and some metals could really condense onto the cool grain surfaces, however, it was too high for the condensation of hydrogen. Oort and van de Hulst [50] assumed that the impinging hydrogen atoms are involved in chemical reactions with the condensed species and are bonded in hydrogen-saturated compounds like H_2O, CH_4, NH_3, and metal hydrides. Thus, the resulting ice conglomerate roughly reflected the interstellar gas composition. A critical point remained the question of the seeds for the start of the grain growth according to this concept.

A further innovation in this dust model was the elegant explanation of the grain size distribution. The authors [50] assumed that the continuous grain growth by gas accretion and hydrogenation of condensed species was controlled by mutual grain destructions due to grain-grain collisions as a consequence of interstellar cloud encounters. From the assumption of the equilibrium between both processes the average particle size distribution function could be derived that together with Mie cross sections of the icy grains allowed satifactorily representing the observed interstellar reddening law and other observed dust properties. It also gave a plausible explanation for the apparent universality of interstellar dust. Deviations of the observed extinction curve towards certain galactic directions, e.g. the Orion region, could be attributed to peculiar local effects.

The theoretical elegance of this model and its obvious success led to honestly name it "classical dust model". The model had a great lobby within the Commission 34 ("Interstellar Matter and Galactic Nebulae") of the International Astronomical Union, which had been formed in 1938 on the IAU General Assembly at Stockholm. For about two decades, the cold way to the formation of interstellar grains in HI clouds dominated the astrophysical considerations on interstellar dust in the textbooks. Speculations on mineral grains in interstellar dust and dust formation in stellar atmospheres seemed to be untimely.

3   Heuristic Dust Modelling with Refractory Grains

3.1   Interstellar Polarization and the Come-Back of Minerals

In the 1950's the discussion on the material properties of interstellar dust was considerably influenced by the detection of the interstellar polarization. In 1949 independently of each other Hiltner [55], Hall [56], and Dombrovski [57] discovered that the light of reddened stars shows a slight linear polarization that could be explained as a result of the interaction of interstellar dust grains with the starlight.

The discovery had two far-reaching consequences:

1. The assumption that all interstellar dust grains are roughly spherical and optically isotropic had to be considerably modified. In order to polarize the transmitting light, a significant part of the particles must be non-spherical or optically anisotropic or both, and, in addition, the grain axes had to get a preferential orientation. This requires the action of a mechanism that permanently exerts a torque damping the irregular tumbling of the grains due to impinging gas atoms and, thus, providing the necessary degree of alignment.

2. Special requirements of the orientation mechanism and/or the fact that non-spherical dirty ice grains of the prevailing classical dust model proved to be very inefficient polarizers focussed again the attention to refractory grains that are much better polarizers. After the era of meteoritic grains in the 1930s, this was the second time that minerals aroused the interests of astrophysicists.

Accurately measuring interstellar polarization was a great challenge to the observers because the small intensity differences to be measured through a polarization filter at different position angles required outstanding photometric accuracy. This especially held for the determination of the wavelength dependence, the "polarization law"; such data became available not until 1959 [58, 59].

The challenge for the theorists was similarly great. One important handicap was the lack of a general light scattering theory for non-spherical grains. Up to now, the only exact solution of light scattering by non-spherical particles is the case of infinite circular cylinders (needles) of abitrary diameters. Light scattering by spheroids could and can be treated only by approximations for very small particles (cf. [53]). In 1937 Wellmann [60] had the right feeling that the interstellar absorption coefficient should be influenced not only by the grain radii and the refraction of the dust material, but also by the shape of the grains. For this reason, he solved the Maxwell equations for the case of light scattering by cylindrical particles for oblique incidence. He was aware that the absorption coefficient depends on the orientation of the cylinder axis relative to the electrical light vector, but his efforts were strictly directed to answering the question if metal cylinders mixed with spheres could reproduce the observed λ^{-1}-law of interstellar extinction. He got a positive result, even if cylinders were much more abundant than spheres, and was contented with it. That aligned cylinders could produce interstellar polarization was not imaginable at this time.

The second challenge of theorists was the grain alignment. This task was first tackled by Spitzer and Tukey [61]. They proposed a static alignment of ferromagnetic grains in a large-scale galactic magnetic field in analogy to compass needles. Concerning the ferromagnetic behaviour of the grain material, already in 1949 Spitzer and Schatzman [62] had suggested that such grain properties could be expected as a consequence of mutual collisions of dirty ice grains. Collisions should support the formation of elongated aggregates as well as the accumulation of Fe and Mg in the refractory collisional remnants. The energy release of chemical reactions should support the formation of ferromagnetic mineral domains in iron oxides of spinel type (maghemite: γ-Fe_2O_3; magnetite: Fe_3O_4) as well as mixed Mg-Fe-oxide (magnesio ferrite: $MgFe_2O_4$).

The dust model [61] had the advantage that ferromagnetic grains would be effective polarizers, and the alignment mechanism the disadvantage that a very strong magnetic field (10^{-3} gauss) was necessary, having lines perpendicular to the spiral arms contrary to the expectations. Some years later, laboratory experiments by Fick [63] with elongated γ-Fe_2O_3-smoke particles, which were produced by an iron arc in a chamber and had average dimensions of 4 μm in length and 0.5 μm in diameter, demonstrated that the mechanism by Spitzer and Tukey [61] also operates with fields of about 10^{-5} gauss. With these grains Fick could also reproduce the interstellar λ^{-1}-law and the observed ratio between polarization and extinction. However, these experiments did apparently not meet much interest in the dust community, mainly because in the meantime the discussion on grain alignment had stroke the other path pointed by the dynamical mechanism proposed by Davis and Greenstein [64] (D-G mechanism). Nevertheless, Fick's experiments are among the "early birds" in the field that today is called "laboratory astrophysics" and does not deserve to be completely ignored.

The D-G mechanism was superior to that of Spitzer-Tukey. It used paramagnetic grains and a magnetic field of only 10^{-5} gauss with lines parallel to the spiral arms of the Galaxy. The paramagnetism of the particles was explained by ferromagnetic impurities in the ice conglomerate of the classical grains (see, also, [62]; later the term "dirty ice" was introduced). The energy dissipation due to paramagnetic relaxation of the grains in the galactic magnetic field resulted in gradually swivelling the angular momentum vectors of the spinning and tumbling grains in the direction of the magnetic field. This way, the rotating grains should be beaded along the magnetic field lines with their long figure axes perpendicular to the field lines.

Later improvement and generalization made the D-G mechanism to an astrophysical standard theory that could be applied to paramagnetic and ferromagnetic as well as diamagnetic grain materials [65, 66]. The expectations that polarization measurements would clearly define the magnetic grain properties did not fulfil, but instead of this the improved D-G mechanism turned out to be an effective tool to trace the magnetic field lines in the Galaxy by polarization measurements.

However, the discovery of the interstellar polarization presented a basic problem to the classical ice model: the inefficient polarization of ice needles. In order to reproduce the observed ratios of polarization to extinction improbable conditions were necessary, namely infinite cylinders (the case of maximum elongated shape) in perfect alignment. Already in 1950, before the problem of the grain orientation was promisingly solved, van de Hulst [67] on the base of his profound knowledge of grain optics had foreseen this weakpoint of the ice model: *"The conclusion is that interstellar grains of the ordinary size as indicated by reddening measurements, may give a barely sufficient amount of polarization... We hope that partly absorbing grains will leave a better margin by giving a stronger polarization".* However, the interstellar extinction sets narrow limits for the enlargement of the imaginary part of the refractive index of ice. Abandoning ices at all and using dust models of refractory solids should become the final solution of this problem. However, for this radical procedure, the time was not yet ripe.

It is, however, noteworthy that already in 1954 Cayrel and Schatzman [68] tentatively considered interstellar graphite grains as a potential candidate that could help out of the polarization worry. Because of their strong optical anisotropy aligned graphite particles would be a very efficient interstellar polarizer. The authors demonstrated this by laboratory measurements of extinction and polarization of oriented coloidal graphite grains smaller than 0.1 μm. They further pointed out that the Davis-Greenstein mechanism, in principle, should also operate with diamagnetic grains like graphite. Concerning the formation of graphite grains, the authors were less enterprising. Apparently they did not know the paper by Rosen and Swings [69], who in 1953 had stated *"The atmosphere of a late N-star should be pictured as containing, probably on its outskirts, solid carbon particles. There is some kind of smoke veil around the star causing a reddening by absorption of the ultraviolet. A veiling effect by smoke has been occasionally envisaged to interpret various astronomical phenomena, even in the case of novae. A late N-star would be a striking example. The smoke veil would vary in variable N-stars.".* Although in [68] a comment by P. Swings on soot grains in N stars was quoted the authors could not link these ideas to propose stars as the proper source of graphite dust. Not even Struve's [70] extended review paper on "Dust and related phenomena in stars", presented at the same congress (6th International Astrophysics Colloquium in Liège), on which Cayrel and Schatzman [68] discussed their new idea, was suggestive in this direction. The time was not yet ripe for a synthesis of the observational evidence of "stardust" that confirmed the old speculations ([46, 48, 49]) with the new requirements of dust astrophysics. On the other hand, at that time the attractivity of the ice model was unbroken and could not be seriously dangered by the polarization problems.

3.2   From Cold to Hot Dust Formation

Apart from the objections due to inefficient polarization ([67, 61, 68, 71]) the ice model met additional criticism. Important counterarguments accumulated during more than two decades have been:

1. The grain formation in usual HI clouds turned out to be much less efficient than it was originally suggested [72]. Additional problems were presented by the search for plausible explanations of the formation of long ice needles.

2. A lot of processes eroding and chemically altering ice grains in the diffuse interstellar medium (ISM) have turned out to be much more destructive than the original mechanism of cloud collisions assumed by Oort and van de Hulst [50], e.g. suprathermal comic rays [73], sputtering in HII regions and shock fronts [74, 75, 76], enrichment of radicals [72, 77] leading to grain explosions [78], photolytic processing transforming volatiles to refractories [79].

3. Some of the observed extinction curves significantly deviated from the theoretical expectations (curve No. 15 in [51]) in the infrared range (see, e.g. [80]).

4. The search for the strong vibrational band of H_2O ice at 3 μm in the diffuse ISM was not successful [81, 82]. When it had finally been detected [83] the carrier ice turned out to occur only in special environments (young stellar objects in molecular clouds and star-forming regions) representing a dust population different from that in the diffuse ISM.

5. Generally, all observed UV extinction curves strongly deviated from the model predictions of ices. Extraterrestrial observations [84, 85, 86] revealed continuing rise of the extinction toward UV, superposed by a strong solid-state band at 217 nm (the extinction "bump"), while curve No. 15 in [51] showed a broad flat plateau there and decreasing extinction in the far UV (cf. Fig.2).

6. Shortly after its detection, the band at 217 nm was attributed to graphite ([87]).

Affected by the growing number of objections against the ice model, in the decade 1960-70 intensive theoretical efforts were undertaken to study alternative dust models consisting of refractory grains. The idea was to reach better reproduction of the observations and to find suitable environments in which such grains could be efficiently formed and ejected in a sufficiently large quantity to the interstellar space. These environments could be the expanding outer layers of evolved stars, e.g. giant stars and cataclysmic variables, but also planetary systems. In any case, the formation place of the grains was relocated from the cold interstellar clouds to much hotter environments, and, consequently, the new models defined the "hot way" of dust formation.

The new approach did not follow a uniform strategy. At the beginning of the 1960s dust modelling was very ambitious, and the model "designers" claimed to offer the perfect alternative to the classical dust. The expectation was to reproduce, with only one type of dust material, as many observations as possible better and without meeting the objections against the cold way. However, in the course of this decade, the condition of chemical uniformity was more and more dropped, and the strategy only aimed to guarantee the best possible compatibility with the observations. The new compromises consisted in the proposal of layered grains (core-mantle up to multi-layer grains) and also grain mixtures. The latter case considered the ISM clouds as "catchment basin" of all possible dust sorts coming from the different circumstellar sources and forming a more or less uniform mixture. Some authors restricted their contribution to propose

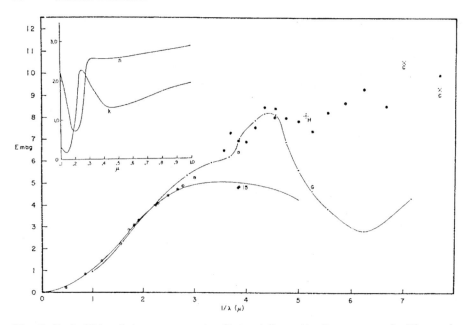

Fig. 2. Early UV rocket measurements of interstellar extinction compared with van de Hulst's curve No. 15 and calculated absorption coefficients of graphite grains (optical constants in the insert). The figure was published in 1965 by Stecher and Donn [87]

only single components that could be significant constituents of the interstellar mixture.

3.3   The Revolutionizing Graphite Model

The hot way of dust formation was opened in 1962/63 with the graphite model by Hoyle and Wickramasinghe [88] and N.C. Wickramasinghe [89]. The authors, who explicitly referred to the old ideas [49, 68], could reproduce the main observational facts on interstellar dust, which were known at that time, with graphite grains having diameters of 10^{-6} to 10^{-5} cm. As source of these particles N stars were assumed, in which the excess carbon should condense to graphite grains in the low-temperature phase of the pulsation cycle of these stars. The grains should be subsequently driven out of the atmospheres by the radiation pressure. Estimates showed that the number of N stars in the Galaxy could account for the necessary quantity of graphite dust. The details of this new dust model, its comparison with the observations and its astrophysical advantages, e.g. in connection with molecule formation, can be found in Wickramasinghe's monograph [74].

The new dust model [88] greatly impacted the theoretical discussion on interstellar dust. However, contrary to the early optimism of the authors, the model did not solve all of the problems and, in particular, could not completely rule out ices. For improving the combatibility with the observations, mainly to get higher grain albedo, composite grains (graphite core + dirty-ice mantle [90, 91]) were introduced. Friedemann and Schmidt [92] pointed to a possible problem of graphite grains with observable consequences. Because of the large grain velocities ($\approx$1000 km/s!) sputtering by impinging interstellar gas atoms should destroy many of them and increase the carbon abundance in the vicinity of N stars.

However, an obvious advantage of the graphite dust was that this model could explain the UV extinction observed by Boggess and Borgman [84] and by Stecher [85]. The interstellar extinction curve showed continuing rise for for $\lambda^{-1} > 3.0$ μm^{-1}, which, however, was interrupted by a big "bump" at about $\lambda^{-1} = 4.6$ μm^{-1}, the first diagnostic solid-state band of interstellar dust. Stecher and Donn [87] assigned it to graphite grains according to the model [88] (Fig. 2).

3.4   The Revitalisation of Iron Grains

In two papers in 1965 and 1967 Cernuschi et al. [93, 94] focused the attention to the adiabatically expanding supernova explosion shells as a possible grain-forming environment. In the first paper [93], they studied the condensation of iron grains. Due to the relatively strong magnetic fields in the SN-shell, the grains should consist of single magnetic domains and stick together, forming elongated particle aggregates that could act as effective interstellar polarizers. The theory had some flaws, as the authors self-critically conceded in their second paper [94]. However, the idea that supernova explosion shells are a dust-forming medium has found confirmation, e.g. by presolar grains (cf. Sect. 5.2).

Independent of the papers [93, 94] Schalén [95], too, revived his old idea and included iron grains in his extinction analysis. At the end of this creative dust modelling phase Hoyle and Wickramasighe [96] seized the supernova origin of dust grains and extended their suggestions not only to iron grains, but also to graphite and silicates. Iron grains were later occasionally included in the dust models (see, e.g., [97]), and today there are more arguments in favour of than against their interstellar occurrence.

3.5   The Disregarded Case: Silicate Grains

For the first time, silicon compounds in the context of interstellar dust emerged in the literature in 1963. Like Hoyle and Wickramasinghe, Kamijo [98] reverted to old stellar condensation speculations [47, 46] and included them in modern models. In the third part of a series of papers on long-period M variables, Kamijo came to the conclusion that nanometre-sized droplets or particles of fused silica (vitreous SiO_2) should be the most abundant species that condensed primarily in the extended envelopes of these stars. The blown out particles were suggested to be condensation seeds in the interstellar gas, i.e., Kamijo [98] understood his result as a contribution to improved condensation conditions for ice grains rather than as a new refractory dust model. Anyhow, mineral grains as potential interstellar dust components must have been in the air at that time. Without giving any reasons, Gaustad [99] in the same year enclosed grains of enstatite

((Mg,Fe)SiO$_3$), hematite (Fe$_2$O$_3$) and α-quartz (SiO$_2$) in his paper "The opacity of diffuse cosmic matter and the early stages of star formation".

In 1968 the present author [100] suggested that all planetary systems present in the Galaxy could be subjected similar disintegration processes of interplanetary bodies like the solar system. This way, fine-grained debris of Mg-Fe-silicates grains mixed with minor portions of, e.g., the meteoritic Fe-Ni-alloy and troilite (FeS) would be permanently released into the interstellar space. Consequently, meteoritic silicates should be a substantial constituent of interstellar dust. Working out in detail the dust model of meteoritic silicates formed the base of the author's Ph.D. thesis [101].

The innovational character of this approach was not only that silicates like olivine and pyroxenes with their favourable optical properties (FUV extinction rise, efficient polarizer, high grain albedo) were included in the dust modelling, but mainly that the evolutionary concept *"that at least part of the interstellar dust is a by-product of the phenomenon of stellar evolution"* (formulation by Herbig [102] on the 16th Astrophysics Colloquium in 1969 in Liège) entered definitely the stage of dust theory. Herbig [102] improved this concept by adding, among others, the conclusion that young planetary systems in the clear-off phase of their "solar nebulae" should release much more dust than the "evolved systems" like ours, on which the first estimates in [100, 101] were based.

In reference to a preprinted abstract of [100], distributed on the IAU General Assembly in Prague in 1967, Knacke [103] calculated extinction cross-sections for grains of quartz and vitreous silica and concluded that particles with the radius 0.2 μm were within 15% in agreement with the extinction measurements by Boggess and Borgman [84]. However, Knacke's [103] paper contained a very important hint to IR observers that such grains should show strong spectral features in the 8–14 μm atmospheric window. It does not matter that he misinterpreted Dorschner's [100] *silicates* as silica, for both solids show IR features in this window.

Still in 1968 Gillett et al. [104] had found a broad spectral feature in some giant stars in the 8-13 μm atmospheric window, but did not give any assignment because they were not sure if it was an absorption band at 8 μm or an emission band at 10 μm (Fig.3). This problem was settled on the 2nd IAU Colloquium on Interstellar Dust (Jena 1969). From the comparison of the observed spectra [104] with the silicate spectra contained in the atlas by Moenke [106] Dorschner [105] concluded that Gillett et al. [104] had observed the 10μm emission band due to the stretching vibrations of the SiO$_4$ tetrahedra of circumstellar silicates. As possible carriers enstatite (MgSiO$_3$) and related pyroxenes as well as mixtures of pyroxenes and olivines ((Mg, Fe)$_2$SiO$_4$) were suggested, whereas quartz was ruled out. As a crucial test for the silicate identification the search for the second fundamental band at about 20μm was recommended.

Since the manuscripts of the Proceedings of the Jena IAU Colloquium were very ill-fated in the hands of the American editor, they could be published only in 1971 after their return from the USA. In the meantime, the identification discussion around the 10 μm band was successfully closed. Already in 1969,

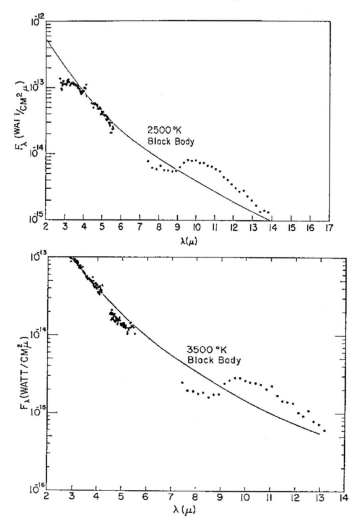

Fig. 3. First spectral indications of an unknown solid state band in the 10μm-region detected by Gillett et al.[104], which in 1969 was identified as the silicate dust fingerprint. Bottom: μ Cep, top: o Cet

Woolf and Ney [107] had published an analog proposal, and, since that time, are commonly quoted in the literature for the successful identification of the circumstellar silicate dust.

3.6 Silicon Carbide and Diamond

In 1969 two more minerals have been suggested to contribute to the interstellar dust: silicon carbide (SiC) and diamond. Friedemann [108], independently of Gilman's [109] thermochemical conclusions published in the same year, calculated the growth of SiC grains in the atmospheres of carbon stars and showed that such grains could leave the star. Carrying out estimates of the total mass of SiC that could be injected this way in the interstellar space, Friedemann [108] was aware that SiC grains should be only a minor dust component.

The Jena dust group at that time favoured the mixture concept of interstellar dust (see, e.g., [110, 100, 108]) that paved the way to the later picture of the "multi-component" interstellar dust. In perfect analogy to the silicate story, SiC grains have been detected already few years later as a common circumstellar dust species in C-rich evolved stars.

In the same year 1969, Saslaw and Gaustad [111] came to the conclusion that *"none of the classical models for the interstellar dust grains fit in with all the modern observations"*, and they presented diamonds as an alternative worth of being studied. They argued that diamond in contrast to graphite could reproduce the steep extinction rise in the FUV, for which first indications had just been found. A second supporting point was the occurrence of diamonds in some types of meteorites. However, at that time only the diamonds in shock-metamorphosed achondrites were known, not to be mixed up with the "presolar" nanodiamonds detected in 1987 [112] in chondrites .

The recommendation of optical studies given by Saslaw and Gaustad [111] was seized by Wickramasinghe [113], who came to the disillusioning conclusion that the diamond hypothesis was in serious conflict with the observations because the calculated extinction curves showed too little IR extinction and a much too high ratio of total to selective extinction, $R = A_V/E_{B_V}$. However, his conclusion was weakened by Landau [114], who showed that a modification of Wickramasinghe's [113] particle radii distribution would much improve the match. Landau commented his realization with the warning that often the choice of a suitable size distribution function can formally provide apparent agreement between calculations and observations for different materials. This way, he had rediscovered the basic "uncertainty principle" of interstellar extinction theory of which already Öpik [27] had taken notice 40 years ago.

Shortly after this debate, diamonds have been forgotten and did not play any longer a role in theoretical dust models. It is one of the wryly traits in the development of interstellar dust research that about twenty years later diamonds [112] turned out to be by far the most abundant ($\approx$1000 ppm per mass!) "presolar" grains preserved in primitive meteorites (see contributions by Bradley and Ott in this book).

3.7   New Positioning of the Points in Dust Astrophysics

Although the dust models proposed in the decade 1960–70 were mostly based on speculative assumptions, they were very helpful in formulating important principles of dust research and provided deep insights into the role of dust in stellar as well as galactic evolution. The dust researchers learned that it is "forlorn hope" to strive for a universal "standard dust model" accounting for all observational facts and resting on a unique type of material formed in one cosmic environment.

Table 1. Refractory dust materials proposed in the models during 1962–69, which later found confirmation by spectral observations and/or detection in meteorites and IDPs

Material	Year / Lit.	Formation environment	Spectral evidence	Fossil dust
Graphite	1962/ [88]	C stars	SD*, IS*, V*	MET
Silica	1963/ [98]	M stars	-	-
Iron	1965/ [93]	Supernovae	-	IDP
Silicate	1968/ [100]	Planet. Systems	SD, IS, DY, V	IDP
SiC	1969/ [108]	C stars	SD	MET
Diamond	1969/ [111]		IS?	MET

The symbols mean:

SD stardust, IS interstellar dust, DY disks of young stars, V Vega objects;

* soot-like carbon, not exactly graphite, as in the original model adopted;

MET isolated presolar grains in meteorites, IDP components from IDPs

The mixture concept more and more gained ground and prepared the modern "multi-component" or "multi-modal" picture of interstellar dust.

From the standpoint of theoretical modelling, dust mixtures simplified the procedure of reproducing the observations, if the constituents and their size distributions were suitably chosen. Further, regionally different extinction and polarization curves could easily be explained by variations of the mixture ratios. The disadvantage of the mixture philosophy was that a larger number of free parameters entered the problem and this meant increasing arbitrariness, at least, as long as for the processes fixing the parameters no plausible theories were available.

An important realization was that the whole supply of different types of dust observations, not only the extinction curve, must be included in dust modelling. Such comprehensive models were often denoted as "unified dust models" [115, 91]. A consequent (and very eloquent) proponent of such improved dust modelling has been the late Mayo Greenberg (see, e.g., the reviews [77, 75, 116]). However, since continuous dust spectra, e.g. extinction, polarization and albedo, as a matter of principle, cannot result in non-ambiguous identification of dust components, striking progress in dust theory could occur not until spectral "fingerprints" of the components were discovered. Progress towards this direction was connected with IR and UV spectroscopy. It is most impressive to see how near the heuristic dust modelling in the 1960s came to reality, and how short the temporal distance between the model proposal and the detection of the respective spectral evidence was. For graphite [88], silicate [100] and silicon carbide [108] this distance amounted only 2-3 years each.

The observational detection of these three dust species confirmed the mixture concept. In 1971 Gilra [117] draw a balance and optimistically concluded *"a mixture of meteoritic silicate, silicon carbide, and graphite particles can explain all the observed properties of interstellar grains"*. However, this mixture could not claim to be the only one that is compatible with the observations (see, e.g., [118]).

The exclusion of ices was a correct decision that has permanently endured (see Table 2). However, the ice problem transiently provided some confusion, after in 1973 strong H_2O ice bands in IR sources had been found [83]. During the 1970s, it became clear that this ice belonged to a special dust population typical of molecular clouds and the dense dust cocoons around young massive stars. In contrast to the diffuse ISM, in cool regions that are screened by strong dust extinction against dissociating UV radiation, H_2O ice can form grain mantles. The existence of different dust populations and evolutionary connections between them represented the next step in modifying the cosmic dust picture in the 1980s (see [5, 6]).

Because of the exciting developments in dust astrophysics in the late 1960s Greenberg agitated for holding an international meeting of higher rank than the Jena Colloquium and found resonance in the IAU administration as well as in the dust community. The first IAU symposium completely devoted to interstellar dust was the very successful Symposium No. 52, held 1972 in Albany, N.Y. Its proceedings [119] positioned the points for the next decade of astrophysical dust research.

4   Dust Mineralogy via Spectral Analysis

4.1   The Impact by the Interstellar UV Band

Even if carbon was the most abundant cosmic element that could form solids, graphite as dust material according to the model [88] looked somewhat too exotic in order to be promptly accepted. It was an unprecedented case of luck in the dust research that useful observational evidence followed the provocative proposal within only three years. Similarly as in 1930 [22] spectral observations of early-type stars anew played the role of the messengers of key information. The UV extinction derived from the spectra of these stars exhibited indications of a "bump" at $\lambda^{-1}=4.6$ μm^{-1}. Being the first true solid-state band, this feature opened the way towards spectral analysis of interstellar dust. Stecher and Donn [87] did not hesitate to assume a connection with graphite grains (cf. Fig. 2), even if the absorption curves calculated with the optical constants of graphite available at that time did not show striking similiarity with the measured UV extinction curve. Anyway, graphite grains of the right size provided an absorption peak at the observed position.

The extinction curves derived from the spectra of the first astronomical satellite observatory, OAO-2, over the whole UV range up to $\lambda^{-1}=9$ μm^{-1} (see Fig. 4) convincingly showed that the "bump" with its peak at $\lambda_0= 217.5$ nm was

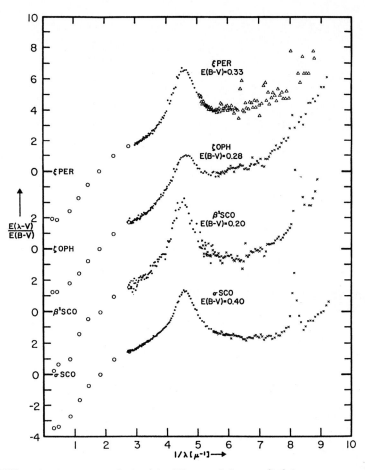

Fig. 4. UV extinction curves derived by Bless and Savage [86] from spectra obtained with the Orbiting Astronomical Observatory (OAO-2)

the result of the superposition of a strong solid-state band over a monotonously rising continuous extinction, which continued the optical extinction curve. The band could be exactly reproduced by a Lorentzian profile, the position, oscillator strength and damping constant pointed due to an electronic transition of an abundant solid carrier. From the measurements of the diffuse galactic light obtained by the same satellite OAO-2 the UV albedo curve of the interstellar dust could be derived. It had a deep minimum at the bump peak, indicating that the 217.5 nm feature was due to almost pure absorption ([120]; see Fig. 5). This pointed to carrier particles very small compared with the wavelength. The equivalent widths of the 217.5 nm band towards different lines of sight turned out to be tightly correlated with the visual extinction E_{B-V}, however, not with the FUV extinction slope [122]. This fact, too, could be interpreted as an indication

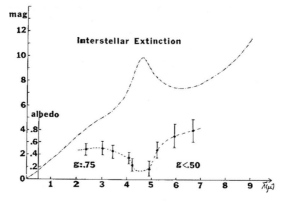

Fig. 5. The band at $\lambda_0 = 217.5$nm is due to pure absorption. The upper curve is the average interstellar extinction curve [86], the lower curve shows the dust albedo derived from observations of the diffuse galactic light [121]

that different components are responsible for different parts of the extinction curve.

An important step was Greenberg's [123] attempt to decompose the extinction curve. The simple sketch, first presented on IAU Symposium No. 52 (see Fig. 6), became a kind of trade mark of the new era of multi-component dust modelling. It is noteworthy that this procedure could be done without making special assumptions on the chemical nature of the components. Of course, it was tempting to attribute the partial curves (2) through (4) (Fig. 6) to hitherto used model components. The visual part (1) was attributed to dielectric grains with radii in the order of 0.1 μm (first ice grains according to van de Hulst's curve No. 15 and/or composite grains with refractory cores and ice mantles; later, after abandoning the ice model, silicate grains). The superimposed band profile (3) was assumed to be due to very small graphite grains. For the FUV extinction nm-sized dielectric grains (mostly silicates) were proposed.

Much observational and laboratory work was concentrated to studies of the UV band in order to get it correctly identified. On the observational side, further space observatories (OAO 3 (1972–81), TD-1A (1974–78), IUE (1978–1996!)) confirmed the band's interstellar ubiquity, its invariant position (λ_0=217.5 $\pm$ 1 nm) in contrast to the remarkable scatter of the width along different lines of sight (FWHM = 48 $\pm$ 12 nm). It turned out to be a typical interstellar solid without a circumstellar counterpart. Spectroscopic evidence for graphite (in the mineralogical sense) was not found in the spectra of carbon stars [124, 125]. From the chemical standpoint, it was not implausible that in a hydrogen-dominated environment pure graphite grains could not be formed. Spectral evidence for the presence of carbon grains in hydrogen-deficient stars were found, e.g. in WC stars [126] and in R CrB stars [127]. However, the strong UV absorption band of the latter was positioned at about 240 nm instead of 217 nm! This was in agreement with spectral data of different soots.

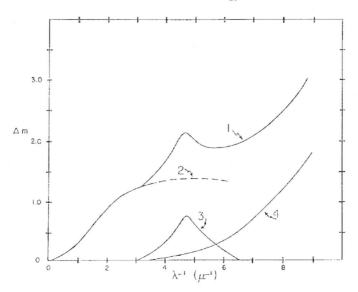

Fig. 6. The decomposition of the extinction curve (1) in three partial curves responsible for the visual and NIR extinction (2), the superimposed UV band (3) and the FUV extinction (4). After [123]

From the theoretical standpoint (including laboratory experiments for obtaining optical constants of different forms of carbonaceous solids) an interesting splitting of opinions occurred in the time between the two large IAU dust symposia (No. 52 in 1972 and No. 135 in 1988). Even if graphite was the initial point for the interpretation of the UV bump, more sophisticated modelling and aspects of grain formation suggested that "graphite" as used in the astrophysical literature is not identical with the mineral graphite. Apart from the intensive search for the right soot type as the band carrier, some dust experts entertained suspicion that carbon could be the wrong track, at all, and searched for alternative explanations of the UV band. On IAU Symposium 135, Draine [128], critically summarizing the identification proposals for the UV band, came to the conclusion that two hypotheses *"stand out as being well-defined, based on at least some laboratory data, and not obviously ruled out by the known observational constraints: 1. graphite grains; and 2. OH^- absorption on the surface of small silicate grains (Steel and Duley, 1987 [129])"*.

The widespread graphite-silicate dualism that had emerged after the silicate discovery and the general acceptance of "graphite" (see, e.g., [130, 124]) was transiently replaced by a "silicate monism" represented by a minority on IAU Symposium No. 135, who pleaded in favour of a silicate origin of the UV band. However, this idea was quickly abandoned, and the same fate had Draine's [128] exclusion of "nongraphitic carbonaceous solids". Amorphous (a-C) and hydrogenated amorphous carbon (a-C:H, HAC) found increasing interest among dust model designers (for review see [131], and cf. also Table 2). One decade later, hydrogenation of soot-like carbon turned out to play the key role in shifting the

UV band of small soot grains towards the interstellar position. The multi-face interstellar carbon and its various problems are reviewed in [132, 6].

4.2   Silicates – the Ubiquitous Dust Constituent

In 1989 Gürtler et al. [133] started their review article on circumstellar silicate dust with the opposite statement *"In the hindsight, it seems rather surprising that silicate particles entered the stage only 20 years ago, because meteorites, asteroids, and the terrestrial planets should have been clear enough evidence for the condensation of silicates taking place whenever and wherever stars and planetary systems are being formed. However, the solar system was generally treated as a special case and links between it and the chemical evolution of the Galaxy or interstellar matter were then not envisaged"*. With the interstellar meteors, every vision of a potential relationship between meteorites and the ISM died away. The reanimation up to the general acceptance of solar system solids as as guide to interstellar dust composition [134, 135] was a procedure of several decades, in spite of the meteoritic dust model [100], the detection of interstellar silicate bands [136], and stimulating suggestions to learn to know interstellar mineralogy by the investigation of primitive meteorites (cf. [137, 138, 139].

In what follows we briefly report the most important events in the discovery story of the silicate bands and some consequences of this second step towards quantitative spectral analysis of cosmic dust.

In 1969 Woolf and Ney [107] correctly considered the feature observed by Gillett et al. [104] in the 8–13 μm atmospheric window as emission band due to circumstellar silicate grains and compared its profile with that of olivine grains in Gaustad's [99] old mineral mixture. As we already mentioned, analogous conclusions drawn at about the same time on the IAU Colloquium on Interstellar Dust at Jena found its way in the literature only in 1971 [105] (see Sect. 3.5).

In the early 1970s impressive progress in circumstellar dust research was reached. The detection of the second vibrational band at about 20μm [140] (bending vibrations of the SiO_4 group) strikingly confirmed the silicate identification. Copious evidence for dust formation in evolved stars [141] established the prevailing view that such stars are the main sources of virgin dust in the Galaxy. Simultaneously, dust formation offered a plausible explanation of the dynamical background of stellar mass loss (see [142]). Mass loss was already a well-known, but insufficiently understood phenomenon (cf. [143]). Now an interesting feedback became obvious: Via momentum coupling between radiation-pressure driven grains and the gas, dust formation supports stellar mass loss, and the outflowing and cooling gas fuels the dust production. The ultimate fate of the "stardust" was to become interstellar dust. The figurative term "stardust" was successfully introduced in the literature by Ney [144]. Already in 1969, Gilman [109] by simple equilibrium calculations of the condensation of solids in late-type giants had shown that the O/C abundance ratio in the stellar atmospheres was the key parameter for the dust production: oxygen-rich stars (O/C>1) supply silicates, carbon-rich (O/C<1) carbon (soot), and stars with O/C$\approx$1 silicon carbide.

Silicate bands were not restricted to circumstellar environments. Still in 1969 Stein and Gillett [145] discovered the 10 μm emission band in the Orion Trapezium nebula, an infrared nebulosity discovered by Ney and Allen [146]. Few months after this first interstellar evidence for silicate dust, Hackwell et al. [136] found the silicate bands in absorption along the line of sight towards the IR sources in the Galactic Centre.

In the first time after the discovery of the 10 μm band, the silicate dust was naturally considered to be crystalline, i.e. consisting of mineral grains (cf. [107, 105]). At that time, this idea appeared obvious rather than devious, since the silicates even in the most primitive meteorites were also crystalline and could be mineralogically classified. In case of need, the observed wide and structureless profiles could be explained as the result of the superposition of many different mineral profiles resulting in wiping out diagnostic profil structure. In this "mineralogically saturated" period, silicate minerals were even used to solve the longstanding problem of the diffuse interstellar bands (DIBs). Several authors, among them also mineralogists, proposed crystal-field transitions of 3d-elements (Ti, Cr, Mn, Fe) as potential cause of the strongest DIBs [147, 148, 149, 150]. However, most of the proposed transitions showed oscillator strengths too low to be compatible with the observed DIBs and the cosmic abundance constraints (see, e.g., [151]). The idea had to be dropped entirely after the structure of the silicates had been recognized to be amorphous. The realization that the circumstellar and interstellar silicates should be far away from the state of regular minerals with crystal structure slowly gained ground in the second half of the 1970s, when laboratory work aiming at the preparation of amorphous silicates started (for review see [152], cf. 5.1).

Stardust silicates did not only exhibit strong deviations from crystalline order visible in the band profiles, but they also showed different absorption behaviour outside of the vibrational bands, e.g. in the visual and NIR, as compared with silicate minerals or glasses. In order to get better agreement between calculated model spectra of dust shells around evolved stars "dirty silicates" were proposed [153], which should be characterized by an increased imaginary part of refractive index in the spectral regions in question. Problems to reach the adequate representation of circumstellar dust opacity have outlived still today.

In 1983 the low-resolution spectrometer data (LRS, spectral range 8–22 μm) of the IRAS satellite drastically increased the number of circumstellar silicate sources. The yield of the IRAS spectroscopy included more than 1800 spectra of evolved stars with the silicate emission features at 10 and 18 μm and about 300 sources with the same features in absorption [154].

As in the case of the "graphite" story, also with the silicates between the publication of the speculative dust model proposal and the discovery of confirming spectral evidence only few years elapsed. However, both stories also showed remarkable differences that are worth of mentioning:

1. The silicate interpretation, which was based on two observed bands, let less space for other identifications. A certain alternative explanation of some tran-

sient significance was the oxide model by Duley and coworkers [155], which, however, did not go beyond the chemical system.

2. The silicates grains revealed a striking environmental ubiquity. They were detected almost simultanously in circumstellar envelopes, in planetary nebulae, in HII regions, in interstellar HI clouds and in molecular clouds, and, also via IR spectroscopy, in interplanetary dust (still in 1969 the 10μm band was found in the spectrum of comet Bennett [156]).

3. The silicate bands, except the "trough" between them, could be observed from the ground.

Unfortunately, the IR silicate bands have also one big problem in common with the UV carbon feature: Like the great variety among the soots, silicates, too, exhibit greatest structural variety nullifying the possibility of accurate "mineralogical diagnosis" of the observed bands. The bands merely prove that there are IR-active Si–O bonds in a spatial (tetrahedral) arrangement that allows bending vibrations and, thus, indicate short-range order. However, their wide and smooth profiles point to lacking translational symmetry in the arrangements of the SiO$_4$ groups, i.e. long-range order typical of crystallinity is absent. These silicates, therefore, are considered amorphous; basic mineralogical information is not available. For instance, we do not know

– whether the "amorphous silicate" contains complete SiO$_4$-tetrahedra or whether there are oxygens deficient;

– to which extent silicon is substituted by other ions;

– how large is the average number of non-bridging oxygens (NBO number);

– which cations with which average coordination are incorporated, etc.

The much evoked "striking spectral resemblance" of laboratory analogue spectra that has been emphasized in many experimental papers of the 1970s was certainly *per se* very impressive, but unfortunately not of ultimate conclusiveness because of lacking unambiguity (cf. [152]).

The amorphousness concept had to be somewhat modified for stardust in evolved stars and also for dust in disks around young and Vega phenomenon stars. The IRAS LRS spectra displayed profile variations from source to source, indicating structural diversity among the silicates and potential contributions of crystalline silicates. Such variations were used to introduce classification schemes (cf., e.g., [157, 158, 159, 160, 161]). "Fine-structure" peaks at constant positions within the 10 μm emission profiles gave reason to look for the presence of particular mineral "fingerprint" bands superposed the amorphous background profile. Interesting cases are the peaks at 11 and 13 μm of the "three-component feature" studied by [157, 158, 159]. The 13-μm peak will be discussed in the next section, here we consider the 11-μm peak tentatively assigned to crystalline olivine.

Support for the identification of a secondary peak at 11.2 μm came from cometary spectra [162]. The same subfeature was also found in main-sequence stars with the so-called Vega phenomenon [163, 164]. The latter is a FIR excess pointing to the thermal emission due to a dust disk around the star, which points more or less to the presence of debris comparable to interplanetary dust (see, e.g., [165]). Some of the Vega phenomenon stars, e.g. β Pictoris, are highly suspicious to be surrounded also by extended clouds of cometary planetesimals.

Such crystallinity peaks superposed to the features of amorphous silicates have been the first weak indications for the presence of circumstellar minerals, justifying the speech of "stardust mineralogy" (for review see [166]). The "ISO revolution" has strikingly confirmed these cautious expectations by the discovery of many highly diagnostic mineral bands in the hitherto not available spectral range beyond the silicate fundamentals (see the contribution of Waters and Molster in this book). It is, however, noteworthy that the "ISO revolution" did not include the interstellar dust. This could mean, that the crystallized stardust ejected into the interstellar space is completely amorphized and/or that amorphous silicate grains are effectively formed in interstellar clouds themselves.

4.3   Other Oxygen-Bearing Dust Minerals

Already in the era of heuristic dust modelling, iron oxides were discussed as interstellar dust constituents, e.g. maghemite (γ-Fe_2O_3, magnetite (Fe_3O_4) and magnesio-ferrite ($MgFe_2O_4$) in the context of interstellar polarization [61] and hematite (α-Fe_2O_3) in Gaustad's [99] opacity mixture for protostars. Huffman [167] counted magnetite (Fe_3O_4) to the *"solids of possible interstellar importance"*. The detection of the very broad structures (VBS) in the interstellar extinction curve [168] formed an important application field for magnetite grains [169, 170, 171]. More recently, Cox [172] called attention to IRAS LRS spectra of HII-regions showing a band-like excess beyond 15 μm. He proposed iron oxides (Fe_3O_4), γ-Fe_2O_3) as carrier material and pointed to the presence of magnetite in carbonaceous chondrites, the most primitive meteorites, to its possible significance for the VBS explanation and the advantages of these oxides in connection with grain alignment.

In the context of this section, we also point to the aforementioned mixed oxide dust model by Duley and coworkers [155]. It was conceived in order to give an alternative explanation of the vibrational bands commonly attributed to silicates. The grains were adopted to consist of diatomic oxides with NaCl-type lattice structure (except SiO, which was considered amorphous), i.e., each metal ion M^{2+} (M = Mg, Fe, Ca, Ni...) is coordinated by six O^{2-} ions. They are formed by kinetic processes in the interstellar environment and can concomitantly explain the observed dust properties as well as the depletion pattern of elements in the interstellar gas. According to this model, the UV feature at 217.5 nm is produced by coordinatively unsaturated O^{2-} ions at the grains' surfaces as proposed in an earlier paper[174]. The 10 μm feature is assumed to be due to the vibrational band of SiO grains, whereas the 18 μm band is caused as a cumulative effect of MgO and FeO. Today the oxide model is no longer of significance in competition to silicates. However, metal oxides, especially those of Fe, still play a role as potential minor ingredients of the multi-component dust (see Table 2).

As a matter of fact, so far non-ambiguous spectral evidence for refractory oxides in one of the dust populations is still lacking, anyhow, there is, at least, one debatable feature in oxygen-rich stars. In IRAS LRS spectra of Miras and SR variables, a weak, but distinct narrow emission band at 13.1 μm was found, which rides on the "red wings" of the strong 10-μm silicate emission bands. As mentioned in the preceding section, in their IR-classification of M stars Little-Marenin and coworkers [157, 158, 159] used the "three-component feature" consisting of two main peaks at 10 and 11 μm and a third one at 13.1 μm. In the literature, Vardya et al. [175] are commonly quoted as the discoverers of the 13-μm band and Onaka et al. [176] as those who identified it as being due to Al_2O_3 grains. Both of these statements deserve some clarification. The paper [175] contains only the lapidary statement *"Among the LRS spectra those of M-type Mira variables are particularly interesting because some show strong silicate emission features at 9.7 and 20 μm, usually attributed to silicate dust, while others show a rather flat spectrum with weaker broad emission features at 12 and 20 μm, whose origin is unknown at present."* The band that Onaka et al. [176] assigned to Al_2O_3 grains was this "weaker broad emission", which can hardly be identical with the much narrower 13.1 μm feature. The identification in [176] was based on optical constants of partially amorphized γ-Al_2O_3 measured by Eriksson et al. [177], which has indeed a band around 12 μm. Corundum (α-Al_2O_3) was excluded in [176] because the authors erroneously assumed the position of the corundum band at 15 μm. Anyhow, the authors of [176] had focussed the attention to aluminium oxide.

Glaccum [178] attributed the 13-μm feature to grains of sapphire, the blue gemstone variety of corundum (α-Al_2O_3). This identification also found some indirect support by the just managed isolation of presolar corundum grains from meteorites [179]. Moreover, aluminium-containing minerals play a basic role in the high-temperature condensates CAI (Ca-Al inclusions) in primitive meteorites. However, the derivation of the accurate 13-μm profile from IRAS LRS spectra met some difficulties resulting in different shapes (cf. [180, 181]). The profile problems settled when the much better resolved ISO-SWS spectra became available. However, the ISO-spectra did not show the second band of corundum grains expected at about 21 μm. Based on ISO-SWS spectra and extended laboratory work, Posch et al. [182] and Fabian et al. [183] have recently shown that the Al–O vibrations of spinel grains ($MgAl_2O_4$) would give a better reproduction of the observed 13-μm band than corundum; they also got additional support for this attribution.

Among the potential oxygen-bearing dust constituents two additional identification proposals should be mentioned: carbonyles and carbonates. Metal carbonyles like $Fe(CO)_5$ have been tentatively proposed by Tielens et al. [184] in order to explain some interstellar absorption bands toward the Galactic Centre (cf. Table 1 in [6]). Carbonate grains have been repeatedly proposed as a dust component. However, convincing spectral evidence for them could not be detected. Recently, a band at about 90 μm in ISO-LWS spectra of some planetary nebulae was tentatively attributed to calcite ($CaCO_3$) or dolomite ($CaMgCO_3$) grains [185].

4.4   Non-oxidic Dust Minerals

The main species of non-oxidic interstellar dust consists of carbon grains, for many years simply designated as "graphite"(cf. 4.1). During the 1990s, "graphite" was more and more replaced by more adequate carbon solids, and, finally, the profile of the 217.5-nm feature could be satifactorily reproduced by laboratory work (cf. the reviews [132, 6]. According to this successful experimental simulation, the interstellar carbon grains should consist of a special type of hydrogenated soot that contains structural units of graphene sheets (aromatically bonded carbon) embedded in an amorphous carbon network structure containing hydrogen heteroatoms. Although carbon grains are assumed to originate as stardust and although presolar graphite grains have been found in meteorites, the final structure of the interstellar extinction bump carrier must have been formed in interstellar conditions.

Carbon probably is not the only elemental solid in interstellar dust. A recent approach to explain the so-called Extended Red Emission (ERE), a galactic photoluminescence phenomenon, is based on nanocrystals of silicon in the ISM. ERE is a band-like excess radiation (FWHM≈100 nm) in the red and NIR, whose peak shifts with increasing radiation field density from 610 to 820 nm. It was detected in reflection nebulae, planetary nebulae, HII regions, dark clouds, galactic cirrus clouds, and finally in the diffuse galactic light (see [186, 187]). The prevailing interpretation was, for a long time, based on a carbonaceous carrier material (see [188]). In some strong ERE sources sharp features have been detected [189, 190]; it is unclear if they are connected with the ERE carrier. In recent years, Ledoux et al. [191] and Witt et al. [192] independently proposed Si nanograins as the alternative hypothesis. If confirmed silicon crystals could represent a hitherto unknown mineral species spread over the whole ISM.

Another non-oxidic dust component is silicon carbide, which has been proposed first as a possible constituent of the multi-component dust (cf. [108, 109]) and, in amazing analogy to "graphite" and silicate dust, has been identified via spectroscopy few years later, however, only as a stardust solid. The SiC-discovery story was initiated by the PhD. thesis of Hackwell [193], who performed a comparative study of IR spectra of M, S, and C stars. Deviating from M and S stars with their typical two silicate peaks, C stars contained only one hump in the 10 μm region, which was distinctly different from the 10-μm silicate band. More resolved spectra of the carbon stars V Hydrae and CIT 6 ([194], Fig. 7) showed that the peak position was 11.3 μm. Having been aware that he had detected a different circumstellar dust species, Hackwell did not offer any assignment of the new band.

Gilra [195] correctly identified this band with the fundamental vibration of circumstellar SiC grains. His paper contained an utmost important conclusion: *"For about a tenth of a micron radius particles the shape of the particle is the most important parameter...Depending on the shape, the emission band(s) should appear between about 10.2 μ and 12.8 μ"*. Based on high-resolution spectra of carbon stars and detailed profile calculations for different shapes, Treffers and Cohen [196] could confirm Gilra's [195] conclusions on the sensitive shape depen-

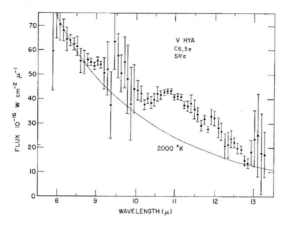

Fig. 7. Infrared excess of V Hya, a C6.3e variable observed by Hackwell [194], the first spectral indication of SiC dust

dence of the SiC profile due to surface modes (Fröhlich modes). They occur at the wavelength λ_F fulfilling the condition $\epsilon(\lambda_F) = -2\epsilon_m(\lambda_F)$, where $\epsilon(\lambda_F)$ and ϵ_m are the dielectric functions of the grain material at λ_F and of the surrounding medium, respectively. SiC is a paradigmatic case of the influence of such modes (see [197]).

At that time, many representatives of the dust community were not yet very familiar with grains shape effects on band profiles, and Gilra's proposal met disbelief. Because of the apparent wavelength disagreement of calculated Mie cross-sections with the observations, Woolf [141] rejected SiC and proposed instead of this silicon nitride (Si_3N_4) as the carrier material. In the following time, much work of the beginning laboratory astrophysics – mainly with size-separated grains obtained from commercial mixtures of α-SiC polytypes (hexagonal or rhombohedral), more rarely on β-SiC (cubic) – was devoted to the spectrum of submicrometre-sized SiC grains in the band region [198, 199, 200] and generally supported Gilra's identification with SiC. Critically must be annotated that the experimenters' conclusions often disregarded shape and matrix effects on the band's position and shape. This explains why different authors came sometimes to different conclusions which SiC modification dominates. The whole problem α- vs. β-SiC has been discussed from the observational and the experimental standpoint by Speck et al. [201].

The IRAS LRS spectral atlas [154] contained more than 500 stars with the SiC band. The profiles showed variations of peak positions and widths from source to source, which impacted first classification work [202]. In this time of commencement, mixing up the SiC feature with one of the UIR bands at about the same wavelength played a certain role. However, since the Unidentified IR (UIR) bands commonly attributed to polycyclic aromatic hydrocarbon molecules always occur as a series with characteristic wavelengths, mix-up could be excluded later.

The few SiC absorption features [201] are probably due to the optical thickness of the corresponding circumstellar shells rather than to interstellar SiC absorption. Whittet et al. [203] have conjectured that SiC could be destroyed in interstellar space by oxidation. A very weak SiC absorption feature could also be hidden within the strong silicate absorption. Nevertheless interstellar SiC must be present, otherwise no presolar SiC grains could be present in meteorites.

In meteorites also fossil interstellar silicon nitride (Si_3N_4) particles have been found, pointing to the fact that they indeed exist. Already before their isolation from meteorites, Russell et al. [204] pointed to this potential dust component by presenting emissivity measurements of crystalline (mixture of α- and β-Si_3N_4) and amorphous samples of silicon nitride. According to these results, grains of Si_3N_4 would be expected to produce observable effects by slightly modifying the SiC profile (cf. [166]). Anyhow, Si_3N_4 deserves further attention as stardust component.

A long-standing question concerns the occurrence of sulfides among the circumstellar solids. In 1981, in their 16–30 μm spectrometry of carbon-rich objects, Forrest et al. [205] could confirm early IR observations by Low et al. [206], who in 1973 had found an excess radiation of the carbon star IRC +10 216 at 34 μm. Kuiper Airborne Observatory spectra [207, 208] finally created clarity that the excess is due to a wide solid-state band peaking at about 30 μm. Following the early MgS assignment [209, 210], Begemann et al. [211] could show that the observations are indeed compatible with magnesium sulfide (MgS). Moreover, based on laboratory work, they suggested the existence of a whole series of mixed Mg-Fe sulfides. In contrast to Mg-rich sulfide grains, the Fe-rich members and pure FeS have sufficient chemical stability to survive the stay in interstellar space and to be included in primitive Solar System solids. The sulfide-bearing GEMS (see section 5.2) give some support to this expectation.

A connection with sulfides was also supposed for the strong emission band at 21 μm detected in 1989 by Kwok et al. [173] in IRAS LRS spectra of four carbon-rich post-AGB stars. Goebel's [212] proposal that silicon disulfide (SiS_2) could be a candidate for the identification has got experimental support by laboratory analogues of amorphous SiS_2 prepared by Begemann et al. [215]. KAO observations by [213] confirmed that the 21-μm band occurs only in the post-AGB stadium, probably as a transient phenomenon. The authors [213] joined the opinion by Buss et al. [214] who pleaded in favour of a carbonaceous carrier material.

Among recent tentative identification proposals of the 21 μm band, diamonds considered by Hill et al. [216] should be mentioned. Following a qualitative suggestion by Koike et al. [217] and the relatively large abundance of presolar diamonds (see section 5.2), the authors [216] reproduced the 21-μm profil by nitrogen-rich diamonds with lattice defects due to irradiation by fast neutrons. However, at present, sulfur- as well as carbon-based proposals can be considered only as hypotheses waiting for support or falsification.

5   The Laboratory Base of Cosmic Dust Mineralogy

5.1   The Approach by Laboratory Astrophysics

The preceding chapter has shown that laboratory measurements are the inevitable condition of reliable identifications of cosmic solids. After 1970 laboratory work, aiming at experimental simulation of cosmic dust materials and processes, became more and more a new institutional astrophysical branch called "laboratory astrophysics". Experimental work in astrophysics was, of course, not restricted to solids, but dust problems have been the first to demonstrate its inevitability for reaching progress. The search for optical properties of many of the "exotic" materials, which have been supposed as dust constituents, in the data collections of the physical, chemical, mineralogical, and technological literature often ended with a "nil return", or they were published only for an insufficient wavelength range. To gradually overcome this *material problem* astrophysicists initiated the determination of optical constants of solids of "possible interstellar importance" [167] over the whole wavelength range of interest in astrophysics.

The wide and smooth band profiles of cosmic dust silicates, which revealed heavily disturbed crystal structure, virtually excluded the direct comparison with mineral spectra. An exception formed indications of "fine structure" in the profiles of stardust bands as discussed in section 4.2. In the early years of laboratory work on silicate problems, many different attempts have been undertaken to gain sufficiently amorphized laboratory analogues (cf. [152]). Despite the often emphasized "striking resemblence" of the laboratory spectra with the observations, there was little conclusiveness that the structure of the laboratory products was exactly the same as that of the carriers of the observed features. Nevertheless, in this phase of heuristic experimentalizing important insights on structure and composition of the interstellar silicates were gained (for overview, see the Proceedings of the OAC2-workshop [218]). The title of the latter in 1987 successfully introduced the term "cosmic dust analogues" for laboratory products prepared for better understanding chemistry and structure of dust solids.

Exept the material problem, dust research met other serious problems. There was a basic *morphology problem*: The real dust grains could not be expected to exhibit such canonical shapes as spheres, infinite cylinders, and small spheroids, for which the light scattering theory offered exact solutions. It is an interesting fact that just the experimental approach to overcome this restriction formed the background for the early development of what gave rise to coin the term "laboratory astrophysics". Its first user probably was J. M. Greenberg in connection with his microwave analogy experiments that he started at the Rensselaer Polytechnic Institute in Troy, N. Y., in 1960 [219, 220]. Similar experiments were also carried out at about the same time in the Tübingen University Observatory by Giese and Siedentopf [221]. Later such experiments were continued with improved equipment at the University of Bochum, Germany, where light scattering by the fluffy cometary grains detected among the IDPs was simulated (see [222, 223], Fig. 8). A more sophisticated version of such analogy techniques working with millimetre waves was developed at the University of Florida [224].

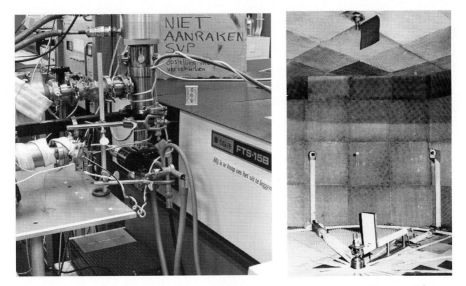

Fig. 8. Laboratory astrophysics in the beginning. Left picture: the experimental setup for the photolysis experiments in Leiden transforming ice mixtures to refractory organic material (RO in Table 2, photo credit: J. M. Greenberg, 1979). Right picture: Microwave scattering setup of the Bochum group. The "particle" is suspended on a nylon threat, transmitter (right) and swivel-mounted receiver (left) are positioned on columns (photo credit: R. H. Giese 1979).

The designation "laboratory astrophysics" was also used in a wider sense. On the first "International Conference on Laboratory Astrophysics", held 1968 in Lunteren, Holland, gas experiments, too, were included in the program. However, the term only slowly gained ground, for in 1972 on IAU Symposium No. 52 [225], where 14 percent of the contributions concerned experimental work, and also on the first European (sponsored by the EAS) dust symposium, held in 1974 in Cardiff, U.K., with the interesting title "Solid-state Astrophysics" [226], the term "laboratory astrophysics" did not yet find official emphasis.

Optical measurements of real-sized particulates produced by grinding raised a *consistency problem* since the measured absorption coefficient in many cases was not compatible with the absorption coefficient calculated with the optical constants of bulk material. Special awareness of this problem and its causes was repeatedly recollected on dust meetings by Huffman and collaborators [167, 227], who formulated important (but often insufficiently observed) warnings of the *"pitfalls in calculating scattering by small particles"* [228]. These concerned the role of surface modes, shape and anisotropy effects, and clumping of particles in matrices. The ample laboratory experience gained by Huffman is reflected by the monograph "Absorption and Scattering of Light by Small Particles" [197] that became a standard work successfully replacing [53]. The method of CDE calculations used there turned out to be a useful criterion whether shape effects play a role or not.

Experimental simulation was also inevitable for the investigation of grain formation and modification by environmental influences. Classical nucleation theory and equilibrium condensation could give, at the best, rough orientation. Experimental simulation and comparison with kinetic calculations proved to be the adequate tools for treating dust formation. One of the first advocates of this type of approach to the *formation problem* was Bertram Donn (for review and literature see [229].

Table 2. Advanced grain modelling for the diffuse ISM

Authors, Reference	Grain type (Species)	Size	UV band
Draine & Lee [236]	b (SIL, GRA)	p	GRA
Chlewicki &	c-m (SIL-RO) +	exp, d	GRA
Laureijs [97]	b (GRA, FE, PAH)		
Greenberg [237]	c-m(SIL-RO)+b(GRA)	exp, d	GRA
Williams [238]	c-m(SIL-HAC)+b(SIL)	p, vs	SIL
Mathis &	co (SIL, GRA,	p	GRA
Whiffen [239]	HAC, VAC) + b(GRA)		
Désert et al. [240]	c-m (SIL-RO)+	p, vs	CAR
	b (AC, PAH)		
Sorrell [241]	b (SIL, AC, GRA)	d	GRA
Rowan-Robinson [242]	b (SIL, AC, GRA)	d, g	GRA
Siebenmorgen &	b (SIL, AC,	p	GRA
Krügel [243]	GRA, PAH)		
Mathis [244]	co (AC, HAC, GRA	p	GRA
	SIL, MOX, VAC) +		
	b (GRA, SIL, MOX)		
Li & Green-	c-m (SIL-RO) +	exp, p	CAR
berg [245]	b (CAR, PAH)		
Zubko et al. [246]	c-m (SIL-RO) +	d	GRA
	ml (SIL,RO,WI) +		
	b (GRA, SI)		

Meaning of the abbreviations:

Grain types: c-m core-mantle, co composite, ml multi-layered, b bare

Chemical species: AC amorphous carbon, HAC hydrogenated amorphous carbon, GRA graphite, CAR carbonaceous material, PAH polycyclic aromatic hydrocarbons, RO refractory organics, MOX metal oxides, SIL silicates, SI silicon, FE iron, VAC vacuum, WI water ice

Size distribution function: d discrete size or narrow intervall, exp exponential law ($n(a) \propto \exp(-Ca^3)$ [247]), g giant grains $\geq 10 \ \mu$m, p power law (MRN-distribution $n(a) \propto a^{-3.5}$ [130]), vs very small grains

Finally, much laboratory simulation work has been and will be in future devoted to the *galactic ecology problem*, i.e. dust evolution and its significance in the galactic ecosystem. Here, the manifold questions of the close connections between cloud and grain evolution ([230, 231], Fig. 9) of the 'life-cycle of dust" (cf. [232]) and the "dust metamorphosis" [5] occurring when the dust passes through different dust populations and acts as an agent in the galactic "dust metabolism" [233]. It attests Greenberg's visionary forsight, when he tackled such problems already in 1975, the foundation year of the "Laboratory Astrophysics Group" at the Huygens Laboratory of der University of Leiden, Holland. This group had a lasting effect to the further development of laboratory astrophysics in Europe and in USA; wellknown activists, e.g. L. Allamandola and L. d'Hendecourt, started their scientific carreer in the Leiden laboratory and were involved in the famous experiments simulating dust evolution by photolytic processing of mantle ices in the molecular cloud dust, transforming C-bearing ices to the refractory organic material of the diffuse ISM dust (for details see [5], Fig. 9). The experiments played a basic role in the development of the coupled evolutionary concept of clouds and grains which characterize the galactic dust "metabolism".

In the modern population concept of the cosmic dust (see [5]) the term "interstellar dust" lost its original universality, which existed still at the time of the laboratory-based dust research. In the strict sense, each population has its own characteristic dust model. Nevertheless, the diffuse ISM dust plays a central role since it represents the dusty debris coming from the different sources. This interstellar "catchment basin" contains stardust from evolved stars, disk dust from protoplanetary discs, "interplanetary dust" from planetary systems, and refractory dust remnants and modified solids from molecular clouds dissolved by star formation. The discovery avalanche of planetary systems around main-sequence stars (cf. [234, 235]) shifts the weights of the contributors. The contributions of planetary systems must no longer be neglected in comparison with the stardust. We conclude this laboratory-devoted section by a listing of the dust models of the last two decades that also reflect the laboratory efforts on understanding galactic particulates inclusive minerals (Tab. 2).

5.2   Interstellar Dust "Fossils" in the Laboratory

After its death in 1930s, the idea of a relationship between interstellar and interplanetary solids resurrected, when silicates entered the stage, first as heuristic model, then by observational evidence (cf. [100, 137, 138]). However, hardly anybody among the "early birds" in dust laboratory simulation work would have dared to hope that, some day, real interstellar dust grains would be available for laboratory analyses. Today this possibility is almost a matter of course. Space missions are underway to collect grains of the local interstellar medium during their passage through the solar system (see the contribution by Mann and Jessberger in this volume).

Moreover, interstellar grains are already now available in terrestrial laboratories. Since 1987 meteoriticists succeeded in isolating presolar grains from primitive meteorites. Their isotope signatures prove the extrasolar origin. They are indeed "fossil" interstellar grains embedded 4.6×10^9 years ago in the meteorite parent bodies of the forming solar system. Being sufficiently resistant, they survived almost unchanged the formation of the solar system and the stay within

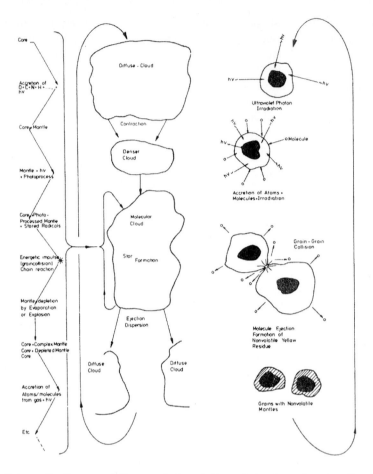

Fig. 9. Schematic diagram of the coupled evolution of clouds and grains. For detailed explanation see Greenberg [231]

the meteoritic rock . The present state of the art in this new exciting research field is reported in the contribution by U. Ott in this volume. These presolar grains in meteorites have broadened the empirical base of dust astrophysics and strengthened its connection with cosmic mineralogy.

The discovery of presolar grains is the second "grand encounter" between astrophysics and meteoritics. In the first one in the 1950s the new table of accurate meteoritic elemental abundances by Suess and Urey [248] turned out to be the challenge that triggered the pioneering work by Burbidge et al. [250] (B^2FH-theory) and Cameron [249], who founded the modern theory of nucleosynthesis. In the second "grand encounter" about twenty years later, astrophysicists began to decode the "cosmic chemical memory" [251] of meteoritic matter (Fig. 10).

"The consequence of this is that today meteoriticists realize that they are also doing astronomy, that they are finding in these falling stones unique memories

of events that predated our solar system formation..." (D.D. Clayton [252] in the "Leonard Medal Address" presented in 1991). In 1982 Clayton [251] in an idiosyncratic but highly original paper with the title "Cosmic chemical memory: a new astronomy" drew a symbolic picture of this "new astronomical science" that he preferentially called "astrochemistry". The "telescopes" are the technological means of the laboratory work providing the "astrochemical" data, i.e. mass spectrometry (investigating the isotopic composition), microanalytics (studying single meteoritic grains, e.g., micron-sized Fremdlinge entirely encapsulated in a different mineral environment), and acid-dissolution techniques (providing insoluble residues, from which the carriers of the extreme isotopic anomalies can be extracted).

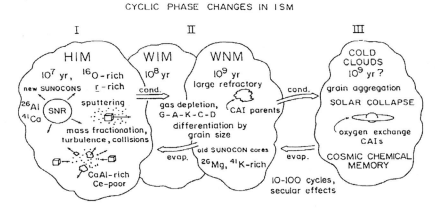

Fig. 10. The cyclic phase changes of ISM, the galactic "generator of chemical memory" [251]. The historical witnesses are the preserved solids (Stardust, SUNOCONS, CAIs, NEBCONS). SUNOCONS (supernova condensates) and NEBCONS (solar nebula condensates) are terms by Clayton that have not generally been accepted by the community. The abbreviations HIM (hot ionized matter), WIM (warm ionized matter), and WNM (warm neutral matter) are the ISM phases according to the generally accepted and meanwhile much improved model by McKee and Ostriker [253].

Clayton's astrochemical "telescopes" have indeed revealed most surprising things. The detection story of astrophysically highly relevant isotope anomalies began already in the 1960s, when Reynolds and Turner [254] became attentive to isotopically anomalous Xe in the Renazzo meteorite. Continuing the "xenology" (coined by Reynolds) showed that not only the heavy (H: ^{134}Xe, ^{136}Xe), but also the light (L: ^{124}Xe, ^{126}Xe) Xe-isotopes were greatly overabundant relative to the solar system values. Since Xe-H and Xe-L were correlated with each other, later the designation was contracted to Xe-HL. Black and Pepin [255] in their isotopic study of carbonaceous chondrites found another noble gase anomaly. Hidden among large amounts of isotopically normal Ne, they detected a small component greatly enriched in the isotope ^{22}Ne. The anomalous neon component was later called neon-E and splitted in two components, Ne-E(H) and Ne-E(L).

Table 3. Presolar grains in meteorites and possible pre-accretional grains in IDPs

Species	Formula Structure	Abund. (ppm)	Size μm	Proposed origin	Detected by (year), Ref.
Meteorites:					
Diamond	β-C fc-cubic	≈1400	0.0026	SN	Lewis et al. (1987) [112]
Silicon carbide	β-SiC fc-cubic	≈14	0.3–20	AGB, SN	Bernatowicz et al.(1987) [262]
Graphite	graphitic spherules [1]	≈10	0.8–20	AGB, SN,WR	Amari et al. (1990) [263]
Titanium carbide (carbide alloys)	TiC (Ti,Zr,Mo)C	[2]	0.005-0.2		Bernatowicz et al.(1991) [264]
Corundum	α-Al$_2$O$_3$ rhomboh.	≈0.01	0.5–5	RGB, AGB	Nittler et al., Hutcheon et al. (1994) [179, 265]
Silicon nitride	Si$_3$N$_4$	> 0.002	0.8–5	SN	Nittler et al. (1995) [266]
Spinel	MgAl$_2$O$_4$	[3]		AGB	Nittler et al. (1997) [267]
Hibonite	CaAl$_{12}$O$_{19}$	[3]		RGB, AGB	Choi et al. (1999) [268]
Rutile (?)	TiO$_2$	[3]			Nittler & Alexander (1999) [269]
IDPs:					
Silicate Glass	GEMS	[4]			Bradley (1994) [270]
Forsterite	Mg$_2$SiO$_4$	[5]	0,3	ISM	Messenger et al. (2002) [271]

Notes:

[1] composite structure: core of nanocrystalline C, surrounded by a mantle of well-graphitized C

[2] nanometre-sized crystallites within μm-sized graphite spherules (*"grains within grains"* [264]), illustrating details of the grain formation

[3] only a few particles

[4] Glass with Embedded Metal and Sulfide: non-stoichiometric silicate glass (depleted in Mg and Si relative to O) with inclusions of FeNi metal and Fe-Ni-sulfides

[5] one particle only; the other ones of this cluster IDP were too tiny for mineral diagnostics

In this historic overview, it is not possible to represent the whole development of unraveling these and other isotopic anomalies and their nucleosynthetic background. We point to review articles by the pioneers of this exciting field [256, 257, 258, 259]. In the last decade, the collaboration between astrophysicists and meteoriticists became very successful. Astrophysicists could establish the base for understanding the observed isotopic patterns by providing mechanisms of nuclear astrophysics in late stages of stellar evolution (RGB-, AGB-stars, supernovae, novae, WR-stars). This way, the stellar origin of the carriers of these exotic isotopic patterns, e.g. Ne-E and Xe-HL, could be proved beyond any doubt. On the other hand, the meteoriticists and cosmochemists improved their painstaking laboratory techniques up to the successful identification and isolation of the carrier materials. The latter turned out to be, as a rule, grains of highly refractory minerals.

With the direct accessibility of IDPs by laboratory methods an analogous development is in the offing (for details see the contribution by Bradley in this volume). However, in this field only single particles were available for the investigation, and the chemical dissolution approach of *"burning down the* [meteoritic] *haystack to find the* [interstellar] *needle"* ([259]) could not be applied. Nevertheless, in IDPs, too, evidence has been found pointing to components of potential interstellar origin. From the astrophysical side, the expectation to find interstellar material in IDPs had been implied by modelling comets as samples of interstellar dust. Greenberg [260] first elaborated this idea in form of a cometary nucleus model and presented it on high ranked meetings in the run-up of the comet Halley encounter (see, e.g., [230]). The problem was that the comet dust mineralogy and morphology derived from the IDPs (that, at that time, were accepted as mainly coming from the disintegration of cometary nuclei) seemed to be very different from that derived from an interstellar dust model ([230]). Nevertheless, the GEMS (see Table 3) show properties that match well with interstellar silicates, and the metallic and sulfidic inclusions are favourable for the magnetic behaviour of the grains in connection with the interstellar polarization [261].

In Table 3 we give a complete list of the presolar grain species detected up to now and also include some IDP components that show properties pointing to interstellar origin.

The discoveries of components of solar system material that point to their interstellar prehistory, which are listed in Table 3, make stardust a new entrant to cosmic mineralogy and, simultaneously, mark the genuinely mineralogical aspect of astromineralogy. Before these grains have been incorporated in the planetesimals of the forming solar system, they spent much more time in the interstellar space than in the stellar envelope. It is, therefore, justified to count them also to the minor interstellar dust components. Unfortunately, several selection effects (interstellar metamorphosis, conditions in the protoplanetary accretion disk, and, finally, the acid-dissolution techniques) effectuate incisive selection and restrict the detectable extrasolar grains to thermally and chemically utmost resistant materials. In particular, the interstellar silicate, sulfide, iron... grains are dissolved in the isolation procedure.

6   Interstellar Dust and the Classical Mineral Definition

Since more than twenty years, terms like "mineral" and "mineralogy" have been applied to interstellar and circumstellar solids (cf. [139, 272, 273]). We finally ask the crucial question: *Is this linguistic usage in astrophysics compatible with the mineral definition in mineralogy?* Improving collaboration will also require standardization of the terminology.

The IMA Commission on New Minerals and Mineral Names (CNMMN) published "Procedures and Guidelines on Mineral Nomenclature" [274] and, in this connection, the following definition is given (based on [275]): *"A mineral substance is a naturally occurring solid formed by geological processes, either on earth or in extraterrestrial bodies. A mineral species is a mineral substance with well-defined chemical composition and crystallographic properties, and which merits a unique name".* This definition is, of course, formulated in terms of traditional mineralogy working with (macroscopic) material samples that could be subjected to all available analytical techniques. However, the mineral classification has also been extended to the micrometre-sized presolar grains, since the modern micro-analytics admits classification of such tiny samples.

For cosmic dust problems some modifications of the mineral definition would be desirable. The term "geological processes" needs some upgrading to mineral-forming processes typical of dust grains, i.e. condensation out of the gas phase and subsequent modifying of the grains. Such a widened scope of the mineral-forming processes is also significant towards an extension of the term "mineral substance". The adaptation of the temperature scale to low interstellar values means that a lot of substances that are volatiles in the conventional mineralogical sight exists as crystalline solids within a definite compositional range. In this sense, they are minerals according to the general definition.

No problems in the conventional sense are raised by the paradigmatic cases presented by the ISO "crystalline revolution" [4]. The multitude of sharp MIR bands coinciding with those in laboratory spectra of forsterite and enstatite allowed the identification of these minerals with a high degree of certainty. However, the proportion of these identified dust minerals is only in the order of a few percent, at the best, of the total galactic dust mass.

Few problems should be connected with metal grains (iron, silicon) among the analogues of Table 2. They are minerals. In a restricted sense, this can also apply to the carrier of the 217.5-nm dust band, i.e. nano-crystalline (graphitic) structural units in the hydrogenated interstellar "soot". However, here a size-limit could be crossed. It is apparently a matter of debate among mineralogists, if there is a lower size-limit in the application of the term "mineral". The CMNNM-paper [274] expresses the following view: *"...with the development of modern analytical techniques, it is now possible to perform complete chemical and crystal-structure analyses on nanometric volumes... Should such submicro-*

scopic domains be accepted as valid mineral species? There is a wide range of opinions on this subject. On the one hand it is argued that if a mineral substance can be characterized in terms of composition and crystallography, then it should be regarded as a valid mineral species. On the other hand, it is contended that the other properties traditionally reported for minerals such as colour, hardness, optical properties, etc., cannot be determined on an area of that size, and that the description is therefore incomplete. Furthermore, the size of the described particle should be sufficiently large so that sufficient type material can be retained to enable a later independent examination to confirm the original characterization.... It has not been possible to reach agreement on a minimal acceptable size for a mineral substance to be regarded as a species, and therefore each case must be decided on its own merits."

Amorphous solids seem to be beyond the limits of the mineral definition, however, a clear demarcation line separating minerals from amorphous solids has not been drawn, so far. Since amorphous substances in interstellar context are of utmost importance (cf. Table 2) we quote this long passages in [274]: *"Amorphous substances are non-crystalline and therefore do not meet the normal requirements for mineral species. The term "crystalline", as generally used in mineralogy, means atomic ordering on a scale that can produce a regular array of diffraction spots when the substance is traversed by a wave of suitable wavelength (X-ray, electrons, neutrons, etc.). However, some geologically derived substances such as gels, glasses and bitumens are non-crystalline. Such substances can be divided into two categories:*

amorphous – substances that have never been crystalline and do not diffract; and metamict – those that were crystalline at one time, but whose crystallinity has been destroyed by ionizing radiation.

Some mineralogists are reluctant to accept amorphous substances as mineral species because of the difficulty of determining whether the substance is a true chemical compound or a mixture, and the impossibility of characterizing it completely; the term "mineraloid" is sometimes applied to such substances.... With modern techniques it is possible to study amorphous phases more effectively than was possible in the past. Spectroscopic methods associated with a complete chemical analysis can often identify an amorphous phase unequivocally. In fact, appropriate spectroscopies (e.g., IR, NMR, Raman, EXAFS, Mössbauer) can reveal the three-dimensional short-range structural environment (chemical bonds) of each atom in the structure. Of course, without the possibility of obtaining a complete crystal structure analysis, which can give the coordinates and the nature of the atoms, the need for a complete chemical analysis is more stringent with amorphous material than with a crystalline phase.

The basis for accepting a naturally occurring amorphous phase as a mineral species could be a series of complete quantitative chemical analyses that are sufficient to reveal the homogeneous chemical composition of a substantial number of grains in the specimen, and physico-chemical data (normally spectroscopic) that prove the uniqueness of the phase.

Metamict substances, if formed by geological processes, are accepted as mineral species if it can be established with reasonable certainty that the original substance (before metamictization) was a crystalline mineral of the same bulk composition. Evidence for this includes the restoration of crystallinity by appropriate heat treatment and the compatibility of the diffraction pattern of the heat-treated product with the external morphology (if any) of the original crystal..."

In principle, these statements offer the possibility to consider amorphous cosmic solids, e.g. interstellar dust silicates, as minerals if their composition is clearly defined and sufficiently uniform. The basic problem is that, at present, for the wide-spread interstellar silicates only insufficient structural and compositional characterization is available by the astronomical observations (see section 4.2). The mineralogical information contained in the spectral "fingerprints" is hopelessly "blurred". However, the situation becomes much less hopeless, if attainable interplanetary solids, e.g. GEMS, should turn out to be indeed typical representatives of interstellar silicates. Their analyses are important not only for the characterization of the glass component, but also of the mineralic inclusions. However, many open questions remain: How, for instance, is the status of the organic (carbonaceous) materials among the dust analogues in Table 2?

More clarity concerning all of these basic questions at the modern interface between astrophysical and mineralogical dust research will surely be reached if in future unchanged interstellar solids are successfully collected by spacecraft, recovered, and subjected to the improved methods of mineralogical and chemical microanalytics.

References

[1] E. F. F. Chladni: Über den Ursprung der von Pallas gefundenen und anderer Eisenmassen und über einige damit in Verbindung stehende Naturerscheinungen. Wittenberg 1794; Neues Verzeichnis der herabgefallenen Stein- und Eisenmassen: in chronologischer Ordnung. Gilberts Annalen der Physik 1815; Über Feuer-Meteore und über die mit denselben herabgefallenen Massen. Wien 1819

[2] W. v. Engelhardt: Probleme der kosmischen Mineralogie. In: *Tübinger Universitätsreden* No. 16 (1963)

[3] J. J. Papike (ed.): Planetary Materials. *Reviews in Mineralogy, Vol. 36* (Mineralogical Society of America, Washington 1998)

[4] C. Jäger, F. Molster, J. Dorschner, Th. Henning, H. Mutschke, and L. B. F. M. Waters: Astron. Astrophys. **339**, 904 (1998)

[5] J. Dorschner and T. Henning: Astron. Astrophys. Rev. **6**, 271 (1995)

[6] J. Dorschner: Interstellar Dust and Circumstellar Dust Disks. In: *Interplanetary Dust*, ed. by E. Grün, B. Å. S. Gustafson, S. Dermott and H. Fechtig (Springer, Berlin 2001), pp. 727–786

[7] J. P. Loys de Chéseaux: *Traité de la Comete qui a paru en Decembre 1743 & en Janvier, Fevrier & Mars 1744* (Lausanne and Geneve 1744), p. 213ff. (1744)

[8] W. Olbers: Über die Durchsichtigkeit des Weltraumes (1823). In: *Wilhelm Olbers. Sein Leben und seine Werke*, Bd. 1, ed. by C. Schilling (Springer, Berlin 1894) pp. 133–141

[9] W. Herschel: Phil. Trans. R. Soc. London **85**, 213 (1785)

[10] J. G. Hagen: Die Geschichte des Nebels "Barnard 86". In: *Specola Astron. Vaticana. Miscellanea Astronomica* Vol. 2 (1929) 61–64

[11] E. E. Barnard: Astron. Nachr. **108**, 370 (1884)

[12] E. E. Barnard: Astrophys. J. **31**, 8 (1910)

[13] E. E. Barnard: Astrophys. J. **49**, 1 (1919)

[14] G. L. Verschuur: *Interstellar Matters. Essays on Curiosity and Astronomical Discovery.* (Springer-Verlag, New York 1989)

[15] M. Wolf: Monthly Not. R. Astron. Soc. **64**, 838 (1904)

[16] M. Wolf: Astron. Nachr. **219**, 109 (1923)

[17] H. Kienle: Die Absorption des Lichtes im interstellaren Raume. In: *Jahrbuch der Radioaktivität und Elektronik* **20**, H. 1, pp. 1–46 (1922)

[18] A. Pannekoek: Proc. Kon. Akad. Amsterdam **23**, No.5, (1920)

[19] H. N. Russell: Astrophys. J. **67**, 83 (1928)

[20] C. Schalén: Astron. Nachr. **236**, 249 (1929)

[21] C. Hoffmeister: Astrophys. J. **69**, 159 (1929)

[22] Y. Öhman: Meddel. Astron. Obs. Upsala No. 48 (1930)

[23] R. J. Trümpler: Lick Obs. Bull. **14**, 154 (No. 420; 1930)

[24] B. P. Gerasimovič: Harvard Obs. Circ. No. 339 (1929)

[25] A. Unsöld: Zeitschr. Astrophys. **1**, 1 (1930)

[26] J. Stebbins, C. M. Huffer, and A. E. Whitford: Astrophys. J. **90**, 209 (1939)

[27] E. Öpik: Harvard Obs. Circ. No. 359 (1931)

[28] G. Mie: Ann. Physik **25**, 377 (1908)

[29] H. Blumer: Zeitschr. Physik **32**, 119 (1925); **38**, 304 (1926); **38**, 920 (1926); **39**, 195 (1926)

[30] C. Schalén: Meddel. Astron. Obs. Upsala No. 58 (1934)

[31] E. Schoenberg and B. Jung: Astron. Nachr. **253**, 261 (1934)

[32] J. H. Oort: Bull. Astron. Inst. Netherlands **6**, 247 (1932)

[33] C. Schalén: Meddel. Astron. Obs. Upsala No. 64 (1936)

[34] E. Schoenberg and B. Jung: Mitt. Univ.-Sternw. Breslau **4**, 61 (1937)

[35] F. L. Whipple: The incentive of a bold hypothesis – hyperbolic meteors and comets. In: *The Collected Contributions of Fred L. Whipple*, Vol. 1. (Smithsonian Astrophys. Obs., Cambridge, Mass. 1972), pp. 3-17

[36] C. Hoffmeister: *Die Meteore, ihre kosmischen und irdischen Beziehungen* (Akadem. Verlagsgesellschaft, Leipzig 1937).

[37] E. Öpik, Monthly Not. R. Astro. Soc. **100**, 315 (1940)

[38] J. G. Porter: Monthly Not. R. Astron. Soc. **104**, 257 (1944)

[39] A. C. B. Lovell: *Meteor Astronomy.* (Clarendon Press, Oxford 1954)

[40] J. L. Greenstein: Harvard Obs. Circ. No. 422 (1938)

[41] L. G. Henyey and J.L. Greenstein: Astrophys. J. **88**, 580 (1938)

[42] L. G. Henyey and J.L. Greenstein: Astrophys. J. **93**, 70 (1941)

[43] A. Güttler: Zeitschr. Astrophys. **31**, 1 (1952)

[44] B. Lindblad: Nature **135**, 133 (1935)

[45] B. Jung: Astron. Nachr. **263**, 426 (1937)

[46] R. Wildt, Zeitschr. Astrophys. **6**, 345 (1933)

[47] P. W. Merrill, Publ. Michigan **2**, 70 (1916)

[48] E. Loreta, Astron. Nachr. **254**, 151 (1935)

[49] J. A. O'Keefe, Astrophys. J. **90**, 294 (1939)

[50] J. H. Oort and H. C. van de Hulst: Bull. Astron. Inst. Netherlands **10**, No. 376, 294 (1946)

[51] H.C. van de Hulst: Rech. Astr. Obs. Utrecht **11**, pt.2 (1949)

[52] H.C. van de Hulst: Rech. Astr. Obs. Utrecht **11**, pt.1, 10 (1946)

[53] H. C. van de Hulst: *Light Scattering by Small Grains* (Wiley, New York 1957)

[54] B. Strömgren: Astrophys. J. **89**, 526 (1939)

[55] W. A. Hiltner: Astrophys. J. **109**, 471 (1949)

[56] J. S. Hall: Science **109**, 166 (1949)

[57] V. A. Dombrovski, Doklady Akad. Nauk Armenia **10**, 199 (1949)

[58] A. Behr: Zeitschr. Astrophys. **47**, 54 (1959)

[59] T. Gehrels: Astron. J. **65**, 470 (1960)

[60] P. Wellmann: Zeitschr. Astrophys. **14**, 195 (1937)

[61] L. Spitzer, Jr., and J. W. Tukey: Astrophys. J. **114**, 187 (1951)

[62] L. Spitzer, Jr., and E. Schatzman: Astron. J. **54**, 195 (1949)

[63] E. Fick: Zeitschr. Physik **138**, 183 (1954) and **140**, 308 (1955)

[64] L. Davis, Jr., and J. L. Greenstein: Astrophys. J. **114**, 206 (1951)

[65] J. Henry: Astrophys. J. **128**, 497 (1958)

[66] V. R. Jones and L. Spitzer: Astrophys. J. **147**, 943 (1967)

[67] H. C. van de Hulst: Astrophys. J. **112**, 1 (1950)

[68] R. Cayrel and E. Schatzman: Ann. d'Astrophys. **17**, 555 (1954)

[69] B. Rosen and P. Swings: Ann. d'Astrophys. **16**, 82 (1953)

[70] O. Struve: Dust and related phenomena in stars. In: *Le particule solides dans les astres*(Université de Liège 1954), pp. 193-222

[71] H. C. van de Hulst: Publ. R. Obs. Edinburgh **4**, 13 (1964)

[72] B. Donn: Some chemical problems of interstellar grains. In: *Le particule solides dans les astres* (Université de Liège 1954), pp. 571-577

[73] H. Kimura: Publ. Astron. Soc. Japan **14**, 374 (1962)

[74] N. C. Wickramasinghe: *Interstellar Grains* (Chapman and Hall Ltd, London 1967)

[75] J. M. Greenberg: Interstellar Grains. In: *Stars and Stellar Systems Vol. VII*, ed. by B.M. Middlehurst and L. A. Aller (Univ. Chicago 1968), pp. 221-364

[76] P. A. Aannestad: Destruction of dirty ice mantles by sputtering. In: *Interstellar Dust and Related Objects. Proc. IAU Symp. No. 52*, ed. by J. M. Greenberg and H. C. van de Hulst (Reidel, Dordrecht 1973), pp. 341–344

[77] J. M. Greenberg: Annu. Rev. Astron. Astrophys. **1**, 267 (1963)

[78] J. M. Greenberg and A. J. Yencha: Exploding interstellar grains and complex molecules. In: *Interstellar Dust and Related Objects. Proc. IAU Symp. No. 52*, ed. by J. M. Greenberg and H. C. van de Hulst (Reidel, Dordrecht 1973), pp. 369–373

[79] W. Hagen, L. J. Allamandola, and J. M. Greenberg: Astrophys. Space Sci.
      65, 215 (1979)
[80] H. L. Johnson: Astrophys. J. **141**, 923 (1965)
[81] R. E. Danielson, N. J. Woolf , J. E. Gaustad: Astrophys. J. **141**, 116
      (1965)
[82] R. F. Knacke, D. Cudaback, and J. E. Gaustad: Astrophys. J. **158**, 151
      (1969)
[83] F. C. Gillett and W. J. Forrest: Astrophys. J. **179**, 483 (1973)
[84] A. Boggess III and J. Borgman: Astrophys. J. **140**, 1636 (1964)
[85] T. P. Stecher: Astrophys. J. **142**, 1683 (1965)
[86] R. C. Bless and B. D. Savage: Astrophys. J. **171**, 293 (1972)
[87] T. P. Stecher and B. Donn: Astrophys. J. **142**, 1681 (1965)
[88] F. Hoyle and N.C. Wickramasinghe: Monthly Not. R. Astron. Soc. **124**,
      417 (1962)
[89] N.C. Wickramasinghe: Monthly Not. R. Astron. Soc. **125**, 87 (1963)
[90] N. C. Wickramasinghe, M. W. C. Dharmawardhana, and C. Wyld:
      Monthly Not. R. Astron. Soc. **134**, 25 (1966)
[91] N.C. Wickramasinghe and K.S. Krishna Swamy: Nature **215**, 895 (1967)
[92] Ch. Friedemann und K.-H. Schmidt: Astron. Nachr. **290**, 233 (1968)
[93] F. Cernuschi, F. R. Marsicano, and I. Kimel: Ann. d'Astrophys. **28**, 860
      (1965)
[94] F. Cernuschi, F. R. Marsicano, and I. Kimel: Ann. d'Astrophys. **30**, 1039
      (1967)
[95] C. Schalén: Medd. Lunds Obs. Ser.1 Nr. 210 (1965)
[96] F. Hoyle and N. C. Wickramasinghe: Nature **226**, 62 (1970)
[97] G. Chlewicki and R. J. Laureijs: Astron. Astrophys. **207**, L11 (1988)
[98] F. Kamijo: Publ. Astron. Soc. Japan **15**, 440 (1963)
[99] J. E. Gaustad: Astrophys. J. **138**, 1050 (1963)
[100] J. Dorschner: Astron. Nachr. **290**, 191 (1968)
[101] J. Dorschner: Ph.D. thesis, University of Jena 1968
[102] G. H. Herbig: Mémoirs Soc. Roy. Sci. Liège **19**, 13 (1970)
[103] R. F. Knacke: Nature **217**, 44 (1968)
[104] F. C. Gillett, F. J. Low, and W. A. Stein: Astrophys. J. **154**, 677 (1968)
[105] J. Dorschner: Astron. Nachr. **293**, 53 (1971)
[106] H. Moenke: Mineralspektren (Akademie Verlag, Berlin 1962)
[107] N. J. Woolf and E. P. Ney: Astrophys. J. Lett. **155**, L181 (1969)
[108] Chr. Friedemann: Astron. Nachr. **291**, 177 (1969)
[109] R. C. Gilman: Astrophys. J. Lett. **155**, L185 (1969)
[110] Ch. Friedemann und K.-H. Schmidt: Astron. Nachr. **290**, 65 (1967)
[111] W. C. Saslaw, J. E. Gaustad: Nature **221**, 160 (1969)
[112] R. S. Lewis, M. Tang, J. F. Wacker, E. Anders and E. Steel: Nature **326**,
      160 (1987)
[113] N. C. Wickramasinghe: Nature **222**, 154 (1969)
[114] R. Landau: Nature **226**, 924 (1970)
[115] J. M. Greenberg and G. Shah: Astrophys. J. **145**, 63 (1966)

[116] J. M. Greenberg: Interstellar Dust. In: *Cosmic Dust*, ed. by J. A. M. McDonnell (John Wiley & Sons, Chichester 1978), pp. 187-294

[117] D. P. Gilra: Nature **229**, 237 (1971)

[118] N. C. Wickramasinghe and K. Nandy: Monthly Not. R. Astron. Soc. **153**, 205 (1971)

[119] J. M. Greenberg and H. C. van de Hulst (editors): *Interstellar Dust and Related Objects. Proc. IAU Symp. No. 52* (Dordrecht Reidel 1973)

[120] A. N. Witt and C. F. Lillie: Astron. Astrophys. **25**,397 (1973)

[121] A. N. Witt: Interstellar dust: observations in the ultraviolet and their interpretations. In: *Interstellar Dust and Related Objects. Proc. IAU Symp. No. 52*, ed. by J. M. Greenberg and H. C. van de Hulst (Reidel, Dordrecht 1973), pp. 53–57

[122] J. Dorschner: Astrophys. Space Sci. **25**, 405 (1973)

[123] J. M. Greenberg: Some scattering problems of interstellar grains. In: *Interstellar Dust and Related Objects. Proc. IAU Symp. No. 52*, ed. by J. M. Greenberg and H. C. van de Hulst (Reidel, Dordrecht 1973), pp. 3–9

[124] B. T. Draine: Astrophys. J. Lett. **277**, L71 (1984)

[125] A. C. H. Glasse, W. A. Towlson, D. K. Aitken, and P. F. Roche: Monthly Not. R. Astron. Soc. **220**, 185 (1986)

[126] M. Cohen, A. G. G. M. Tielens, J. D. Bregman: Astrophys. J. Lett. **344**, L13 (1989)

[127] J. H. Hecht: Astrophys. J. **367**, 635 (1991)

[128] B. T. Draine: On the Interpretation of the $\lambda 2175$Å Feature. In: *Interstellar Dust. Proc. IAU Symp. No. 135*, ed. by L. J. Allamandola and A. G. D. M. Tielens (Kluwer, Dordrecht 1989), pp. 313–327

[129] T. M. Steel and W. W. Duley: Astrophys. J. **315**, 337 (1987)

[130] J. S. Mathis, W. Rumpl, and K. H. Nordsieck: Astrophys. J. **217**, 425 (1977)

[131] W. W. Duley: Carbonaceous grains. In: *Dust and Chemistry in Astronomy*, ed. by T. J. Millar and D. A. Williams (Inst. Physics Publ., Bristol), pp. 71–101

[132] Th. Henning and M. Schnaiter: Carbon – from space to laboratory. In: *Laboratory Astrophysics and Space Research*, ed. by P. Ehrenfreund, C. Kraft, H. Kochan and V. Pironello (Kluwer, Dordrecht 1999), pp. 249–277

[133] J. Gürtler, Th. Henning and J. Dorschner: Astron. Nachr. **310**, 319 (1989)

[134] A. P. Jones, D. A. Williams: Monthly Not. R. Astron. Soc. **224**, 473 (1987)

[135] J. A. M. McDonnell: Solar system dust as a guide to interstellar matter. In: *Dust in the Universe*, ed. by M. E. Bailey, D. A. Williams (Cambridge Univ. Press, Cambridge 1988) pp. 169–181

[136] J. A. Hackwell, R. D. Gehrz, and N. J. Woolf: Nature **227**, 822 (1970)

[137] A. G. W. Cameron: Interstellar grains in museums? In: *Interstellar Dust and Related Objects. Proc. IAU Symp. No. 52*, ed. by J. M. Greenberg and H. C. van de Hulst (Reidel, Dordrecht 1973), pp. 545–547

[138] A. G. W. Cameron: The role of dust in cosmogony. In: *The dusty universe*, ed. by G. B. Field and A. G. W. Cameron (Neale Watson Acad. Publ. , New York 1975), pp. 1-31

[139] R. F. Knacke: Mineralogical similarities between interstellar dust and primitive solar system material. In: *Protostars and Planets. Studies of Star Formation and of the Origin of the Solar System*, ed. by T. Gehrels and M. S. Mathews (Univ. Arizona Press Tucson, Ariz., 1978), pp. 112–133

[140] F. J. Low and K. S. Krishna Swamy: Nature Nature **227**, 1333 (1970)

[141] N. J. Woolf: Circumstellar infrared emission. In: *Interstellar Dust and Related Objects. Proc. IAU Symp. No. 52*, ed. by J. M. Greenberg and H. C. van de Hulst (Reidel, Dordrecht 1973), pp. 485–504

[142] R. D. Gehrz and N. J. Woolf: Astrophys. J. **165**, 285 (1971)

[143] A. J. Deutsch: The mass loss from red giant stars. In: *Stars and Stellar Systems Vol. VI*, ed. by J. L. Greenstein (Univ. Chicago Press 1960), pp. 543–568

[144] E. P. Ney: Science **195**, 541 (1977)

[145] W. A. Stein and F. C. Gillett: Astrophys. J. Lett. **155**, L193 (1969)

[146] E. P. Ney and D. A. Allen: Astrophys. J. Lett. **155**, L197 (1969)

[147] J. Dorschner: Astron. Nachr. **292**, 107 (1970)

[148] P. G. Manning: Nature **226**, 829 (1970)

[149] D. R. Huffman: Astrophys. J. **161**, 1157 (1970)

[150] W. A. Runciman: Nature **228**, 843 (1970)

[151] J. Dorschner: Nature **231**, 124 (1971)

[152] J. Dorschner and Th. Henning: Astrophys. Space Sci. **128**, 47 (1986)

[153] T. W. Jones and K. M. Merrill: Astrophys. J. **209**, 509 (1976)

[154] F. M. Olnon and E. Raimond, E. (eds.): IRAS Catalogues and Atlases. Atlas of Low-resolution Spectra. Astrophys. Suppl. Ser. **65** 607–1065 (1986)

[155] W. W. Duley, T. J. Millar, and D. A. Williams: Astrophys. Space Sci **65**, 69 (1979)

[156] R. W. Maas, E. P. Ney, and N. J. Woolf: Astrophys. J. Lett. **160**, L101 (1969)

[157] I. R. Little-Marenin and S. D. Price: The shapes of circumstellar "silicate" features. In: *Summer School on Interstellar Processes*, ed. by D.J. Hollenbach and H.A. Thronson (NASA Tech. Memo. 88342, Washington, D.C. 1986) pp. 137-138

[158] I. R. Little-Marenin, and S. J. Little: Astrophys. J. **333**, 305 (1988).

[159] I. R. Little-Marenin, and S. J. Little: Astron. J. **99**, 1173 (1990)

[160] J. P. Simpson: Astrophys. J. **368**, 570 (1991)

[161] G. C. Sloan and S. D. Price: Astrophys. J. **451**, 758 (1995)

[162] M. S. Hanner, D. K. Lynch, R. W. Russell: Astrophys. J. **425**, 274 (1994)

[163] R. F. Knacke, S. B. Fajardo-Acosta, C. M. Telesco, J. A. Hackwell, D. K. Lynch, and R. W. Russell: Astrophys. J. **418**, 440 (1993)

[164] S. B. Fajardo-Acosta and R. F. Knacke: Astron. Astrophys. **295**, 767 (1995)

[165] D. E. Backman and F. Paresce: Main-sequence stars with circumstellar solid material: the Vega phenomenon. In: *Protostars and Planets III*, ed. by E. H. Levy, J. I. Lunine, and M. S. Matthews (Univ. Arizona Press, Tucson, Ariz. 1993), pp. 1253–1304

50   Johann Dorschner

[166] J. Dorschner: Stardust Mineralogy. The Laboratory Approach. In: *Formation and Evolution of Solids in Space*, ed. by J. M. Greenberg and A. Li (Kluwer, Dordrecht 1999), pp. 229–264

[167] D. R. Huffman and J. L. Stapp: Optical measurements on solids of possible interstellar importance. In: *Interstellar Dust and Related Objects. Proc. IAU Symp. No. 52*, ed. by J. M. Greenberg and H. C. van de Hulst (Reidel, Dordrecht 1973), pp. 297–301

[168] D. S. Hayes, G. A. Mavko, R. R. Radick, K. H. Rex, and J. M. Greenberg: Broadband structure in the interstellar extinction curve. In: *Interstellar Dust and Related Objects. Proc. IAU Symp. No. 52*, ed. by J. M. Greenberg and H. C. van de Hulst (Reidel, Dordrecht 1973), pp. 83–90

[169] P. G. Manning: Nature **255**, 40 (1975)

[170] D. R. Huffman: Advances in Physics **26**, 129 (1977)

[171] I. G. van Breda and D. C. B. Whittet: Monthly Not. R. Astron. Soc. **195**, 79 (1981)

[172] P. Cox: Astron. Astrophys. **236**, L29 (1990)

[173] S. Kwok, K. Volk, and B. J. Hrivnak: Astrophys. J. **345**, L51 (1989)

[174] W. W. Duley: Astrophys. Space Sci **45**, 253 (1976)

[175] M. S. Vardya, T. de Jong, and F. J. Willems: Astrophys. J. Lett. **304**, L29 (1986)

[176] T. Onaka, T. de Jong, and F. J. Willems: Astron. Astrophys. **218**, 169 (1989)

[177] T. S. Eriksson, A. Hjortsberg, G. A. Niklasson, and C. G. Granquist: Appl. Optics **20**, 2742 (1981)

[178] W. Glaccum: Infrared dust features of late-type stars and planetary nebulae. In: *Airborne Astronomy Symposium on Galactic Ecosystem: From Gas to Stars to Dust*, ed. by M. R. Haas, J. A. Davidson, and E. F. Erickson (A.S.P. Conf. Ser. 73, San Francisco, 1995) p. 395

[179] L. R. Nittler, C. M. O'D. Alexander, X. Gao, R. M. Walker, and Zinner, E.: Nature **370**, 443 (1994)

[180] G. C. Sloan, P. D. LeVan, and I. R. Little-Marenin: Astrophys. J. **463**, 310 (1996)

[181] B. Begemann, J. Dorschner, Th. Henning, H. Mutschke, J. Gürtler, C. Kömpe, and R. Nass, R.: Astrophys. J. **476**, 199 (1997)

[182] T. Posch, F. Kerschbaum, H. Mutschke, D. Fabian, J. Dorschner, J. Hron: Astron. Astrophys. **352**, 609 (1999)

[183] D. Fabian, Th. Posch, H. Mutschke, F. Kerschbaum, and J. Dorschner: Astron. Astrophys. **373**, 1125 (2001)

[184] A. G. G. M. Tielens, D. H. Wooden, L. J. Allamandola, J. Bregman, and F. C. Witteborn: Astrophys. J. **461**, 210 (1996)

[185] F. Kemper, C. Jäger, L. B. F. M. Waters, Th. Henning, F. J. Molster, M. J. Barlow, T. Lim, and A de Koter: Nature **415**, 295 (2002)

[186] A. N. Witt: Visible/UV scattering by interstellar dust. In: *Interstellar Dust. Proc. IAU Symp. No. 135*, ed. by L. J. Allamandola and A. G. D. M. Tielens (Kluwer, Dordrecht 1989), pp. 87–100

[187] K. D. Gordon, A. N. Witt, and B. C. Friedmann: Astrophys. J. **498**, 522 (1998)
[188] S. S. Seahra and W. W. Duley: Astrophys. J. **520**, 719 (1999)
[189] W. W. Duley: Astrophys. Space Sci **150**, 387 (1988)
[190] K. D. Gordon, A. N. Witt, R. J. Rudy, R. C. Puetter, D. K. Lynch, S. Mazuk, K. A. Misselt, G. C. Clayton, and T. L. Smith: Astrophys. J. **544**, 859 (2000)
[191] G. Ledoux, M. Ehbrecht, O. Guillois, F. Huisken, B. Kohn, M. A. Laguna, I. Nenner, V. Paillard, R. Papoular, D. Porterat, and C. Reynaud: Astron. Astrophys. **333**, L39 (1998)
[192] A. N. Witt, K. D. Gordon, and Furton, D. G.: Astrophys. J. **501**, L111 (1998)
[193] J. A. Hackwell: Ph.D. thesis, London University College 1971
[194] J. A. Hackwell: Astron. Astrophys. **21**, 239 (1972)
[195] D. P. Gilra: Dust particles and molecules in the extended atmospheres of carbon stars. In: *Interstellar Dust and Related Objects. Proc. IAU Symp. No. 52*, ed. by J. M. Greenberg and H. C. van de Hulst (Reidel, Dordrecht 1973), pp. 517–528
[196] R. Treffers and M. Cohen: Astrophys. J. **188**, 545 (1974)
[197] C. F. Bohren and D. R. Huffman: *Absorption and Scattering of Light by Small Particles* (Wiley, New York 1983)
[198] J. Dorschner, C. Friedemann, and J. Gürtler: Astron. Nachr. **298**, 279 (1977)
[199] C. Friedemann, J. Gürtler, R. Schmidt, and J. Dorschner: Astrophys. Space Sci. **79**, 405 (1981)
[200] A. Borghesi, E. Bussoletti, L. Colangeli, and C. de Blasi: Astron. Astrophys. **153**, 1 (1985)
[201] A. K. Speck, M. J. Barlow, and C. J. Skinner: Monthly Not. R. Astron. Soc. **288**, 431 (1997)
[202] J. H. Goebel, P. Cheeseman, and F. Gerbault: Astrophys. J. **449**, 246 (1995)
[203] D. C. B. Whittet, W. W. Duley, and P. G. Martin: Monthly Not. R. Astron. Soc. **244**, 427 (1990)
[204] R. W. Russell, M. A. Chatelain, J. H. Hecht, and J. R. Stephens: Si_3N_4 emissivity and the unidentified infrared bands. In: *Interstellar Dust. Contributed Papers*, ed. by A. G. G. M. Tielens and L. J. Allamandola. NASA CP-3036 (1989), pp. 157–162
[205] F. J. Forrest, J. R. Houck, and J. F. McCarthy: Astrophys. J. **248**, 195 (1981)
[206] F. J. Low, G. H. Rieke, and K. R. Armstrong: Astrophys. J. Lett **183**, L105 (1973)
[207] P. Cox: Far-infrared spectroscopy of solid-state features. In: *Astronomical Infrared Spectroscopy: Future Observational Directions*, ed. by S. Kwok (Astron. Soc. Pacific Conf. Ser. Vol. 41, San Francisco 1993), pp. 163–170
[208] A. Omont, S. H. Moseley, P. Cox, W. Glaccum, S. Casey, T. Forveille, K.-W. Chan, R. Szczerba, Loewenstein, R. F., Harvey, P. M., and Kwok, S.: Astrophys. J. **454**, 819 (1995)

[209] J. H. Goebel and S. H. Moseley: Astrophys. J. Lett. **290**, L35 (1985)

[210] J. A. Nuth III, S. H. Moseley , R. F. Silverberg, J. H. Goebel,and W. J. Moore: Astrophys. J. Lett. **290**, L41 (1985)

[211] B. Begemann, J. Dorschner, Th. Henning, H. Mutschke, and E. Thamm: Astrophys. J. Letter **423**, L71 (1994)

[212] J. H. Goebel: Astron. Astrophys. **278**, 226 (1993)

[213] A. Omont, P. Cox, S. H. Moseley, W. Glaccum, S. Casey, T. Forveille, R. Szczerba, K.-W. Chan: Mid- and far-infrared emission bands in C-rich protoplanetary nebulae. In: *Airborne Astronomy Symposium on Galactic Ecosystem: From Gas to Stars to Dust*, ed. by M. R. Haas, J. A. Davidson, and E. F. Erickson (A.S.P. Conf. Ser. 73, San Francisco, 1995), pp. 413–418

[214] R. H. Buss Jr., M. Cohen, A. G. G. M. Tielens, M. W. Werner, J. D. Bregman, F. C. Witteborn, D. Rank, S. A. Sandford: Astrophys. J. **365**, L23 (1990)

[215] B. Begemann, J. Dorschner, Th. Henning, and H. Mutschke: Astrophys. J. **464**, L195 (1996)

[216] H. G. M. Hill, A. P. Jones, and L. B. d'Hendecourt: Astron. Astrophys. **336**, L41 (1998)

[217] C. Koike, N. C. Wickramasinghe, N. Kano, K. Yamakoshi, T. Yamamoto, C. Kaito, S. Kimura, and H. Okuda: Monthly Not. R. Astron. Soc. **277**, 986 (1995)

[218] E. Bussoletti, C. Fusco, and G. Longo (eds.): *Experiments on Cosmic Dust Analogues. Procedings of the Second International Workshop of the Astronomical Observatory of Capodimonte (OAC2), held at Capri, Italy, Septemper 8-12, 1987* (Kluwer: Dordrecht 1988)

[219] J. M. Greenberg: Rensselaer Research **5**, No.1 (1960)

[220] J. M. Greenberg, N.E. Pedersen, J.C. Pedersen: J. Applied Physics **32**, 233 (1961)

[221] R. H. Giese and H. Siedentopf: Z. Naturforschung **17a**, 817 (1962)

[222] R. Zerrull, R.H. Giese and K. Weiss: Appl. Optics **16**, 777 (1977)

[223] R. H. Zerrull, R. H. Giese, S. Schwill, and K. Weiss: Scattering by particles of non-spherical shape. In: *Light Scattering by Irregularly Shaped Particles* ed. by. D. W. Schuerman (Plenum Press, New York 1980), pp. 273-282

[224] B. Å. S. Gustafson: Optical properties of dust from laboratory scattering measurements. In: *Physics, Chemistry and Dynamics of Interplanetary Dust*, ed. by B. Å. S. Gustafson and M. S. Hanner (A.S.P. Conf. Ser. 104, San Francisco, 1995) p. 401–408

[225] J. M. Greenberg and H. C. van de Hulst (eds.): *Interstellar Dust and Related Objects. Proc. IAU Symp. No. 52* (Reidel, Dordrecht 1973)

[226] N. C. Wickramasinghe and D. J. Morgan (eds.): *Solid State Astrophysics*, Astrophysics and Space Science Library Vol. 55 (Reidel, Dordrecht 1976)

[227] D. R. Huffman: Astrophys. Space Sci. **34**, 175 (1975)

[228] D. R. Huffman: Pitfalls in calculating scattering by small particles. In: *Interstellar Dust. Proc. IAU Symp. No. 135*, ed. by L. J. Allamandola and A. G. D. M. Tielens (Kluwer, Dordrecht 1989), pp. 329–336

[229] B. Donn: Experimental investigations relating to the properties and formation of cosmic grains. In: *Interrelationships Among Circumstellar, Interstellar, and Interplanetary Dust*, ed. by J. A. Nuth III and R. E. Stencel.
NASA CP-2403 (1986), pp. 109–134

[230] J. M. Greenberg: What are comets made of? A model based on interstellar
dust. In: *Comets*, ed. L. L. Wilkening (Univ. Arizona Press, Tucson, Ariz.
1982), pp. 131–163

[231] J. M. Greenberg: Occasional Rep. Royal Obs. Edinburgh **12**, 1 (1984)

[232] A. P. Jones: The lifecycle of interstellar dust. In: *From Stardust to Planetesimals*, ed. by Y. J. Pendleton and A. G. G. M. Tielens (Astron. Soc.
Pacific, San Francisco 1997), pp. 97–106

[233] J. Dorschner: Rev. Mod. Astron. **6**, 117 (1992)

[234] M. Perryman: Rep. Progr. Phys. **63**, 1209 (2000)

[235] J. Schneider: The Extrasolar Planets Encyclopaedia,
http://www.obspm.fr/encycl/encycl.html

[236] B. T. Draine and H. M. Lee: Astrophys. J. **285**, 89 (1984)

[237] J. M. Greenberg: The core-mantle model of interstellar grains and the
cosmic dust connection. In: *Interstellar Dust. Proc. IAU Symp. No. 135*,
ed. by L. J. Allamandola and A. G. D. M. Tielens (Kluwer, Dordrecht
1989), pp. 345–355

[238] D. A. Williams: Grains in diffuse clouds: carbon-coated silicate cores. In:
Interstellar Dust. Proc. IAU Symp. No. 135, ed. by L. J. Allamandola and
A. G. D. M. Tielens (Kluwer, Dordrecht 1989), pp. 367–373

[239] J. S. Mathis and G. Whiffen: Astrophys. J. **341**, 808 (1989)

[240] F. J. Désert, F. Boulanger, and J. Puget: Interstellar dust models for
extinction and emission. Astron. Astrophys. **237**, 215 (1990)

[241] W. F. Sorrell: Monthly Not. R. Astron. Soc. **243**, 570 (1990)

[242] M. Rowan-Robinson: Monthly Not. R. Astron. Soc. **258**, 787 (1992)

[243] R. Siebenmorgen and E. Krügel: Astron. Astrophys. **259**, 614 (1992)

[244] J. S. Mathis: Astrophys. J. **472**, 643 (1996)

[245] A. Li and J. M. Greenberg: Astron. Astrophys. **323**, 566 (1997)

[246] V. G. Zubko, T. L. Smith, and A. N. Witt: Astrophys. J. **511**, L57 (1999)

[247] S. S. Hong and J. M. Greenberg: Astron. Astrophys. **70**, 695 (1978)

[248] H. E. Suess and H. C. Urey: Phys. Rev. **28**, 53 (1956)

[249] A. G. W. Cameron: Chalk River Rep. AECL, CRL-41 (1957)

[250] E. M. Burbidge, G. R. Burbidge, W. A. Fowler, and F. Hoyle: Rev. Mod.
Phys. **29**, 547 (1957)

[251] D. D. Clayton: Quart. J. R. Astron. Soc. **23**, 174 (1982)

[252] D. D. Clayton: Meteoritics **27**, 5 (1992)

[253] C. McKee and J. P. Ostriker: Astrophys. J. **218**, 148 (1977)

[254] J. H. Reynolds and G. Turner: J. Geophys. Res. **69**, 3263 (1964)

[255] D. C. Black and R. O. Pepin: Earth Planet. Sci. Lett. **6**, 395 (1969)

[256] E. Anders: Annu. Rev. Astron. Astrophys. **9**, 1, (1971)

[257] E. Anders: Phil. Trans. R. Soc. London **A323**, 287 (1987)

[258] E. Anders, E. Zinner: Meteoritics **28**, 490 (1993)

[259] E. Zinner: Presolar material in meteorites: an overview. In: *Astrophysical Implications of the Laboratory Study of Presolar Materials. CP-402*, ed. by T. J. Bernatowicz and E. K. Zinner (American Inst. Physics 1997), pp. 3–57.

[260] J. M. Greenberg: From interstellar dust to comets. In: *Comets, Asteroids and Meteorites*, ed. A. Delsemme (University of Toledo, Toledo, Ohio 1977), pp. 491–497

[261] P. G. Martin: Astrophys. J. **445**, L63 (1995)

[262] T. Bernatowicz, G. Fraundorf, T. Ming, E. Anders, B. Wopenka, E. Zinner, P. Fraundorf: Nature **330**, 728 (1987)

[263] S. Amari, E. Anders, A. Viraq, and E. Zinner: Nature **345**, 238 (1990)

[264] T. J. Bernatowicz, S. Amari, E. K. Zinner, R. S. Lewis: Astrophysical J. **373**, L73 (1991)

[265] I. D. Hutcheon, G. R. Huss, A. J. Fahey, and G. J. Wasserburg: Astrophys. J. **425**, L97 (1994)

[266] L. R. Nittler, P. Hoppe, C. M. O'D. Alexander, + 7 authors: Astrophys. J. Lett. **453**, L25 (1995)

[267] L. R. Nittler, C. M. O'D. Alexander, X. Gao, R. M. Walker, and E. Zinner: Astrophys. J. **483**, 457 (1997)

[268] B. G. Choi, G. J. Wasserburg, G. R. Huss: Astrophys. J. **522**, L133 (1999)

[269] L. R. Nittler and C. M. O'D. Alexander: Lunar Planet. Sci. **30**, Abstr. 2041 (1999)

[270] J. P. Bradley: Science **265**, 925 (1994)

[271] S. Messenger, L. P. Keller, and R. M. Walker: Lunar Planet. Sci. **33**, Abstr. 1887 (2002)

[272] J. Dorschner, J. Gürtler and Th. Henning: Steps toward interstellar silicate dust mineralogy. In: *Interstellar Dust. Contributed Papers*, ed. by A. G. G. M. Tielens and L. J. Allamandola. NASA CP-3036 (1989), pp. 369–370

[273] A. G. G. M. Tielens: Towards a circumstellar silicate mineralogy. In: *From Miras to Planetary Nebulae*, ed. by M. O. Mennessier and A. Omont (Edition Frontieres, Gif sur Yvette 1990), pp. 186-200

[274] E. N. Nickel and J. D. Grice: The IMA Commission on New Minerals and Mineral Names: Procedures and Guidelines on Mineral Nomenclature (1998)

[275] E. N. Nickel: Canad. Min. **33**, 689 (1995)

Formation and Evolution of Minerals in Accretion Disks and Stellar Outflows

Hans-Peter Gail

University of Heidelberg, Institute for Theoretical Astrophysics, Tiergartenstr. 15, D-69121 Heidelberg, Germany

Abstract. The contribution discusses dust formation and dust processing in oxygen rich stellar outflows under non-explosive conditions, and in circumstellar discs. The main topics are calculation of solid-gas chemical equilibria, the basic concepts for calculating dust growth under non-equilibrium conditions, dust processing by annealing and solid diffusion, a discussion of non-equilibrium dust formation in stellar winds, and in particular a discussion of the composition and evolution of the mineral mixture in protoplanetary accretion discs. An overview is given over the data on dust growth, annealing, and on solid diffusion for astrophysically relevant materials available so far from laboratory experiments.

1 Introduction

The existence of dust particles in the interstellar medium is known since nearly a century. Since some thirty years we also know of the existence of dust in shells around certain highly evolved stellar types which form dust in their stellar outflows, and about fifteen years ago dusty circumstellar discs around young stellar objects have been detected. A lot of observational efforts have been undertaken since the beginning to establish the nature and composition of the dust in these different environments and remarkable progress has been achieved on the observational side, especially during the last nearly two decades since infrared observations from satellites became available.

Parallel to the observational work a lot of theoretical efforts have been undertaken to explain the formation of dust particles and their composition and properties in interstellar space and in circumstellar environments. The success of the theoretical analysis has always been meager, however, since this problem turned out to be particular difficult and unintelligible because of a complex interplay between a lot of different chemical, physical and mineralogical processes, most of which are not well understood even under laboratory conditions. So we do not even now have a really complete picture of which processes are responsible for dust formation, growth and processing, and which processes ultimately determine the chemical composition and mineralogical structure of cosmic dust.

Some details of the general problem of dust formation and its chemical and mineralogical evolution especially in protoplanetary discs can already be solved, however. By laboratory investigations of dust related processes by a number of groups, especially from Japan, many coefficients required for calculating formation, destruction and chemical and physical processing of dust grains have been experimentally determined during the last decade. Most of these investigations had the main purpose to study processes relevant for processing meteoritic material in the early Solar System, but the results can immediately be carried over to processes in the gas-dust-mixtures of protoplanetary accretion discs and of circumstellar dust shells. This allows to put theoretical calculations of dust formation and processing now on a much firmer basis than this was possible till now. This contribution gives an overview over some of the major problems encountered in calculating compositions and properties of cosmic mineral mixtures and how they can be tackled.

2   Dust-Forming Objects and Their Element Abundances

Dust formation in cosmic environments is generally associated with phases of severe mass loss during late stages of stellar evolution. In the cooling outflow of stellar winds or of expelled mass shells conditions become favourable for condensation of the heavy element content of the gas phase into solid particulates. During most of the life of a star, with only minor exceptions, its photospheric element abundances reflects the standard cosmic element abundances of the interstellar matter at the time of birth of the star within a molecular cloud. The surface element abundances change if the stars evolve off from the main sequence and enter advanced stages of their evolution for two reasons:

(1) In low- and intermediate-mass stars, convective mixing between the nuclear burning interior and the stellar photosphere carries some quantities of the freshly synthesized heavy nuclei from the centre to the surface. In rapidly rotating hot stars some mixing from the central region to the photosphere occurs by circulation currents.

(2) In evolved massive stars surface abundances change because the outer stellar zones are peeled off by massive stellar winds and initially deeply embedded zones are exposed to the surface which formerly have been subject to nuclear burning processes.

Dust formation in highly evolved stars, thus, may and in fact often does occur in matter with a composition considerably different from the standard cosmic element abundance.

Matter with unusual chemical composition also is expelled during nova and supernova explosions. Dust formation is observed to occur also in such environments, but this contribution does not consider dust formation in element mixtures and environments originating from explosive processes.

As long as dust forming stars show the standard cosmic element mixture in their photosphere, they mainly form silicate dust. With progressive change of the surface element mixture they start to form completely different dust mixtures in their stellar outflows. The best known example is the change of spectral type from M over S to C for stars of medium mass during their evolution on the thermally pulsing AGB. This results from the mixing of freshly produced carbon from He burning to the surface after each pulse. The change in chemistry and spectral type from being oxygen rich for M stars to carbon rich for C stars is

Table 1. Estimated elemental abundances for the more important dust forming stars with non-explosive mass ejection. Abundances for AGB stars are for models with $2\,M_\odot$, for massive stars for a model with $60\,M_\odot$ ($Z = 0.02$ in both cases). The abundances are particle densities with respect to the total particle density. The carbon rich mixture on the AGB essentially equals the oxygen rich one except for the indicated enhancements. The abundances of the elements heavier than Ne essentially equal solar system abundances in all cases

Element	AGB M-star	AGB C-star	CNO processed	He burning	solar system
H	9.07×10^{-1}		4.34×10^{-1}	consumed	9.09×10^{-1}
He	9.15×10^{-2}		5.66×10^{-1}	8.59×10^{-1}	8.89×10^{-2}
C	2.17×10^{-4}	$1...2 \times \epsilon_O$	2.23×10^{-5}	1.22×10^{-1}	3.23×10^{-4}
N	1.91×10^{-4}		1.17×10^{-3}	destroyed	8.49×10^{-5}
O	6.54×10^{-4}		2.95×10^{-5}	1.27×10^{-2}	6.74×10^{-4}
Ne	1.07×10^{-4}		1.07×10^{-4}	5.68×10^{-3}	1.07×10^{-4}
Mg					3.50×10^{-5}
Al					2.75×10^{-6}
Si					3.23×10^{-5}
S					1.69×10^{-5}
Ca					1.99×10^{-6}
Ti					7.82×10^{-8}
Fe					2.94×10^{-5}
Ni					1.62×10^{-6}
Zr		$...100\times$			3.72×10^{-10}
No.	(1)	(2)	(3)	(4)	(5)

References:(1): Boothroyd & Sackmann [9], Schaller et al. [106] (3)+(4): Meynet et al. [74] (5): Anders & Grevesse [2], Grevesse & Noels [45]

accompanied by a drastic change of the dust manufacturing processes, which turns from silicate to soot production.

Similarly, severe changes of the elemental composition of the stellar outflows occur for later evolutionary stages and for the more massive stars. Such changes in the element mixture also result in drastic, but less well known, changes in the composition of the dust mixture formed in the outflow.

Table 1 lists the abundances of a number of elements, which may be important in some way or the other for the dust formation process, for different element mixtures appearing during different stages of non-explosive stellar evolution at the stellar surface. These are:

1. M stars on the AGB. Reduced C and slightly reduced O abundance, enhanced N abundance due to first and second dredge up of CNO processed material. Gradually increasing C abundance and increased s-Process element abundances after onset thermal pulsing.
2. C stars on the AGB. Carbon enriched by third dredge up of products of He burning. Strongly increased abundances of s-Process elements.

58   Hans-Peter Gail

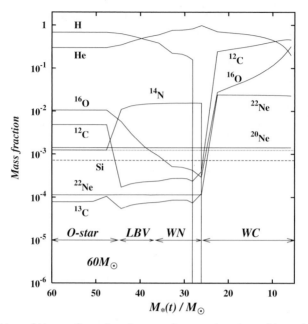

Fig. 1. Evolution of the surface abundances of a massive star with an initial mass of
$60\,M_\odot$ on the main sequence due to mass loss (without rotation) according to Meynet
et al. [74]. The mass scale $M_*(t)$ is the remaining mass of the star. Indicated are
the evolutionary stages where the surface abundances approximately correspond to
the abundances of O-stars, LBV's, WN-, and WC-stars, respectively. The dashed line
shows the initial abundance of the dust forming element Si. The dotted line is the lower
limit for the O abundance where normal silicate dust can be formed

3. CNO processed material. Material exposed by massive stellar winds of massive stars. C and O abundance strongly reduced, in advanced stages reduced
   below the abundances of Si, Mg, and Fe, strongly enhanced N (cf. Fig. 1).
   In its extreme form the material is (nearly) free of H which has been converted into He. Such material is encountered in Luminous Blue Variables
   and especially in WN stars.
4. Products of He burning. If formerly H burning outer layers are peeled off by
   massive stellar winds of massive stars. Nearly pure He, C, and O, and some
   Ne (cf. Fig. 1). This material is encountered in WC stars and in the rare R
   Cor B stars (borne again AGB-stars).
5. Solar system abundances are shown for comparison in Table 1.

The abundances of the elements between Ne and the iron peak are essentially unchanged in these element mixtures since temperatures during H and He burning
are insufficient for processing elements with nuclear charge $Z > 10$.

   The dust formed from these different element mixtures has quite different
chemical compositions. In mixtures (1) and (2) of Table 1 the standard silicate
and carbon dust, respectively, are formed which are usually observed in circumstellar shells of AGB stars or Red Supergiants. In the transition case of S stars

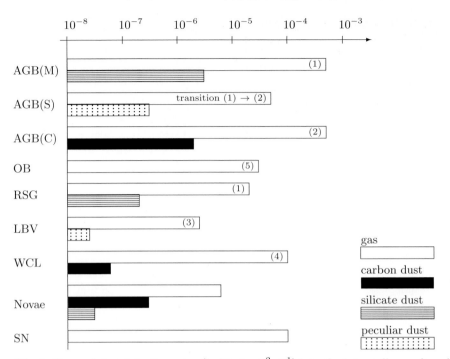

Fig. 2. Gas and dust injection rates (in $M_\odot$ kpc^{-2} a^{-1}) into the interstellar medium by the different stellar sources (modified data from Tielens [123]). The abbreviations for the dust sources are: AGB= asymptotic giant stars of spectral type M, S, or C, OB= massive stars on the upper main sequence, RSG= red supergiants, LBV= Luminous Blue Variables, WCL= Wolf-Rayet stars of spectral type WC8-11. The different compositions of the dust mixture formed in the different dust forming objects are simplified to the three basic types 'silicate', 'carbon', and 'peculiar'. For supernovae, no dust loss rate is given because of the uncertainty with respect to the nature and efficiency of dust production in SN. The numbers refer to the element mixtures in Table 1

with $\epsilon_C \approx \epsilon_O$ neither silicates nor carbon dust can be formed since C and O are blocked in the CO molecule. Some peculiar type of dust is formed in this case (discussed in Sect. 6.1). From mixture (3) of Table 1 also peculiar dust will be formed since only insufficient amounts of C and O are present for silicate or carbon dust formation, like in S stars. From mixture (4) carbon dust is formed, but its properties may be different from carbon dust in AGB star shells because the lack of H requires a different formation mechanism in this case.

Figure 1 shows how mixtures (3) and (4) occur during the course of stellar evolution of massive stars, and Fig. 2 presents a brief overview over the dust forming stellar types and the different kinds of dust produced in the different objects.

3 Equilibrium Condensation

Calculation of chemical equilibrium compositions of solid-gas-mixtures is an important tool for studying dust formation in space. Though in most dust forming objects the gas-dust-mixture is not in a state of chemical equilibrium, the calculation of the hypothetical equilibrium state yields information on the most stable condensates probably existing in the element mixture under consideration and helps to find out the dust species which can be expected to exist in that environment. This concept has been applied with much success to dust formation problems in circumstellar shells and accretion discs.

3.1 Theory: Pure Substances

The problem of determining the chemical equilibrium between several species in a mixture with given total pressure p and temperature T and a given total number of moles of the elements is a standard problem of thermodynamics, which is treated in almost every textbook. It is shown that for this problem the equilibrium state corresponds to the minimum of the Gibbs function $G(p, T, n_i)$ with respect the variations of the mole fractions n_i of all gaseous and solid species i which may be formed from the elements contained in the system. Liquids usually need not to be considered in accretion discs and circumstellar shells because pressures are low.

In thermodynamics it is shown (e. g. Atkins [3]) that a chemical reaction of the type

$$iA \; + \; jB \; + \; kC \; + \; \ldots \; \longrightarrow \; lD \; + \; mE \; + \; nF \; + \; \ldots \tag{1}$$

between reactants A, B, C, ... with products D, E, F, ... is in equilibrium if the following relation holds

$$\sum_i \nu_i \mu_i = 0 \,. \tag{2}$$

The ν_i in (2) are the stoichiometric coefficients i, j, ... in (1) which are usually taken to be positive for the products and negative for the reactants. The μ_i are the chemical potentials of species i defined by

$$\mu_i = \left(\frac{\partial G}{\partial n_i} \right)_{p,T} . \tag{3}$$

For astrophysical applications in most cases one is interested only in the special case of the reaction of formation of some molecular compound or condensate from the free atoms in the gas phase. We consider only this case.

For an ideal mixture the Gibbs function is given by

$$G(p, T, n_j) = \sum_i n_i G_i(p, T) \tag{4}$$

where $G_i(p, T)$ is the partial free enthalpy of one mole of species i at temperature T and pressure p, which equals the chemical potential μ_i in this case. We denote by $\mu_0 = \mu(p_0, T)$ the chemical potential at temperature T and at the standard pressure p_0 of one bar. The equilibrium condition (2) then can be written as

$$\sum_i \nu_i \left(\mu_i - \mu_{0,i} \right) = - \sum_i \nu_i G_i(p_0, T) . \tag{5}$$

One defines the activity a_i of component i by

$$RT \ln a_i = \mu_i(p, T) - \mu_i(p_0, T) \tag{6}$$

which changes (5) into

$$\sum_i \nu_i \ln a_i = -\Delta G/RT . \tag{7}$$

where

$$\Delta G = \sum_i \nu_i G_i(p_0, T) . \tag{8}$$

Equilibrium Conditions for Gases and Solids. Consider a mixture of ideal gases and solids at the rather low pressures in circumstellar shells or protoplanetary accretion discs. The activities can be determined from the thermodynamic relation (see Atkins [3])

$$\partial G/\partial p = V . \tag{9}$$

For n moles of an ideal gas at pressure p we substitute at the rhs. of (9) the ideal gas equation of state $pV = nRT$ and integrate from p_0 to p

$$G(p, T) = G(p_0, T) + nRT \ln p/p_0 . \tag{10}$$

From definition (3) we obtain

$$\mu(p, T) = \mu(p_0, T) + RT \ln p/p_0 \tag{11}$$

and from definition (6):

$$a = p/p_0 \quad \text{(ideal gas)} . \tag{12}$$

For n moles of a solid we have to substitute nV_{mol} at the rhs. of (9) where V_{mol} is the molar volume of the solid. At low pressures V_{mol} can be considered as constant which yields

$$G(p, T) = G(p_0, T) + nV_{\mathrm{mol}} \left(p - p_0 \right) \tag{13}$$

and (3) shows

$$\mu(p, T) = \mu(p_0, T) + V_{\mathrm{mol}} \left(p - p_0 \right) . \tag{14}$$

The molar Gibbs function $G(p_0, T)$ of a solid is typically of the order of several hundred kJ/mole, while the product $V_{\mathrm{mol}} \left(p - p_0 \right)$ is of the order of only a few J/mol at low pressures. The p-V-work can be completely neglected in this case and (6) shows

$$a = 1 \quad \text{(pure solids)} . \tag{15}$$

The equilibrium condition (7) then reads as follows

$$\prod_{\substack{i \\ \text{all gases}}} \left(\frac{p_i}{p_0}\right)^{\nu_i} \prod_{\substack{i \\ \text{all solids}}} a_i^{\nu_i} = e^{-\Delta G/RT} \,. \tag{16}$$

At low pressure the activities a_i of pure solids equal unity. (16) is the well known law of mass action in case that gases and solids are present in a mixture.

We need two special cases: If a molecule of composition $A_i B_j C_k \ldots$ is formed from the free atoms A, B, C, $\ldots$ from the gas phase, then according to (16) its partial pressure in chemical equilibrium is

$$p_{A_i B_j C_k \ldots} = p_A^i p_B^j p_C^k \ldots e^{-\Delta G/RT} \quad \text{(for molecules)}\,, \tag{17}$$

where all pressures are in units of the standard pressure $p_0 = 1\,\text{atm}$. If a solid with composition $A_i B_j C_k \ldots$ is formed from the free atoms A, B, C, $\ldots$ from the gas phase it is in chemical equilibrium with the gas phase if

$$1 = a_{A_i B_j C_k \ldots} = p_A^i p_B^j p_C^k \ldots e^{-\Delta G/RT} \quad \text{(for solids)}\,. \tag{18}$$

While for gas phase species (17) yields the partial pressure of the molecule, for solids relation (18) does not provide direct information on the abundances of the solids in the mixture but instead defines for each solid a condition for the partial pressures of the free atoms which has to be satisfied if the solid has to exist in equilibrium with the gas phase.

If we start with some arbitrary initial state of a solid gas-mixture and if we have a state of chemical equilibrium between all the gas phase species, then the partial pressures of the free atoms p_A, p_B, p_C, $\ldots$ can readily be determined for this state. The quantities $a_{A_i B_j C_k \ldots}$ can take values $a > 1$, $a = 1$, or $a < 1$ if the solids are not in chemical equilibrium with the gas phase.

If we have $a > 1$ for some solid compound in the mixture, the number of moles of this solid in the mixture can be increased by condensing material from the gas phase into this solid until the associated reduction of the partial pressures of the gas phase species decreases the value of a determined by the lhs. of (18) to unity, from which point on the solid would be in equilibrium with the gas phase.

If we have $a < 1$ for some solid this solid can vaporise or can decompose by reactions with gas phase species until (i) either the solid disappears from the mixture or (ii) the associated increase of the partial pressures of the gas phase species increases a to unity from which point on the solid exists in chemical equilibrium with the gas phase.

This shows: In a state of chemical equilibrium we have

- $a = 1$ for each of the solids present in the gas-solid-mixture, or
- $a < 1$ for all solids which do not exist in this mixture.

Constraints set by Element Abundances. The partial pressures of the free atoms in a mixture and the fraction of the elements condensed into solids is determined by the element abundances ϵ_k in the mixture. For each solid one can define a degree of condensation f_j for instance as the fraction of the least abundant of the elements forming this solid, which actually is condensed into this solid. If a solid is present in the mixture then $f_j > 0$ and the corresponding a_j equals unity. If it is absent from the mixture then $f_j = 0$ and $a_j < 0$. For each element k one has the following condition for the partial pressures of the molecules and the solids containing this element

$$\epsilon_k P_H \left(1 - \sum_{\substack{\text{all solids } i \\ \text{with element } k}} f_i \right) = \sum_{\substack{\text{all molecules } i \\ \text{with element } k}} \nu_{i,k} p_i , \tag{19}$$

in order to satisfy the condition set by the abundance of the element in the mixture (stoichiometric conditions). P_H is the fictitious partial pressure of all H nuclei if they where present as free atoms. The $\nu_{i,k}$ at the rhs. are the number of atoms of element k contained in molecule i. For the molecules one introduces (17) into the rhs. of (19). If there are totally K elements in the mixture from which L solids might be formed, one has a system formed by the K equations (19) and the L equations (18) for the K partial pressures of the free atoms and the L degrees of condensation of the solids. A solution of this system has to be found in such a way that for each of the solids either $a = 1$ and $f > 0$ or $f = 0$ and $a < 1$.

Minimizing the Gibss Function. For given p and T in chemical equilibrium the Gibbs free enthalpy

$$G = \sum_i n_i \mu_i \tag{20}$$

takes its absolute minimum subject to the constraints

$$\sum_i n_i \nu_{ik} = n_k , \tag{21}$$

where μ_i is the chemical potential of species i, n_i its number of moles in the mixture, ν_{ik} the number of atoms of element k in species i and n_k the number of moles of the elements k forming the mixture. The composition of a mixture can also be determined by solving this minimization problem.

3.2   Sources of Data

The free enthalpy ΔH_f° of formation of one mole of a substance in the standard state (usually $T = 298.15\,\text{K}$ and $p = 1\,\text{bar}$) from the elements in their reference states and the entropy S° can be obtained from a number of compilations. From these data

$$\Delta G_f^\circ = \Delta H_f^\circ - T S^\circ \tag{22}$$

is calculated for a substance and for the free atoms from which it is formed. The ΔG for the formation of the substance from free atoms follows by (8).

Extensive tabulations for ΔH_f° and S° are the JANAF tables[1] (Chase et al. [20]), Barin [4] and Saxena et al. [105] which cover all the data presently needed for calculations of gas-solid-mixtures in circumstellar shells and protoplanetary accretion discs. Additional data on solids may be found in Kubaschewski & Alcock [62]. Convenient polynomial approximations for the temperature dependence of ΔG for the formation of many solids and gases of astrophysical interest from free atoms are published in Sharp & Huebner [113].

3.3   Solution of the Problem

The problem can be solved numerically in different ways. For pure molecular equilibria this is a standard problem of stellar physics, the calculation of complex solid-gas-mixtures, however, is a rather new problem for astrophysics, which only recently raised in the context of modeling Brown Dwarf atmospheres and protostellar accretion discs. In planetary physics and in the context of laboratory investigations of meteorites the problem of calculating the composition of solid-gas-mixtures in matter with a cosmic element mixture has been studied already much earlier. The method of choice depends on the degree of information required. General multi-purpose methods for calculating equilibria in gas-solid-mixtures are mostly based on the minimization of the Gibbs free energy function $G(P, T, n_j)$ of the whole mixture (cf. Sect. 3.1). Such methods are extensively described in Smith & Missen [115]. A good description of the Gibbs energy minimization method is given in Saxena et al. [105]. Such methods have to be preferred for complex mixtures where a big number of solids have to be checked for their possible condensation.

In many astrophysical problems one only needs to know the most abundant condensates either because only these are of importance, for instance for the purpose of model calculations for accretion discs or circumstellar shells, or because less abundant solid components are (presently) not observable. Some preliminary information on the possible condensates can readily be obtained by first assuming no condensates to be present and calculating the molecular composition of the gas phase. The partial pressures of the free atoms taken from this solution can be used to calculate the quantities a by means of (18). The resulting numerical values of a satisfy either $a < 1$, in which case the corresponding solids cannot be formed, or they satisfy $a \geq 1$. The solids with the highest calculated values of $a \geq 1$ are the possible candidates for condensation.

Figure 3 shows an example of such a calculation for solar system abundances (cf. Tab 1) for a sequence of temperatures and a fixed total pressure of $P = 5 \times 10^{-9}$ bar representative for the dust formation zone of stellar winds (cf. Fig. 4). Obviously solid Fe, forsterite Mg_2SiO_4 and enstatite $MgSiO_3$ are the possible candidates for condensation. Enstatite will form despite of its lower activity because olivine alone cannot take up all the available Si and Mg in

---

[1] also available at: http://webbook.nist.gov/chemistry/form-ser.html

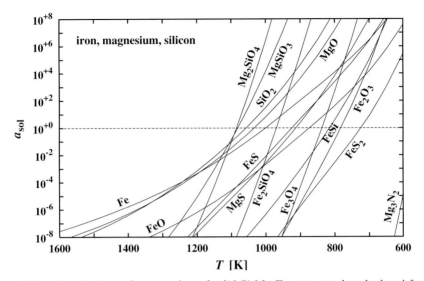

Fig. 3. Pseudo-activities a for a number of solid Si-Mg-Fe compounds calculated from (18) for a sequence of temperatures and a pressure of $P = 5 \times 10^{-9}$ bar representative for the dust formation zone in stellar winds

the standard cosmic element mixture. Other Si bearing compounds will not be formed because of low activities. At about 700 K the activity of FeS becomes higher than the activity of solid Fe, at low temperatures part of the Fe will take up all the available S. Of course, such preliminary conclusions require verification by explicit calculations of equilibrium compositions.

Another possibility is to determine in advance the stability limits of the condensates (cf. Fig. 4), for instance by the kind of considerations in Gail [35], Gail & Sedlmayr [39], Sharp & Wasserburg [114]. Then for each P-T-combination one knows the mixture of solids present in this state and one knows the particular system of equations which has to be solved for this state.

3.4   Results for Some Element Mixtures

Calculations of the composition of solid-gas-mixtures for different purposes and different element mixtures have frequently been performed in the past. The first such calculation for determining the possible condensates in circumstellar dust shells covering the observed C/O abundance variations between M and C stars was that of Gilman [43] while for the primitive solar nebula the first complete calculations of the equilibrium mixture where the seminal papers of Larrimer [64] and Grossman [47]. There have been a number of such calculations preceding that of Gilman, Larimer and Grossman and there followed numerous other calculations. Some papers of a more general type are for instance Lattimer, Schramm & Grossman [66], Saxena & Eriksson [104], Sharp & Huebner [113], Sharp & Wasserburg [114], Lodders & Fegley [70, 71, 72] ([71] contains a bibliography on

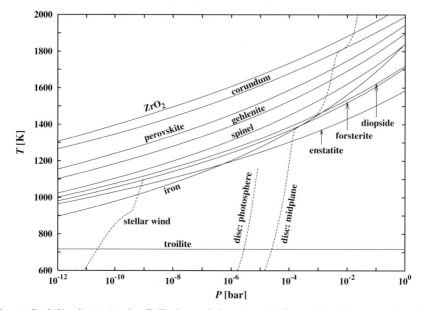

Fig. 4. Stability limits in the P-T plane of the minerals formed by the most abundant refractory elements Si, Mg, Fe, Al, Ca and of Ti and Zr in a Solar System element mixture. The dotted lines are P-T trajectories corresponding to a stellar wind (left) of an AGB-star with a mass-loss rate of $10^{-5}\,\mathrm{M}_\odot\,\mathrm{yr}^{-1}$ and to the photosphere at $\tau = \frac{2}{3}$ (middle) and the midplane of a stationary protoplanetary accretion disc (right) of a solar like protostar with an accretion rate of $10^{-7}\,\mathrm{M}_\odot\,\mathrm{yr}^{-1}$

condensation calculations for stellar problems), Glassgold [44], Ebel & Grossman [26], Krot et. al [61].

As an example for condensation in the standard cosmic element mixture Fig. 4 shows stability limits for the most important minerals formed from Mg, Si, Fe, Al, and Ca. These are the main dust components which are expected to be formed in chemical equilibrium. Additionally the upper stability limits of Ti and Zr compounds are shown, which are not abundant enough to form abundant dust components, but which may be important for the formation of seed nuclei for growth of dust grains formed by more abundant elements.

As a second example Fig. 5 shows the upper stability limits of the most abundant condensates in a mixture where the elements C, N, O evolved into isotopic equilibrium according to the CNO-cycle (mixture (3) in Tab. 1). The abundances of C and O drop below the abundances of Mg, Fe, and Si (cf. Fig. 1), but the mixture stays oxygen rich in the sense that $\epsilon_O > \epsilon_C$. Such material is encountered in the outflow of WN stars. Less extreme versions of this element mixture are ejected during the outbursts of LBV stars. If such stars form dust, the dust will have a rather unusual composition since neither the oxygen rich minerals of M stars nor the mixture of carbon dust and carbides as in C stars can be formed. Instead solid iron and iron-silicon alloys are probably the dominant

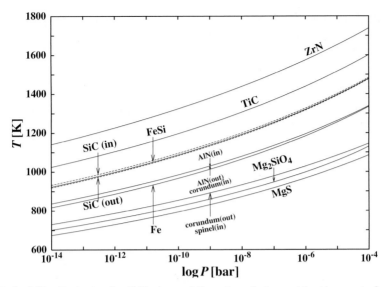

Fig. 5. Stability limits in the P-T plane of the minerals formed by the most abundant refractory elements C, Si, Mg, Fe, Al, Ca and of Ti and Zr in a CNO-cycle processed material

dust components. The composition of the dust is similar to that expected to form in S stars where ϵ_C nearly equals ϵ_O (Ferrarotti & Gail [32]).

Results for the composition of carbon rich mixtures may be found in may of the papers on equilibrium condensation, especially in Lodders & Fegley [70, 71], Sharp & Wasserburg [114] and Lattimer et al. [66].

3.5 Solid Solutions

Some of the minerals existing in circumstellar shells and in protoplanetary accretion discs form solid solutions. Especially this holds for the important silicate dust components and the iron in the standard cosmic element mixture:

- Olivine with composition $Mg_{2x}Fe_{2(1-x)}SiO_4$ is a solid solution formed from the end members forsterite Mg_2SiO_4 and fayalite Fe_2SiO_4.
- Pyroxene with composition $Mg_xFe_{1-x}SiO_3$ is a solid solution formed from enstatite $MgSiO_3$ and ferrosilite $FeSiO_3$.
- Iron forms a solid solution with Ni

From the less abundant Al-Ca-minerals the minerals

- gehlenite ($Al_2Ca_2SiO_7$) and åkermanite ($MgCa_2Si_2O_7$) form a solid solution which is known as melilite.

There are a number of other solid solutions bearing less abundant elements and the abundant solid solutions contain additional minor solution components (e.g.

Table 2. Parameters for the Margules formulation (26) of the activity coefficients for some astrophysical important binary solid solutions

Component 1	Component 2	W_{12} kJ/mole	W_{21} kJ/mole	Source
forsterite	fayalite	4. 500	4. 500	[105],[128]
enstatite	ferrosilite	0 for $T > 873$	0 for $T > 673$	[105]
		$13.1 - 0.015\,T$ else	$3.37 - 0.005\,T$ else	
gehlenite	åkermanite	24. 2881	0. 5021	[105]
iron	nickel	-10. 135	-3. 417	[104]

Saxena & Eriksson [104], Ebel & Grossman[26]). In view of the rather limited accuracy of present day model calculations only the most abundant solid solutions and their most abundant solution components need to be considered. The theory of solid solutions with applications to minerals of interest for astrophysics is described for instance in Saxena [103, 104], Saxena et al. [105]. Useful information can be obtained also from eg. Schmalzried & Navrotsky [107], Schmalzried [108], Philpotts [95], Putnis [98].

In an ideal binary solid solution $A_x B_{1-x}$ of A and B with mole fractions x and $1 - x$, respectively, there is no change in the enthalpy due to mixing, but the entropy changes by the entropy of mixing (e. g. Atkins [3])

$$\Delta S = -nR\left(x \ln x + (1 - x)\ln(1 - x)\right) . \tag{23}$$

n is the number of moles of mixing sites per mole of the solid solution. For a solid solution like that of Fe and Ni one has one mole of lattice sites over which the Fe and Ni atoms may be distributed in one mole of the solid solution, i.e. $n = 1$ in this case. For olivine there are two moles of mixing sites per mole of SiO_4 tetrahedrons in the solid solution over which the Mg^{2+} and Fe^{2+} cations may be distributed, i.e. $n = 2$ in this case.

Since for the solid solution one has $\Delta G = \sum_i x_i G_i$ according to (8), Eq. (23) implies that the chemical potential of a component in an ideal solid solution is (e.g. Putnis [98])

$$\mu_i(p, T) = \mu_i(p_0, T) + nRT \ln x_i , \tag{24}$$

where x_i is the mole fraction of the solution component i. The activity a_i of component i in an ideal solid solution, as defined by (6), equals x_i. For non-ideal solid solutions one defines

$$\mu_i(p, T) = \mu_i(p_0, T) + nRT \ln \gamma_i x_i \tag{25}$$

with some activity coefficient γ_i describing deviations from ideality. γ_i has to be determined experimentally. The results for binary solutions often can be fitted by an expression of the type (unsymmetric Margules formulation)

$$RT \ln \gamma_1 = x_2^2 \left[W_{12} + 2x_1 \left(W_{21} - W_{12} \right) \right] \tag{26}$$

and the same for γ_2 with indices 1 and 2 interchanged. The interaction parameters W_{ij} for some astrophysical important solid solutions are given in Table 2.

For the astrophysically important silicates olivine and orthopyroxene the deviations from ideality are quite small and they may be assumed to form ideal solid solutions, but this not admitted for nickel-iron or melilite.

Formation of Nickel-iron. As example consider the formation of nickel-iron (first considered for Solar System conditions by Grossman [47]). For the reactions

$$\mathrm{Ni(g)} \longrightarrow \mathrm{Ni(in\ nickel-iron)}, \quad \mathrm{Fe(g)} \longrightarrow \mathrm{Fe(in\ nickel-iron)}$$

of formation of solid Ni and Fe as components of the solid solution nickel-iron from the free atoms one has according to (18) the equilibrium conditions

$$a_{\mathrm{Ni}} = x_{\mathrm{Ni}}\gamma_{\mathrm{Ni}} \qquad = p_{\mathrm{Ni}}\, e^{-\Delta G(\mathrm{Ni})/RT} \tag{27}$$
$$a_{\mathrm{Fe}} = (1 - x_{\mathrm{Ni}})\gamma_{\mathrm{Fe}} = p_{\mathrm{Fe}}\, e^{-\Delta G(\mathrm{Fe})/RT}. \tag{28}$$

x_{Ni} is the mole fraction of Ni in the solid solution. $\Delta G(\mathrm{Ni})$ and $\Delta G(\mathrm{Fe})$ are the change of free enthalpy in the formation of the pure solids Ni and Fe, respectively, from free atoms, and p_{Ni} and p_{Fe} are the partial pressures of the free atoms in units of standard pressure p_0. Condensation of nickel-iron essentially is independent of all other condensation processes. Since Ni and Fe exist in the gas phase only as free atoms the stoichiometric equations (19) can be written as

$$\epsilon_{\mathrm{Ni}}\,(1 - f_{\mathrm{Ni}})\,P_{\mathrm{H}} = p_{\mathrm{Ni}} \tag{29}$$
$$\epsilon_{\mathrm{Fe}}\,(1 - f_{\mathrm{Fe}})\,P_{\mathrm{H}} = p_{\mathrm{Fe}} \tag{30}$$

where f_{Ni} and f_{Fe} are the fractions of Ni and Fe, respectively, condensed into nickel-iron. In terms of f_{Ni} and f_{Fe} one has

$$x_{\mathrm{Ni}} = f_{\mathrm{Ni}}\epsilon_{\mathrm{Ni}} / \left(f_{\mathrm{Ni}}\epsilon_{\mathrm{Ni}} + f_{\mathrm{Fe}}\epsilon_{\mathrm{Fe}} \right). \tag{31}$$

The relation of P_{H} to the total pressure P is

$$P = (\tfrac{1}{2} + \epsilon_{\mathrm{He}})P_{\mathrm{H}}, \tag{32}$$

since hydrogen is completely associated to H_2. For given T and P Eqs. (27) ... (31) uniquely determine the composition and abundance of nickel-iron.

A representative result is shown for conditions typically encountered in circumstellar dust shells ($P = 10^{-10}$ bar) in Fig. 6. Note that adding some Ni to the iron increases the condensation/vaporization temperature of nickel-iron by about 20 K as compared to that of pure iron. Results for protoplanetary discs, where typically one has $P = 2 \times 10^{-3}$ bar in the region of dust vaporization, are similar, except that nickel-iron vaporizes at about $T = 1\,505$ K.

Note that below $T \approx 720$ K part of the iron forms FeS (cf. Fig. 4) in chemical equilibrium.

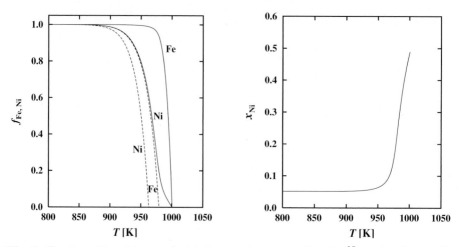

Fig. 6. Condensation of iron and nickel at total pressure $P = 10^{-10}$ bar (stellar wind). *Left part:* Fraction of the total Fe and Ni condensed into nickel-iron (full lines) and separate condensation of the pure substances (dashed lines) if the formation of the solid solution would not be possible. *Right part:* Mole fraction x_{Ni} of Ni in the nickel-iron

Iron Content of Silicates. A particular important example for solid solutions as dust components in circumstellar dust shells and in accretion discs are the solid solutions olivine and pyroxene. These solid solutions seem to have first been considered for Solar System conditions by Blander & Katz [6] and in more detail by Grossman [47]. See also Saxena & Eriksson [104]. The mole fractions of the iron bearing end members in these solution series has presently become of special observational relevance for circumstellar dust (e.g. Molster [76], Molster et al. [77, 78, 79]).

The condensation of olivine, pyroxene, and iron is calculated for instance as follows: The equilibrium conditions for the formation of the solution components of olivine and pyroxene according to (18) are

$$a_{\mathrm{fo}} = (x_{\mathrm{fo}}\gamma_{\mathrm{fo}})^2 = p_{\mathrm{Mg}}^2 p_{\mathrm{Si}} p_{\mathrm{O}}^4 \, e^{-\Delta G_{\mathrm{fo}}/RT} \tag{33}$$

$$a_{\mathrm{fa}} = (x_{\mathrm{fa}}\gamma_{\mathrm{fa}})^2 = p_{\mathrm{Fe}}^2 p_{\mathrm{Si}} p_{\mathrm{O}}^4 \, e^{-\Delta G_{\mathrm{fa}}/RT} \tag{34}$$

$$a_{\mathrm{en}} = x_{\mathrm{en}}\gamma_{\mathrm{en}} = p_{\mathrm{Mg}} p_{\mathrm{Si}} p_{\mathrm{O}}^3 \, e^{-\Delta G_{\mathrm{en}}/RT} \tag{35}$$

$$a_{\mathrm{fs}} = x_{\mathrm{fs}}\gamma_{\mathrm{fs}} = p_{\mathrm{Fe}} p_{\mathrm{Si}} p_{\mathrm{O}}^3 \, e^{-\Delta G_{\mathrm{fs}}/RT} \tag{36}$$

ΔG_{fo}, ΔG_{fa}, ΔG_{en}, and ΔG_{fs} are the partial free enthalpies of the formation of one mole of the pure components forsterite, fayalite, enstatite, and ferrosilite from free atoms, respectively. The γ's are the activity coefficients, which for most purposes can be set to unity. Additionally one have the mass action laws for the formation of nickel-iron as just before since at higher pressures solid iron coexists with the silicates (cf. Fig. 4).

At the pressures and temperatures of interest where condensed silicates exist all H is associated to H_2, all C is bound in CO, nearly all Si not bound in solids

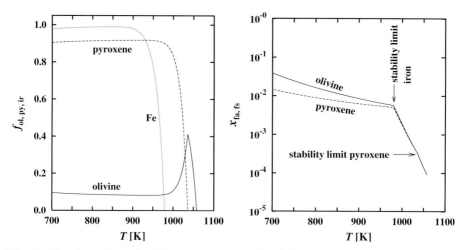

Fig. 7. Condensation of olivine, pyroxene and solid iron at total pressure $P = 10^{-10}$ bar, representative for the condensation zone in stellar winds. *Left part:* Fraction of the Si condensed into olivine and pyroxene and fraction of Fe condensed into solid iron. *Right part:* Mole fractions x_{fa} and x_{fs} of fayalite and ferrosilite in olivine and pyroxene, respectively

is bound in SiO, and all O not bound in CO, SiO and solids is bound in H_2O. The set of stoichiometric equations (19) in its simplest form for the condensation of olivine, pyroxene and solid iron is

$$\left(\epsilon_O - \epsilon_C - (4f_{ol} + 3f_{py})\,\epsilon_{Si}\right) P_H \qquad = p_{H_2O} + p_{SiO} \qquad (37)$$

$$\left(\epsilon_{Mg} - \left[2(1 - x_{fa})f_{ol} + (1 - x_{fs})f_{py}\right]\epsilon_{Si}\right) P_H = p_{Mg} \qquad (38)$$

$$\left(\epsilon_{Fe}\,(1 - f_{ir}) - \left[2x_{fa}f_{ol} + x_{fs}f_{py}\right]\epsilon_{Si}\right)P_H \qquad = p_{Fe} \qquad (39)$$

$$\epsilon_{Si}\left(1 - f_{ol} - f_{py}\right) P_H \qquad = p_{SiO}\,. \qquad (40)$$

The equilibrium conditions are

$$(1 - x_{fa})^2 = p_{Mg}^2\, p_{Si}\, p_O^4\, e^{-\Delta G(fo)/RT} \qquad (41)$$

$$x_{fa}^2 \qquad = p_{Fe}^2\, p_{Si}\, p_O^4\, e^{-\Delta G(fa)/RT}\,. \qquad (42)$$

$$1 - x_{fs} \qquad = p_{Mg}\, p_{Si}\, p_O^3\, e^{-\Delta G(en)/RT} \qquad (43)$$

$$x_{fs} \qquad = p_{Fe}\, p_{Si}\, p_O^3\, e^{-\Delta G(fs)/RT}\,. \qquad (44)$$

$$1 \qquad = p_{Fe}\, e^{-\Delta G(ir)/RT}\,. \qquad (45)$$

Here f_{ol} denotes the fraction of Si condensed into olivine, x_{fa} the mole fraction of fayalite Fe_2SiO_4 in the solid solution $Mg_{2(1-x_{fa})}Fe_{2x_{fa}}SiO_4$, f_{py} the fraction of Si condensed into pyroxene, x_{fs} the mole fraction of ferrosilite $FeSiO_3$ in the solid solution $Mg_{1-x_{fs}}Fe_{x_{fs}}SiO_3$, and f_{ir} the fraction of Fe condensed into solid iron. Additionally one has the laws of mass action for the molecules H_2O and

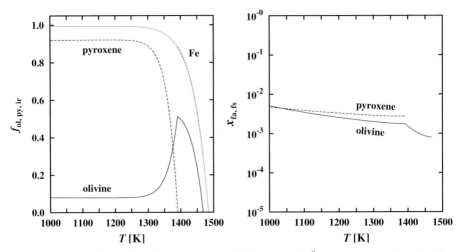

Fig. 8. Same as Fig. 7, but for a pressure of $P = 2 \times 10^{-3}$ bar representative for the dust vaporization zone in protoplanetary accretion discs

SiO and Eq. (32) for the total pressure. An extension of the system to a more complex gas phase chemistry is straight forward.

Equations (37) ... (45) form a set of nine equations for the nine unknowns p_O, p_{fe}, p_{Mg}, p_{Si}, f_{ol}, x_{fa}, f_{py}, x_{fs}, and f_{ir} which may be solved for instance by Newton-Raphson iteration, or one uses for the solution of the system one of the multi-purpose methods described in Smith & Missen [115]. The solution is subject to the restrictions that

$$0 \le f_{ol} \le 1, \quad 0 \le f_{py} \le 1, \quad 0 \le f_{ir} \le 1. \tag{46}$$

Representative results for the composition of olivine and pyroxene are shown in Fig. 7 for conditions encountered in the condensation zone of circumstellar dust shells, and in Fig. 8 for conditions in the dust vaporization region in protoplanetary accretion discs. The iron content of the silicates is very low for temperatures $T > 700$ K in chemical equilibrium, in fact they are essentially pure forsterite and enstatite. At much lower temperatures ($T \lesssim 600$ K) the iron content of the silicates in a chemical equilibrium mixture rapidly increases with decreasing temperature, but equilibration is prevented in circumstellar shells and protoplanetary discs at such low temperatures by slow cation diffusion (cf. Sect. 5.2).

4   Dust Growth Processes

4.1   Characteristic Scales

The evaporation and growth of an ensemble of dust grains is determined by two basically different processes: By the processes which remove or add material from the gas phase from or to the surface of a grain, and by transport processes in the gas phase by which material for the growth of particles is supplied or the material evaporated from grains is carried away. They are characterized by a number of length and time scales and dimensionless numbers.

Characteristic Length Scales. There are three different length scales for the problem. The first one is the mean free path length of the gas particles with respect to collisions with H_2

$$\lambda_p = 1/n_p\sigma_p . \tag{47}$$

σ_p is an average collision cross-section for gas phase species which for order of magnitude estimates may be approximated by $\sigma_p = 10^{-15}\,\mathrm{cm}^2$. n_p is the number density of H_2. The second important length scale is the particle size a. From a and λ_p one defines the dimensionless Knudsen number

$$Kn = \lambda_p/a . \tag{48}$$

$Kn \to 0$ or $Kn \to \infty$ correspond to the limit cases where either the interaction of a grain with the ambient gas can be treated in the hydrodynamic limit or has to be treated by the methods of the theory of rarefied gas dynamics, respectively. Numerically we have

$$Kn = 10^9 \; \left(10^{10}\mathrm{cm}^{-3}/n_p\right) \left(1\,\mu\mathrm{m}/a\right) . \tag{49}$$

In the dust formation zone of circumstellar shells we have $n_p \approx n_{H_2} = 10^{10} \dots 10^5$ and $a < 1\,\mu\mathrm{m}$. The Knudsen number always is very large in this case. In protoplanetary accretion discs we have $n_p \approx 10^{15} \dots 10^{13}$ in the regime between the region of dust evaporation and ice evaporation. The Knudsen number satisfy $Kn \gg 1$ only for small particles with $a \lesssim 1\,\mathrm{mm}$.

The third important length scale is the inter-grain distance

$$d_{\mathrm{gr}} = 1/n_{\mathrm{gr}}^{1/3} , \tag{50}$$

where n_{gr} is the particle density of grains. From d_{gr} and λ_p we define a second dimensionless Knudsen number

$$Kn_{\mathrm{gr}} = \lambda_p/d_{\mathrm{gr}} . \tag{51}$$

The limit cases $Kn_{\mathrm{gr}} \to 0$ or $Kn_{\mathrm{gr}} \to \infty$ correspond to the cases where either one has only one grain within the sphere with radius λ_p around this grain (case B in Fig. 9) or there are a lot of other grains within this sphere of influence (case A in Fig. 9), respectively. Either the grains are well isolated from each other and only weakly interact by the slow transport processes (e.g. for matter, momentum) via the gas phase (case B), or the particles strongly interact by means of direct particle exchange (case A), respectively. Figure 9 illustrates the two limit cases.

Obviously, in case A all grains "see" changes in the gas phase concentration of species consumed or liberated by particle growth or evaporation, respectively,

Case A   **Case B**

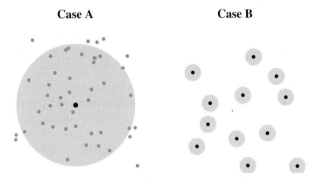

Fig. 9. The two limit cases for grain growth and evaporation. Case A: $Kn_{gr} \gg 1$, the sphere of influence (indicated by a gray circle) with radius λ around one grain (indicated by the big black dot) contains many other dust grains. Case B: $Kn_{gr} \ll 1$, the spheres of influence of the different grains are well separated

in the same way and immediately. There are no local concentration reductions or enhancements around a particle. In case B, however, growth or evaporation results in locally reduced or increased concentrations of growth species around a dust grain, which are only slowly adjusted to the average concentrations in the gas phase by the slow diffusion process.

For an order of magnitude estimate we assume the dust grains to have uniform sizes a and to consume all atoms of some element with abundance ϵ (by number, relative to H). Then we have

$$\tfrac{4\pi}{3} a^3 n_{gr} = V_0 \epsilon N_H . \tag{52}$$

The quantity V_0 denotes the volume of one formula unit of the condensate in the solid

$$V_0 = A m_H / \rho_{gr} , \tag{53}$$

where A is the atomic weight of one formula unit and ρ_{gr} the mass density of the solid. N_H is the fictitious number density of H nuclei, if all hydrogen atoms are present as fee atoms. At temperatures where dust exists, we have $N_H = 2n_{H_2}$. It follows

$$Kn_{gr} = \frac{1}{a n_p^{2/3} \sigma_p} \left(\frac{3V_0 \epsilon}{2\pi} \right)^{1/3} . \tag{54}$$

Using numbers for olivine grains from Tab. 1 and 3 we obtain numerically

$$Kn_{gr} = 2.3 \times 10^3 \ (1\,\mu m/a) \left(10^{10}\,cm^{-3}/n_p \right)^{2/3} . \tag{55}$$

In the growth zone of circumstellar shells we usually have $n_p < 10^{10}\,cm^{-3}$ and $a < 1\,\mu m$, hence $Kn_{gr} \gg 1$, which corresponds to case A in Fig. 9. In protoplanetary discs, however, particle densities are high ($10^{13} \ldots 10^{15}\,cm^{-3}$) and dust particles may become much bigger than $1\,\mu m$. The limit case $Kn_{gr} \gg 1$, case A in Fig. 9, is valid in protoplanetary discs only if grains remain smaller than about $1\,\mu m$, for bigger grains case B of Fig. 9 applies.

Characteristic Time Scales. There is a characteristic timescale associated with the diffusion of gas phase species through the gas phase

$$t_{\mathrm{diff}} = \lambda_p^2/D \qquad (56)$$

where D is the diffusion coefficient which can be approximated by $D \approx \lambda_p v_{p,\mathrm{th}}$, where $v_{p,\mathrm{th}} = (kT/m_p)^{1/2}$. This time is of the order of magnitude of the time required for the development of a stationary sphere of concentration enhancement or reduction by particle evaporation or growth, respectively, around a particle at rest with respect to the gas. For longer times, particle transport towards or away from a grain is determined by a stationary diffusion current. Numerically we have

$$t_{\mathrm{diff}} = 0.5 \ \left(10^{10}\,\mathrm{cm}^{-3}/n_p\right) (1000\,\mathrm{K}/T) \ [\mathrm{s}] \,. \qquad (57)$$

In many cases external forces drive a motion of dust grains through the gas with a drift velocity v_{dr}. The characteristic time scale associated with this process is

$$t_{\mathrm{dr}} = \lambda_p/v_{\mathrm{dr}} \,. \qquad (58)$$

From the two time scales t_{diff} and t_{dr} one defines the dimensionless Mach number

$$Ma = t_{\mathrm{diff}}/t_{\mathrm{dr}} = v_{\mathrm{dr}}/v_{p,\mathrm{th}} \,. \qquad (59)$$

In the limit case $Ma \to 0$ the drift of the grain through the gas is so slow that diffusion of vapour species towards or away from the grain relax to a stationary state. In the opposite limit $Ma \to \infty$ vapour transport towards or away from the grain is determined by the motion of the grain through the gas. For circumstellar shells one has $Ma \gtrsim 1$ (except for the smallest, just nucleated grains) but in this case also case A of Fig. 9 holds, in which case diffusion anyhow does not play any role. For protoplanetary discs the grain drift velocity relative to the gas remains smaller than $v_{p,\mathrm{th}}$ for grain sizes $a \lesssim 10\,\mathrm{cm}$, cf. Fig. 3 of Weidenschilling & Cuzzi [127], for instance. Thus, for grain sizes between $\approx 1\,\mu\mathrm{m}$ and $\approx 10\,\mathrm{cm}$ the transport of vapor species during growth or evaporation towards or away from a particle may be regulated by diffusion through the gas phase. Bigger particles are ventilated by their motion through the gas and diffusion is unimportant.

4.2   Growth of Dust Grains

The simplest growth process for dust grains is that where the chemical composition of the condensate is the same as that of the vapour, as for instance in the case of the condensation of iron atoms into solid iron or of H_2O molecules into ice. This is *homomolecular* growth, in contrast to *heteromolecular* growth where the vapour has a different chemical composition from that of the solid. Homomolecular growth simply requires a step by step addition of the basic molecule to the solid, heteromolecular growth necessarily is a complicated process which requires the reaction of several gas phase species at the surface of the solid to form the basic building block of the condensate. From the abundant cosmic minerals only iron forms by homomolecular growth. All silicate compounds and all aluminium and calcium compounds decompose on evaporation and, thus, grow by heteromolecular growth processes.

Homomolecular Growth. We start with homomolecular growth as the most simple case and assume that the growth is not diffusion limited and the particles are at rest with respect to the gas. The rate at which particles hit the surface per unit area and time according to a classical result of the kinetical theory of gases is

$$R_{\text{coll}} = n\sqrt{kT/2\pi m}\,. \tag{60}$$

n is the particle density of the growth species and m its mass. If α denotes the probability that a collision results in the deposition of the growth species onto the grain, the growth rate per unit surface area is given by

$$J^{\text{gr}} = \alpha R_{\text{coll}} = \alpha n\sqrt{kT/2\pi m}\,. \tag{61}$$

The growth or sticking coefficient α has to be determined by laboratory experiments.

The evaporation rate of a grain is given by

$$J^{\text{vap}} = \alpha' R_{\text{coll}} = \alpha' n'\sqrt{kT_{\text{d}}/2\pi m} \tag{62}$$

with an evaporation coefficient α'. n' refers to the particle density of the growth species in chemical equilibrium between the gas phase and the condensate and is given by the vapour pressure p_{eq} as

$$n' = p_{\text{eq}}/kT_{\text{d}}\,. \tag{63}$$

From detailed balancing arguments one finds $\alpha = \alpha'$, but for non-equilibrium growth this does not necessarily hold. Since usually only one, if ever, of these coefficients is known one alway assumes $\alpha = \alpha'$. Note that the temperature T_{d} in (62), the internal temperature of the grains, needs not to be the same as the temperature T in (61), the kinetic temperature of the gas phase.

The number N of basic building blocks forming a grain increases with time as

$$\frac{d\,N}{d\,t} = \oint_{\mathcal{A}} dA\,(J^{\text{gr}} - J^{\text{vap}})\,. \tag{64}$$

The integration is over the surface of a grain. For particles at rest with respect to the gas the growth rate J^{gr} does not depend on the local position and orientation of the grain surface and can be taken out of the integral. The rates J^{vap}, J^{gr} usually are different for different crystal facets. The present state of the theory of cosmic dust does not allow to treat the detailed growth of crystals with their specific crystal shapes. Thus, one introduces average evaporation and growth rates and writes

$$\frac{d\,N}{d\,t} = \mathcal{A}\,(J^{\text{gr}} - J^{\text{vap}})\,, \tag{65}$$

where $\mathcal{A}$ is the total surface area of the grain. If the grains are spherical with radius a then $NV_0 = \frac{4\pi}{3}a^3$ and $\mathcal{A} = 4\pi a^2$. V_0 is defined by (53). Then

$$\frac{d\,a}{d\,t} = V_0\,(J^{\text{gr}} - J^{\text{vap}})\,. \tag{66}$$

If the growth and evaporation rate are determined by (61) and (62) the equation takes the form

$$\frac{da}{dt} = V_0 \alpha \sqrt{\frac{kT}{2\pi m}} \left(n - \frac{p_{eq}}{kT}\right) = V_0 \alpha \left(p - p_{eq}\right)/\sqrt{2\pi m kT} \qquad (67)$$

which is a variant of the Hertz-Knudsen equation.

Heteromolecular Growth. Most substances of astrophysical interest grow by heteromolecular processes since the nominal molecule corresponding to the chemical formula of the solid does not exist. The magnesium silicate forsterite Mg_2SiO_4, for instance, in a hydrogen rich environment requires for its formation a net reaction of the kind

$$2Mg + SiO + 3H_2O \longrightarrow Mg_2SiO_4(s) + 3H_2O, \qquad (68)$$

since Mg, SiO and H_2O are the only abundant species in the gas phase from which forsterite can be formed. In such cases there often exist a *rate determining reaction* which rules the whole growth process. For forsterite experimental results (see below) indicate that the addition of SiO from the gas phase determines the rate of forsterite growth. In such cases (67) also describes the growth of dust grains by heteromolecular growth where p refers to the gas phase partial pressure of the least abundant molecule involved in the rate determining reaction and p_{eq} is its partial pressure in the vapour.

Consider the formation of some solid D from gas phase species A, B, C, ...

$$iA + jB + kC + \ldots \longrightarrow lD + mE + nF + \ldots \qquad (69)$$

where E, F, ... are some reaction products, and let the reaction with gas phase species A be the rate determining step for formation of D. The activity of solid D according to (18) is

$$a_{sol} = \frac{p_A^i p_B^j p_C^k \cdots}{p_E^m p_F^n \cdots} e^{\Delta G/RT}. \qquad (70)$$

For the equilibrium pressure p_{eq} in (67) one has to take the value of p_A calculated from (70) by letting $a_{sol} = 1$. Then $p_{eq}/p = p_{A,eq}/p_A = 1/a_{sol}^{1/i}$ and (67) changes into

$$\frac{da}{dt} = V_0 \alpha \sqrt{\frac{kT}{2\pi m}} \left(n - \frac{p_{eq}}{kT}\right) = V_0 \alpha p \left(1 - \frac{1}{a_{sol}^{1/i}}\right)/\sqrt{2\pi m kT} \qquad (71)$$

This shows that the activity a_{sol}, calculated according to (70), rules the growth or the evaporation of the solid in a non-equilibrium mixture of gases and solids: If $a > 1$ the solid condenses, if $a < 1$ the solid evaporates. Calculations of activities of possible candidates for condensation like that shown in Fig. 3 help to find out the possible non-equilibrium condensates.

For homomolecular growth the activity a_{sol} in (71) is equivalent to the conventional supersaturation ratio.

Table 3. Some quantities important for calculating dust growth or evaporation for abundant dust components. Data for ρ and A from the CRC-handbook (Lide [69]). For α see text

Dust species	ρ g cm^{-3}	A	V_0 cm^3	α
quartz	2.648	60.085	3.80×10^{-23}	0.05
iron	7.87	55.845	1.19×10^{-23}	0.9
forsterite	3.21	140.694	7.33×10^{-23}	0.1
fayalite	4.30	203.774		
enstatite	3.19	100.389	5.26×10^{-23}	
ferrosilite	4.00	131.93		
olivine			7.55×10^{-23}	0.1
pyroxene			5.35×10^{-23}	0.1

If the transport of gas phase material to and from the grain is limited by diffusion (case B in Fig. 9), the local concentration at the surface of a particle is fixed by the vapour pressure. One has to solve the stationary diffusion problem for this case in order to determine the rate at which particles arrive at or leave from the surface of the grain. This problem is discussed, for instance, in textbooks on cloud physics (e.g. Pruppacher & Klett [97]).

Laboratory Condensation Experiments. Experimental investigations of condensation of cosmic dust analogs have been performed for Mg-Fe-Si condensates by Rietmeijer & Nuth [99], for Fe-Si condensates, by Rietmeijer et al. [100], and for Al-Si-condensation by Rietmeijer & Karner [101]. These give some hints on the composition and structure of grains forming in cosmic element mixtures.

4.3   Condensation Coefficients

The evaporation or condensation coefficients α have to be determined from laboratory experiments. A number of such experiments have been performed during the last decade and α is now known for the abundant dust minerals, though with an only moderate accuracy in most cases.

Olivine: This material evaporates congruently, which means that the residue has the same composition as the starting material (Nagahara et al. [85]), only the last residues contain some fraction of silica SiO_2 (Nagahara et al. [84]). The evaporation of forsterite in vacuum has been studied by Hashimoto [50]. The evaporation coefficient α derived by Hashimoto for the evaporation reaction

$$Mg_2SiO_4(s) \longrightarrow 2Mg + SiO + O_2 + O \tag{72}$$

is between 0.09 and 0.16 in the experimental temperature regime $1\,600 \ldots 1890°$C. Similar results are obtained in Wang et al. [125]. The composition of the vapour of forsterite has been determined by mass spectrometry of the outflow from a Knudsen cell by Nichols & Wasserburg [89] and from a Langmuir configuration by Nichols et al. [90]. In both cases the composition of the vapour is found to be in accord with the evaporation reaction (72). A new determination of the condensation coefficient by Inaba et al. [54] gave $\alpha = 0.18$ at $1\,700°C$. In all experiments it is assumed that there is no difference between the condensation and evaporation coefficient. A typical value of the condensation/evaporation coefficient is 0.1 which is given in Table 3.

Evaporation of olivine in the presence of hydrogen is studied by Nagahara & Ozawa [86, 87] for pressures of H_2 between 10^{-9} and 10^{-3} bar at a temperature of $1\,700°C$. Below 10^{-6} bar the evaporation rate is found to be independent of the H_2 pressure and similar to the vacuum evaporation rate, above 10^{-6} bar the evaporation rate increases with H_2 pressure (in bar) as

$$J^{\mathrm{vap}} = (1.72\,P^{1.19} + 9.87) \times 10^{-7} \tag{73}$$

in units $\mathrm{g\,cm^{-2}\,s^{-1}}$. They interpreted their results as a superposition of free evaporation according to (72) at low pressures and chemisputtering with H_2 according to the reaction

$$\mathrm{Mg_2SiO_4(s)} + 3\mathrm{H_2} \longrightarrow 2\mathrm{Mg} + \mathrm{SiO} + 3\mathrm{H_2O} \tag{74}$$

at higher pressures. Based on this reaction they determined a condensation coefficient α. At low H_2 pressures α roughly is constant $\alpha = 0.06$. At higher pressure it increases up to about 0.2. A typical value is $\alpha = 0.1$, the same value as determined from vacuum evaporation. The results of Nagahara & Ozawa [86, 87] are criticized by Hashimoto [51].

The evaporation of forsterite in protoplanetary discs and the relation between vacuum evaporation and evaporation in H_2 gas has been discussed by Tsuchiyama et al. [124].

Pyroxene: Enstatite evaporates incongruently with forsterite as residue forming a surface layer on enstatite (Tachibana et al. [119, 120]). Tachibana et al. [120] find that during the early phases the thickness of the forsterite evaporation residual layer grows linear with time, implying that the initial forsterite layer is porous and the evaporation is controlled by surface reactions. During the later phases of evaporation a constant thickness layer evolves which is diffusion controlled. The layer thickness at the transition from surface reaction to diffusion controlled growth of the forsterite residual layer is not reported, but can be estimated from the figures to occur at a layer thickness of about $5\,\mu m$ thickness. Below this layer thickness the growth rate of the forsterite residual layer thickness is given by (Tachibana et al. [120])

$$k = 6.9 \times 10^2\,e^{-43540/T}\ \mathrm{cm\,s^{-1}}\,. \tag{75}$$

Note that part of the forsterite layer first formed from enstatite later evaporates again, i.e. the true increase of layer thickness is given by the difference between the increase by forsterite formation, given by 75) and the decrease by forsterite evaporation.

A detailed study on enstatite evaporation in vacuum and in H_2 gas has been performed by Tachibana et al. [121].

Silica: The evaporation coefficients for SiO_2 reported in the literature are small. Landolt-Börnstein [63] gives 0.02 to 0.04 in the temperature region 1429 ... 1460°C. Hashimoto [50] finds an evaporation coefficient of 0.01 at abut 1 600°C. All these experiments are conducted at much higher temperatures than those encountered in circumstellar shells or protoplanetary discs in the region of dust formation or evaporation. The measurements of Mendybaev et al. [73] on the kinetics of SiO_2 evaporation are conducted at lower temperatures an in the presence of H_2. Ferrarotti & Gail [31] derived a value of $\alpha = 0.07$ at 1 000 K from the results. A value of $\alpha = 0.05$ is probably a typical value for the condensation/evaporation coefficient of SiO_2. This is the value given in Table 3.

Iron: The evaporation coefficient of iron is 0.8 ... 1.0 in the temperature region 1020 ... 1440°C (Landolt-Börnstein [63]). In a new determination Tachibana et al. [118] obtained $\alpha = 0.83 ... 0.92$ at temperatures between 1170 and 1360°C. A typical value of the condensation/evaporation coefficient of iron is 0.9 which is given in Table 3.

4.4   Nucleation Theory

Dust formation in many cases is a two step process which starts with the formation of tiny seed nuclei from the gas phase and proceeds by growth to macroscopic sized grains via precipitation of condensible material from the gas phase onto the seed nuclei. The seed nuclei need not to be formed from the same material as the grains that finally emerge from the process. The formal theory of nucleation and its application to astrophysical problems is discussed e.g. in Salpeter [102], Draine [25], Gail & Sedlmayr [37], Patzer et al. [94], Yamamoto et al. [131]. The main problem in astrophysical problems is to determine the nature of the seed nuclei, since this determines the onset of dust formation.

For the problem of dust formation in carbon rich environments the rich data bases developed for the purposes of modeling terrestrial flames allowed a modeling of the basic mechanisms of carbon dust nucleation. The first attempt to model carbon dust nucleation from the gas phase by Keller [57] was followed by the more detailed treatments of Frenklach & Feigelson [34] and especially by Cherchneff et al. [21]. The investigations of presolar dust grains (Bernatowicz et. al [5]) have shown that carbon dust in some (but not all) cases grows on seed nuclei of metal carbides like TiC, ZrC, MoC or solid solutions of these carbides. An overview over the problem is given by Sedlmayr & Krüger [112]. Recent major progress is the calculation of nucleation in WC stars by Cherchneff et al. [22] and in supernova ejecta by Clayton et al. [23].

Dust nucleation in oxygen rich environments remains an unsolved problem up to now. The present state with respect to nucleation is described in Gail & Sedlmayr [39, 40]. The possible candidates for nucleation seem to be clusters of the types Al_nO_m and Ti_nO_m and at high pressures possibly Fe_n, but no definite conclusion with respect to the nature of the nucleation species is possible at present since the data with respect to cluster structures, bond energies, vibrational frequencies, and reaction rates required for calculating the cluster formation process are not yet available. These have to be determined by quantum mechanical calculations. Some progress for Ti_nO_m clusters has recently been made by Jeong et al. [55] and for Al_nO_m clusters by Chang et al. [18, 19], Patzer et al. [93] and more calculations are currently under way.

5   Dust Processing

For dust formed in oxygen rich environments there arise some complications in determining the composition and properties of the dust mixture because some of the oxides form solid solutions with variable compositions which may change their composition by solid diffusion or they may undergo solid-state chemical reactions. Additionally the optical properties of amorphous and crystalline oxides are considerably different and one has to determine the degree of crystallinity of such materials. For dust material formed in carbon rich or other chemically diverse environments there seems to be presently no need for considering such complications.

5.1   Annealing

The dust formed under astrophysical conditions in most cases has an amorphous lattice structure, where the basic building blocks of the minerals, e.g. the SiO_4–tetrahedra in the silicates, show a disordered arrangement in the solid. In some cases, however, if the material is heated for sufficiently long times to high temperatures, internal rearrangement processes are activated by which the material evolves into the regular crystalline lattice structure (annealing). For instance, annealing is thought to be responsible for the observed crystalline dust component in cometary dust or in meteoritic matrix material (see Sect. 7.7).

Annealing of Amorphous Olivine and Orthopyroxene. Annealing has been studied in the laboratory for the most important dust components formed in an oxygen rich environment by Hallenbeck, Nuth and Daukantas [48], Brucato et al. [13], Hallenbeck and Nuth [49], Fabian et al. [28], Thompson and Tang [122].

Fabian et al. [28] studied annealing of enstatite glass, of smokes obtained by laser ablation of forsterite and enstatite targets, and of nano-sized silica particles. The materials have been annealed at temperatures between 1 000 K and 1 121 K for different durations. The evolution of spectral IR absorption features associated with the development of a crystalline structure have been monitored and the characteristic annealing time τ was determined. From this, the activation energy E_a for annealing was derived from

$$\tau^{-1} = \nu_{\text{vib}} \, e^{-E_a/kT} \, . \tag{76}$$

Table 4. Activation energies for annealing of some silicate dust components

Mineral	Structure	E_a K	Temperature range [K]	Ref.
forsterite	smoke	39 100	1000-1120	[28]
enstatite	smoke	42 040	1000-1120	[28]
enstatite	glass	40 100...42 300	1000-1120	[28]
quartz	smoke	49 000	1000-1120	[28]
pyroxene	smoke	47 300	1070	[13]

ν_{vib} is a characteristic vibration frequency, which is taken to be $\nu_{\text{vib}} = 2 \times 10^{13}\,\text{s}^{-1}$ (Lenzuni et al. [68]). The results obtained for E_a show a marked scatter and some temperature dependence. Average values given by Fabian et al. [28] are shown in Table 4.

Brucato et al. [13] formed amorphous pyroxene grains by laser ablation of natural enstatite and the structural changes during annealing are monitored by the evolution of the IR spectrum. The activation energy for annealing, again, is determined from relation (76). Converting to the same ν_{vib} as used in Fabian et al. [28], they obtained an activation energy $E_a/k = 47\,300\,\text{K}$, significantly higher than that of Fabian et al. [28]. The difference is probably due to the different composition and structure of the samples.

Hallenbeck et al. [48] produced amorphous magnesium silicate smokes by vapour condensation. The dust has been annealed at temperatures between $1\,000\,\text{K}$ and $1\,300\,\text{K}$ and the IR spectra were monitored as function of the annealing time. Contrary to the experiments of Fabian et al. [28] and Brucato et al. [13] they found after initial rapid annealing a stall phase without significant change of the absorption spectrum followed later by further evolution towards crystallinity. Such a stall phase has also been detected in the annealing experiment by Thompson and Tang [122], who prepared their samples as precipitates from reaction of $MgCl_2$ and Na_2SiO_2 in water. Probably the different annealing behaviour is due to the different composition and structure of the samples. Activation energies for annealing have not been derived by Hallenbeck et al. [48], so unfortunately their results can not directly be used in model calculations for the evolution of the dust properties.

Annealing in protoplanetary discs occurs at lower temperatures than that for which laboratory studies are performed because of the much longer timescales available. Application of the laboratory data to astrophysical problems requires extrapolation to much lower temperatures. Whether this is admitted or not is not known.

Growth of Crystallized Regions. Crystallization usually is a two step process which starts with the formation of growth centres and then proceeds by a local rearrangement of the lattice building blocks at the border between the

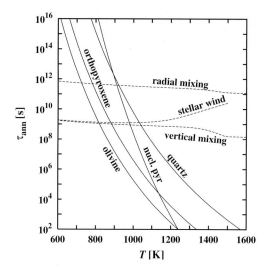

Fig. 10. Characteristic time scales for annealing of a $0.1\,\mu m$ dust grain and time required for formation of growth centres in enstatite (*Full lines*). Data for the activation energy from Fabian et al. [28]. For comparison, the time scales for radial and vertical transport processes (mixing) in protoplanetary discs and the hydrodynamic timescale r/v for a stationary stellar wind model are shown as dashed lines

crystallized and non-crystalline region, by which the crystallized region expands into its amorphous environment.

The growth of a crystallized region starting from some growth centre, can be described in the most simple case by an equation of the type

$$\frac{d V_{cr}}{d t} = 6 V_0^{\frac{1}{3}} V_{cr}^{\frac{2}{3}} \nu_{vib} e^{-E_a/kT} \qquad (77)$$

(Gail [35]). Similar equations have been used by Kouchi et al. [59] and Sogawa and Kozasa [116]. V_{cr} is the volume of the crystallized region. V_0 is defined by (53). The characteristic timescale for the crystallization can be defined by

$$\tau_{cry} = V_{cr} \, |dV_{cr}/dt|^{-1} \,. \qquad (78)$$

The run of τ_{cry} for the silicate dust species olivine, orthopyroxene, and quartz are shown in Fig. 10.

τ_{cry} becomes shorter than the timescale for radial mixing and transport in protoplanetary discs at about $800\,K$ for olivine and orthopyroxene. Above this temperature these two dust species rapidly crystallize, if crystallization starts on pre-formed growth centres. Quartz crystallizes for $T \gtrsim 900\,K$.

In stellar winds τ_{cry} becomes shorter than the timescale for cooling of the outflowing gas between about $900\,K$ and $850\,K$. Dust formed above about $900\,K$ will have a crystalline structure because any initial disorder of whatsoever origin will rapidly be annealed. Dust formed below about $800\,K$ will be amorphous if the initial growth process results in a disordered material, because the disorder in the lattice structure later cannot be healed up.

If growth of crystalline regions within a dust grain starts from more than one growth centre, then the growth regions from some instant on come into contact. Growth then only proceeds into the remaining amorphous part of the grain volume and there result different crystallized regions in the grain. This

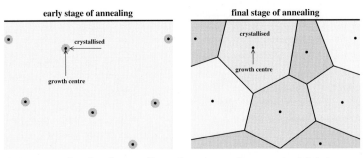

Fig. 11. Schematic sketch of annealing of an amorphous material into a crystalline material. It is assumed that growth centres form at random inside the volume. *Left:* Early stage, where only small crystallized regions have grown around the centres. *Right:* Late stage, when the whole volume is crystallized

situation is sketched in Fig. 11. The simple growth equation (77) then cannot be used for the late phases of crystallization. Also, if growth centres preferentially form at surfaces, as it was observed by Fabian et al. [28], the growth cannot be treated in the simple approximation (77). The average degree of crystallization x_{cr} of a grain can approximately be determined from the solution of (77) by

$$x_{cr} = \min\left(V_{cr}(t)\zeta_{cr}/V_0\,,\ 1\right)\,. \tag{79}$$

Here ζ_{cr} denotes the number of growth centres per Si nucleus. According to this definition of ζ_{cr} the quantity V_0/ζ_{cr} is the average volume of the amorphous material available for each growth centre, into which the crystallized region may expand (Gail [36]).

The growth centres may already pre-exist in the amorphous material or they have to be formed by a nucleation process. The nucleation of growth centres for crystallization of circumstellar dust on the basis of classical nucleation theory was studied by Kouchi et al. [59] and by Sogawa & Kozasa [116], Kozasa & Sogawa [60].

Some experimental hints on the formation of seed nuclei for crystallization in enstatite can be found in Fig. 2 of Fabian et al. [28], from which one reads off an induction time required for the formation of seed nuclei of about 235 hours at $1\,080\,\mathrm{K}$ and about 18 hours at $1\,121\,\mathrm{K}$. This can be fitted by

$$\tau_{nuc} = 2.62 \times 10^{-25}\,\mathrm{e}^{75\,870/T}\,\mathrm{s}\,. \tag{80}$$

The time required for the formation of crystallization centres in amorphous enstatite is shown in Fig. 10. If crystallization requires the nucleation of growth centres, the temperature required for onset of crystallization in protoplanetary discs will be shifted from $800\,\mathrm{K}$ to about $900\,\mathrm{K}$.

5.2 Solid Diffusion

Diffusion is a transport of matter within a mixture (solution) driven by concentration gradients. The temperature dependence of the diffusion coefficient can

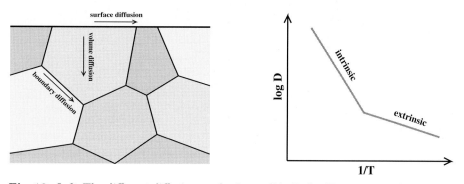

Fig. 12. *Left:* The different diffusion modes in a solid. *Right:* Temperature dependence of diffusion data showing extrinsic and intrinsic regimes

usually be approximated by the Arrhenius form

$$D = D_0 \, e^{-Q/RT} \, . \tag{81}$$

R is the gas constant. The frequency factor D_0 and the activation energy Q have to be determined for each meterial by laboratory experiments.

Solid diffusion of the different constituents within the dust minerals is important because it determines to a large extent the composition of growing or vaporizing dust species, the rate of solid chemical reactions, and the existence and extent of internal composition gradients within dust grains. Counterdiffusion of the main components forming a substance, which changes its local composition, is called *interdiffusion*. Generally interdiffusion of species in a composite material tends to homogenize any internal composition gradients within grains, but diffusion in solids involves significant activation energy barriers and does not operate, or operates only incompletely, at low temperatures.

Diffusive transport of matter in solids may occur by three major path's: by *volume diffusion* through the grain lattice, by *surface diffusion* across a grain surface, and by *grain boundary diffusion*. The slowest process is volume diffusion. It requires the existence of point defects in the solid (e.g. vacant lattice sites or interstitial atoms). Diffusion along grain-grain boundaries, if such exist like in fine-grained polycrystalline material, is much easier. It requires the existence of line or surface defects like e.g. dislocations or internal grain boundaries. Matter transport across free surfaces or via the gas phase usually is fast. Here we consider volume diffusion, since this is responsible for homogenization of composition differences within microcrystallites and for solid reactions.

A brief general discussion of the basic principles of solid diffusion in silicate minerals and glasses is given by Freer [33]. Extensive compilations of data for diffusion in minerals are given by Freer [33] and Brady [11].

Intrinsic and Extrinsic Diffusion. The lattice defects required for solid diffusion have two significantly different contributions: either they are permanently present resulting from impurity ions, dislocations, grain boundaries and so on, or

Table 5. Diffusion coefficients for cation interdiffusion for some important dust materials: $D = D_0 \exp\{-T_0/T - \alpha x\}$

Mineral	Cations	D_0 cm^2/s	T_0 K	α	Temperature range [K]	Ref.
olivine	Fe-Mg	2.0×10^{-2}	27 180	6.91	1250-1570	[17]
pyroxene	Fe-Mg	1.1	43 300	—	1020-1170	[109]
melilite	Al-Si	$9.3 \times 10^{+1}$	50 510	3.34	1470-1620	[88]

they are created spontaneously as pairs of vacancies and interstitial ions by thermal excitation at high temperatures. Diffusion mediated by this kind of defects usually requires higher activation energies than diffusion mediated by impurity ions and other pre-existing defects, since additionally to the activation energy for migration the activation energy for defect creation has to be overcome. Hence, there exist two regions with different temperature dependencies and quite different values of the diffusion coefficient, the region of *intrinsic* diffusion at high temperature and the region of *extrinsic* diffusion at low temperature. In an Arrhenius plot of the diffusion coefficient the transition from intrinsic to extrinsic diffusion is characterized by the occurrence of a kink. This is illustrated in the right part of Fig. 12.

Obviously, the diffusion coefficient corresponding to intrinsic diffusion is a property of the material under consideration and, in principle, at a given temperature is the same for all particles formed by the same material. The extrinsic diffusion coefficient, however, depends on the history and formation conditions of a particle and may by grossly different for different specimens of the same material.

Since timescales available for diffusion both under circumstellar shell or protoplanetary disc conditions are rather long compared to laboratory measurement conditions, diffusion effects in cosmic objects become important already at a much lower temperature ($< 1000\,\mathrm{K}$) than in the laboratory. For astrophysical applications one needs diffusion data for rather low temperatures where possibly diffusion is dominated by extrinsic diffusion. One has therefore to be cautious in applying laboratory measured diffusion coefficients obtained for intrinsic diffusion to astrophysical problems.

Interdiffusion of Mg and Fe in Olivine. The interdiffusion of the Mg^{++} and Fe^{++} cations in the magnesium-iron silicates olivine and orthopyroxene is of particular importance because these minerals form the dominating dust species in the standard cosmic element mixture. The exchange of Mg and Fe between olivine, orthopyroxene, and iron and its diffusional transport within grains determines the chemical composition of the dust mixture in circumstellar shells and protoplanetary discs.

Diffusion of Mg and Fe in olivine with particular emphasis on astrophysical or geophysical applications has been studied in the laboratory several times

(Buening and Buseck [14], Misener [75], Morioka [80, 81], Chakraborty et al. [16], Chakraborty [17]). Especially the older studies of Buening and Buseck [14] seemed to indicate an intrinsic-extrinsic diffusion transition. The most recent and more consistent determination of Mg-Fe-interdiffusion in olivine by Chakraborty [17] yields much lower diffusion coefficients than older determinations and show no indications for a change from intrinsic to extrinsic diffusion. It is argued by Chakraborty that true intrinsic diffusion is unlikely to be observable for olivine and that the observed interdiffusion of Fe and Mg over the whole experimental temperature regime is determined by the presence of Fe^{+++} impurities, i.e. extrinsic diffusion dominates. Extrapolation of the results to lower temperatures seems to be uncritical in this case.

For olivine with composition $Fo_{86}Fa_{14}$ Chakraborty finds $D_0(5.38 \pm 0.89) \times 10^{-5}\,cm^2\,s^{-1}$ and $Q = 226 \pm 18\,kJ/mol$. The diffusion coefficient of olivine depends considerably on the Fayalite concentration and this may be approximated by an exponential law (Buening and Buseck [14], Misener [75])

$$D_0(x) = D_0(0)\,e^{-\alpha x}, \tag{82}$$

where x is the mole fraction of forsterite in olivine. Values of α for a number of cations are compiled in Morioka [81]. Chakraborty [17] finds essentially the same dependence of D_0 on x as the previous studies. For $T = 1000°C$ one finds from Fig. 4 of that paper a value of $\alpha = 3\ln 10$. Since the study of Chakraborty seems to be the most reliable one, we choose this value for α. Then one finds for olivine $D_0(0) = 2.0 \times 10^{-2}\,cm^2\,s^{-1}$. The data are listed in Table 5.

The activation energy for Mg-Fe interdiffusion in olivine is much lower than the activation energy for annealing (cf. Tables 4 and 5). Internal gradients in the Fe/(Mg+Fe) concentration within dust grains are erased already at a lower temperature than disorders in the lattice structure.

The role of Mg-Fe-cation interdiffusion in olivine evaporation is studied in Ozawa & Nagahara [92].

Interdiffusion of Mg and Fe in Pyroxene. The Mg-Fe interdiffusion in orthopyroxene minerals seems to be less well studied. Some laboratory data on Mg self-diffusion are reported by Schwandt et al. [109]. A detailed study on Fe-Mg interdiffusion in natural specimens of orthopyroxene was performed by Klügel [58]. There are no experimental data over an extended temperature regime. The results of Klügel [58] seem to indicate (Fig. 7 of the paper) that the data of Schwandt et al. [109] for Mg tracer diffusion can be taken as representative also for Mg-Fe interdiffusion in orthopyroxene. The frequency factor of the coefficient of Mg self-diffusion in crystallographic (100) direction in Orthopyroxene with composition $En_{88}Fs_{12}$ according to Schwandt et al. [109] is $\log D_0 = 0.04 \pm 2.48$ (D in units $cm^2\,s^{-1}$) and the activation energy is $360 \pm 52\,kJ/mole$. These data are listed in Table 5. The rather low temperature for which D was experimentally determined makes it rather unlikely that the results refer to intrinsic diffusion, i.e. extrapolation of the diffusion coefficient to lower temperatures probably is admitted. Most likely the diffusion coefficient depends on the enstatite-ferrosilite

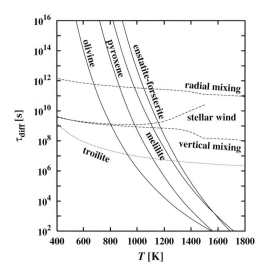

Fig. 13. *Full lines:* Characteristic timescales for solid interdiffusion of cations in $0.1\,\mu$m dust grains. For olivine and orthopyroxene a 10% iron content is assumed, for melilite a 10% content of åkermanite. *Dashed lines:* Time scales for radial and vertical transport processes are as in Fig. 10. *Dotted lines:* Troilite formation from iron by reaction with H_2S at total gas pressure of 10^{-3} bar

mixing ratio in orthopyroxene similar as in the case of olivine, but no experimental results are available.

The activation energy for Mg-Fe interdiffusion in olivine seems to be somewhat higher than the activation energy for annealing (cf. Tables 4 and 5), but the difference is probably within the experimental errors.

Interdiffusion of Al and Si in Melilite. The interdiffusion of Al + Al vs. Mg + Si in the melilite, a solid solution between gehlenite ($Ca_2Al_2SiO_7$) and åkermanite ($Ca_2MgSi_2O_7$) is important for the evolution of the Al-Ca dust component, since melilite is one of the important Ca-Al compounds formed in the oxygen rich element mixture.

The interdiffusion process in this material is quite complex since it requires paired diffusion of $Al^{3+} + Al^{3+}$ versus $Mg^{2+} + Si^{4+}$ because of charge balancing. This results in a complex composition dependence of the diffusion coefficient as found by Nagasawa et al. [88]. For astrophysical applications only the gehlenite rich melilite mixture is of interest. The activation energy Q for mixtures with composition $Ge_{100}\ \dots\ Ge_{50}Åk_{50}$ found by Nagasawa et al. [88] is 420 ± 50 kJ/mole. The frequency factor depends on the mole fraction x of gehlenite in the solution as in (82) if $x > 0.5$. For $x < 0.5$ the dependency is quite different. The frequency factor $D_0(0)$ in (82) can be determined from Fig. 2 of Nagasawa et al. [88] as $D_0(0) = 92.7$ for the gehlenite rich mixture ($x < 0.5$) and a coefficient α for (82) of 3.34. The temperature interval in the laboratory experiments of Nagasawa et al. [88] covers the temperature region where gehlenite may be formed in protoplanetary discs. No substantial extrapolation of the laboratory results is required in this case. The data are listed in Table 5.

Figure 13 shows the characteristic diffusion time

$$\tau_{\mathrm{diff}} = a^2/D \tag{83}$$

Table 6. Rate coefficients for the formation of reaction rims: $k = k_0 \exp\{-T_0/T\}$

Minerals	k_0 cm^2/s	T_0 K	Temperature range [K]	Ref.
enstatite-forsterite	4.0×10^3	60 730	1450-1820	[53]

for dust grains of radius $a = 0.1\,\mu$m for the different materials. For comparison the Figure also shows the characteristic timescales for radial and vertical mixing in protoplanetary discs for a stationary disc model and the characteristic hydrodynamic time scale r/v for a stationary wind model.

5.3   Chemical Reactions

There are a number of chemical reactions between the gas phase and the solids and between solids which are important for the evolution of the dust mixture, especially in the case of protoplanetary discs. Only a few of these reactions seem to have been studied in the laboratory.

Enstatite Formation from Olivine. Since olivine is stable up to higher temperatures than pyroxene (cf. Fig 4), but pyroxene is the dominant component at low temperatures, a solid state chemical reaction is required for conversion of forsterite into enstatite and vice versa. An experimental study of the formation of enstatite rims on forsterite in a Si-rich gas has been performed by Imae et al. [53]. The vapour from thermal vaporization of cristobalite ($SiO_2 \rightarrow SiO + O_2$) reacted with a forsterite crystal and formed a layer of stoichiometric enstatite which covered the forsterite. A parabolic rate law

$$x^2 = kt \tag{84}$$

was observed for the growth of an enstatite layer of thickness x. This indicates that the reaction is controlled by diffusion through the enstatite layers. A model is favoured, that growth occurred by Si^{4+} and $2Mg^{2+}$ counterdiffusion via grain boundary diffusion. The experimental results for the rate constant k are fitted by

$$k = k_0 \, e^{-Q/RT} \tag{85}$$

with $Q = 505\pm188$ kJ/mole and $\log k_0 = 3.6\pm5.6$ where k_0 is in units cm^2/s. The results are obtained at rather high temperatures between 1 450 K and 1 820 K. Since the basic diffusion mechanism controlling the solid reaction seems to be grain boundary diffusion, the data probably can be extrapolated to lower temperatures. The formation of a forsterite layer on entstatite during enstatite evaporation is studied by Tachibana et al. [121].

Formation of Troilite. At temperatures below about $720\,K$ the iron reacts with H_2S molecules from the gas phase to form the sulfide FeS (cf. Fig 4). This reaction has been studied in great detail in the laboratory by Lauretta et al. [67], which also give references to the preceding studies on FeS formation. Lauretta et al. found a linear growth kinetics for FeS layers on Fe up to a layer thickness of $\approx 30\,\mu m$ and a parabolic growth kinetics controlled by diffusion of Fe^{2+} through FeS above $30\,\mu m$. Iron grains in protoplanetary accretion discs prior to the planetary formation stage probably are smaller than $10\,\mu m$ (Kerridge [56]) and the linear growth rate has to be applied. They determined for this case a rate coefficient for the formation of FeS from solid Fe of $k_f = 5.6 \cdot \exp(-3360/T)$ in units gram FeS per cm^2, hour, and atm and $k_r = 10.3 \cdot \exp(-11\,140/T)$ for the reverse reaction. The reaction timescale in a gas of total pressure $P = 10^{-3}\,bar$ and solar abundance is shown in Fig. 13.

A more recent study of FeS reduction by H_2 is Tachibana & Tsuchiyama [117]. Because of the big specimens used in their experiments the results are not applicable for very small dust grains.

Formation of Magnetite. At temperatures below about $380\,K$ in chemical equilibrium the iron metal would be converted under protoplanetary disc conditions into magnetite (Fe_3O_4). The kinetics of the formation of magnetite from iron has been studied by Hong & Fegley [52]. They found much slower conversion of iron into magnetite than of iron into troilite. The oxidation reaction was studied at standard pressure and no dependence of the rate constant on H_2 and H_2O pressures has been determined, so unfortunately it is not possible to use the results in model calculations at the much lower pressures in accretion discs and circumstellar shells.

Reduction of Silicates by H_2. The reduction of silicates by hot H_2 gas was studied by Allen et al. [1].

6   Circumstellar Dust Shells

Dust is formed in late stages of stellar evolution by stars from almost the whole range of initial stellar masses on the main sequence. Here we discuss the formation of the most abundant minerals in circumstellar shells with an oxygen rich element mixture. We do not consider the soot formation problem in carbon rich environments and we do not consider the problem of formation of secondary dust components formed from Al and Ca (see Fig. 4 for the most stable Al-Ca-compounds). Al bearing dust grains have been found as presolar grains (cf. Nittler et al. [91]) but presently nothing seems to be known on the condensation kinetics of such dust materials.

6.1   Predictions of Dust Composition by Equilibrium Calculations

Low and medium mass stars reach the upper part of the AGB as M-stars with the slight modifications of their initial element abundances by first and second dredge

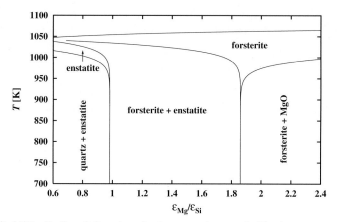

Fig. 14. Stability limits of the abundant magnesium and silicate compounds at $P = 10^{-10}$ bar for varying abundance ratios of magnesium to silicon. The picture shows the regions of existence of the different condensates in chemical equilibrium (Ferrarotti & Gail [31])

up processes (mixture (1) in Tab. 1) where they start to experience severe mass-loss by a dust driven wind. After the onset of thermal pulsing the products of He-burning are mixed during the "third dredge-up" from the core into the stellar envelope and the carbon abundance increases in discrete steps after each pulse (c.f Groenewegen et al. [46]), while the oxygen abundance essentially remains unchanged.

Stars with initial masses $M_* \lesssim 1.5\,M_\odot$ have lost most of their envelope and evolve off from the AGB towards the stage where they excite a planetary nebula (PN) before the carbon abundance in their envelope exceeds that of oxygen (e.g. Groenewegen et al. [46]). They remain M-stars during their whole AGB evolution.

Stars with initial masses in the range $1.5 \lesssim M_* \lesssim 4 \ldots 5\,M_\odot$ suffer several thermal pulses and increase their carbon abundance in the envelope from a C/O abundance ratio of about 0.4 to C/O>1 where carbon is more abundant than oxygen. The star then becomes a C-star and remains to be so until the star evolves off from the AGB towards the PN-stage.

Stars with initial masses between $4 \ldots 5\,M_\odot$ and $8\,M_\odot$ most likely experience 'hot bottom burning'. These stars do not become carbon stars (cf. Lattanzio & Forestini [65]). They develop a strong N overabundance instead.

If during the stepwise increase of the carbon abundance in the envelope the C/O abundance ratio is close to unity, the star appears as of spectral type S (Scalo & Ross [110]) where the chemistry is different from both that of M and of C stars. For stars with initial masses in the region $1.5 \lesssim M_* \lesssim 2\,M_\odot$, however, the C/O ratio increases in a few and rather big steps from C/O< 1 to C/O>1 and the intermediate state of C/O≈ 1 of an S star probably is skipped in most cases (Groenewegen et al. [46]).

The chemistry in the stellar envelope and the spectral appearance of the stars is completely different for each of the three different carbon to oxygen abundance ratios C/O< 1, C/O≈ 1, and C/O> 1. This also holds for the chemical composition of the dust formed in their dust shells.

Massive stars from the region of initial masses $8 \lesssim M_* \lesssim 25\,M_\odot$ reach the RGB (Schaller et al. [106]) before they explode as supernovae and form dust during their evolutionary stage as Red Supergiant. The element mixture in their atmospheres in this stage corresponds to that of M stars on the AGB (e.g. Schaller et al. [106]).

M Stars: From chemical equilibrium condensation calculations it is known that for the standard cosmic element mixture the most abundant refractory elements Si, Mg, Fe form the silicates olivine ($Mg_{2x}Fe_{2(1-x)}SiO_4$) and pyroxene ($Mg_xFe_{1-x}SiO_3$) (cf. Fig. 4) and solid nickel-iron (cf. Fig. 6). The iron content of the silicates in chemical equilibrium at high temperatures is small, as is demonstrated in Fig. 7. Their composition corresponds to nearly pure forsterite Mg_2SiO_4 and enstatite $MgSiO_3$.

The abundance ratio of Mg/Si of Pop I stars shows a considerable scatter between different stars (cf. Edvardsson et al. [27]) and varies between about 1.0 and 1.5 and in some cases becomes as low as 0.6, especially for Pop II metallicities (cf. Fig. 1 in Ferrarotti & Gail [31]). The stability regions of the different equilibrium condensates with varying Mg/Si abundance ratio for a pressure typical for the condensation zone in a stellar wind is shown in Fig. 14. For Mg/Si abundance ratios less than unity quartz (SiO_2) additionally becomes a stable condensate, which is absent from the mixture for $\epsilon_{Mg}/\epsilon_{Si} > 1$ since in the latter case Si (and Mg) can completely be condensed into the magnesium silicates. For other elements of interest for dust formation (Fe, Al, Ca) the scatter in element abundances seems to be lower (cf. Edvardsson et al. [27]) and of no significant influence on the dust mixture.

S Stars: Chemical equilibrium calculations for determining the stable condensates for varying C/O abundance ratios at the transition from M to C stars ($\epsilon_C/\epsilon_O \approx 1$) have been performed by Ferrarotti & Gail [32]. The main characteristics of the chemistry at the transition is the lack of oxygen and carbon which prevents the formation of silicates and oxides at one hand and of carbon dust and carbides on the other hand. The most abundant refractory elements according to equilibrium calculations form solid iron and FeSi (cf. Fig. 15). The magnesium would condense in chemical equilibrium as MgS (not shown in Fig. 15) at temperatures below about 800 K and probably is of no interest for circumstellar condensation in S stars.

C Stars: The condensation of carbon dust is not treated in this contribution. The equilibrium chemistry of the carbon rich mixture is discussed at length by e.g. Lodders & Fegley [70, 71], Sharp & Wasserburg [114].

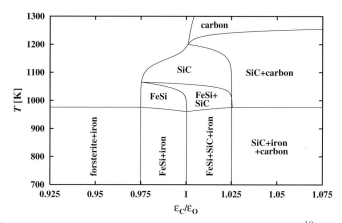

Fig. 15. Stability limits of the abundant condensates at $P = 10^{-10}$ bar for varying carbon to oxygen abundance ratios at the transition from M over S to C stars. The picture shows the regions of existence of the different condensates in chemical equilibrium (Ferrarotti & Gail [31])

CNO Processed Element Mixture: In material having burned hydrogen via the CNO-cycle and having reached equilibrium of C, N, and O isotopes the abundance of C and O drop below that of the abundant dust forming elements Si, Fe, and Mg (cf. Fig. 1), but the mixture remains oxygen rich (mixture (3) in Tab. 1). Insufficient amounts of O and C are available to form oxides or carbides. In this respect the chemistry in this mixture has some similarities with that of S-stars. The stability limits of the most stable condensates formed from this mixture are shown in Fig. 5. Solid iron and Fe-Si-alloys are expected to be the most abundant dust components formed from this mixture. The chemical equilibrium dust composition with respect to Fe and Si for this mixture is shown in Fig. 16.

Such dust may be formed in the rare cases where WN stars form dust or during post-AGB evolution after a late thermal pulse (cf. Blöcker [7]).

6.2   Non-equilibrium Dust Formation

In stellar outflows the dust forms in a rapidly cooling and rapidly diluted environment. Under these conditions the chemistry in such shells does not evolve into an equilibrium state. Dust formation in stellar outflows is a time dependent problem which must be calculated by solving the equations for the time dependent condensation kinetics.

Timescales. The dust formation problem is governed by three characteristic timescales.

Growth timescale: The growth of grains is determined by (71). The essential growth species are a few abundant molecules from the gas phase which usually

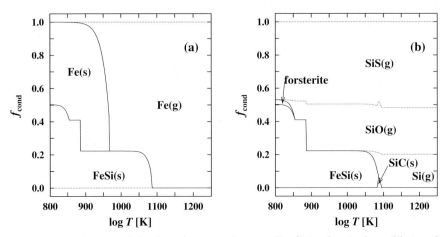

Fig. 16. Distribution of the dust forming elements Fe, Si in chemical equilibrium between solids and the gas phase for CNO-cycle processed material at $P = 10^{-10}$ bar

carry nearly all atoms of an element which is essential for the formation of a dust species. In the oxygen rich element mixture for instance the abundant dust species olivine and pyroxene are formed from SiO, Fe, and Mg, and from H_2O as oxidizing agent. If the growth species are not yet substantially consumed by grain growth one has for the particle density n of the rate determining growth species

$$n = \epsilon \dot{M}/4\pi r^2 v(1 + 4\epsilon_e)m_H \qquad (86)$$

if we assume for simplicity a stationary outflow. ϵ is the abundance of the element forming the essential growth species. The characteristic timescale for particle growth up to size a_d is (Eq. 71)

$$\tau_{gr} = \frac{a_d}{|\dot{a}_d|} = \frac{a4\pi r^2 v(1 + 4\epsilon_{He})m_H}{V_0 \alpha v_{th} \epsilon \dot{M}(1 - a_{sol}^{-1/i})} . \qquad (87)$$

Using the relation $4\pi R_*^2 \sigma T_{eff}^4 = L_*$ und introducing numerical values for forsterite from Table 3 one obtains numerically

$$\tau_{gr} = 1.8 \times 10^8 \frac{1}{1 - a_{sol}^{1/i}} \frac{0.1}{\alpha} \frac{v}{c_T} \frac{r^2}{(4R_*)^2} \frac{L_*}{10^4 L_\odot} \frac{1}{\dot{M}_{-5}} \left(\frac{2500K}{T_{eff}}\right)^4 \text{sec.} \qquad (88)$$

c_T is the isothermal sound velocity of the gas and $\dot{M}_{-5}$ the mass-loss-rate in units $10^{-5} M_\odot$ yr^{-1}. a_{sol} is the activity.

   Cooling timescale: Assuming for simplicity a grey temperature structure $T \propto r^{-1/2}$ in the dust forming region we obtain for the time required by a gas parcel to cool by ΔT

$$\tau_{cool} = 2(r/v)(\Delta T/T) . \qquad (89)$$

For the hydrodynamic expansion timescale we have

$$\tau_{\exp} = \frac{r}{v} = 1.8 \times 10^8 \, \frac{c_T}{v} \, \frac{r}{R_*} \, \text{sec.} \qquad (90)$$

The isothermal sound velocity c_T at $1000\,\text{K}$ is about $2\,\text{km/s}$.

At the onset of condensation ($a_{\text{sol}} = 1$ at this point) the wind enters the condensation zone with a rather low velocity which is of the order of the sonic velocity. The time required for a cooling by $10\,\text{K}$ according to (89) is $3.7 \times 10^6\,\text{s}$ which is short compared to the growth time (88) of $1.8 \times 10^8\,\text{s}$ at $r = 4R_*$ and $v = c_T$. A decrease in T by $10\,\text{K}$ typically increases the activities a_{sol} of the dust materials of interest by factor of about 10 (cf. Fig. 3). Short after the onset of dust growth the evaporation term ($\propto 1/a_{\text{sol}}^i$) in the growth equation becomes negligible because of significant cooling without significant depletion of the gas phase from condensible material.

From (89) one infers that during the time required for dust grains to grow to their final size of about $0.1\,\mu\text{m}$ the gas in the stellar wind cools by several $100\,\text{K}$.

Acceleration timescale: After significant amounts of dust from the most abundant refractory elements are condensed the gas is rapidly accelerated by radiation pressure on grains. Neglecting pressure effects, the velocity increase between two radii r_1 and r_2 is

$$v_2^2 - v_1^2 = 2GM_*(\Gamma - 1) \int_{r_1}^{r_2} dr/r^2 = 2GM_*(\Gamma - 1)\,(r_2 - r_1)/(r_2 r_1) \qquad (91)$$

where

$$\Gamma = f\kappa\, L_*/4\pi c G M_* = 0.77 \times f\,\kappa\, \frac{L_*}{10^4 L_\odot}\, \frac{M_\odot}{M_*} \qquad (92)$$

is the ratio of radiative acceleration to gravitational attraction. κ is the mass extinction of the dust, which is about 20 at $T_{\text{eff}} = 2\,500\,\text{K}$ (e.g. Fig. 10 in Ferarotti & Gail [32]), and f is the fraction of the condensible material condensed into grains. Starting from about sonic velocity $c_T \ll v$ at r_1 yields $v^2 \approx (\Gamma - 1)v_{\text{esc}}^2\, \Delta r/r$. The time τ_{acc} required for a gas parcel to move from r_1 to r_2 can be approximated by $\tau_{\text{acc}} = 2\Delta r/v$ which yields

$$\tau_{\text{acc}} \approx \frac{2}{\Gamma - 1}\, \tau_{\exp}\, v^2/v_{\text{esc}}^2 . \qquad (93)$$

Since Γ usually is of the order of $2\ldots5$ the time required to accelerate the wind material to its final expansion velocity of $v \approx v_{\text{esc}}(4R_*)$ is of the same order as the growth timescale τ_{gr} if $\dot{M} \approx 10^{-5}\,M_\odot\,\text{yr}^{-1}$.

The increase of v from c_T to the supersonic outflow velocity increases the growth timescale by a factor of about 10 compared to its initial value. The further growth of grains is strongly slowed down from this point on. Due to the further increase of $\tau_{\text{gr}} \propto r^2$ (cf. Eq. 88) as a gas parcel moves outwards the grain growth essentially ceases by rapid dilution of the wind material after acceleration of the wind to strongly supersonic outflow velocities has taken place.

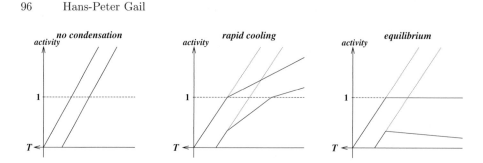

Fig. 17. Schematic evolution of the activities of two substances formed from the same gas phase material during cooling of the gas phase. *Left:* Cooling is so rapid that essentially no condensation occurs. *Middle:* Rapid cooling where the more stable substance has not consumed the condensible material at the instant where the less stable material starts to condense. *Right:* Cooling is so slow that the system always is close to chemical equilibrium. The less stable material never condenses because its activity always remains < 1

For mass-loss rates exceeding about $\dot{M} \approx 10^{-5}\,\mathrm{M}_\odot\,\mathrm{yr}^{-1}$ one has $\tau_{\mathrm{gr}} \leq \tau_{\mathrm{acc}}$ and the condensation of the condensible material into grains will be more or less complete. For mass loss rates $\dot{M} < 10^{-5}\,\mathrm{M}_\odot\,\mathrm{yr}^{-1}$ condensation is incomplete because $\tau_{\mathrm{gr}} > \tau_{\mathrm{acc}}$ and grain growth is suppressed before condensation is complete.

At low mass loss rates $\dot{M} < 10^{-5}\,\mathrm{M}_\odot\,\mathrm{yr}^{-1}$ one has to be aware that only part of the material is condensed and one has bigger to much bigger gas to dust ratios than in the ISM!

In this estimate it is assumed that sufficient dust condenses to accelerate the wind material to supersonic velocities which requires a sufficiently high mass-loss rate. This problem is discussed in Gail & Sedlmayr [38].

Non-equilibrium Condensation. As we have seen, in a stellar wind the gas cools by several 100 K until either all condensible material is consumed or further growth is suppressed by rapid dilution. This has the consequence that for certain solids, formed from just the same growth species as those dust species which are just growing, the activity a_{sol} of somewhat less stable materials raises above unity. From this point on such less stable materials may condense from the gas phase, if they either can nucleate from the gas phase or on the surface of already existing grains.

The evolution of the activities of two possible condensates competing for the same growth species from the gas phase during cooling schematically is shown in Fig. 17. Only one of the two components is formed in chemical equilibrium (right part) because the activity of the thermodynamic less stable material remains less than unity. During rapid cooling, growth cannot take pace with the shift of the equilibrium state with decreasing temperature. The activity of the more stable material exceeds unity since not all material from the gas phase, which can be condensed in a chemical equilibrium state, is really condensed. For this reason the activity of the less stable material also exceeds unity from some point on and this material also starts to condense.

An inspection of Fig. 3 shows that for the standard oxygen rich element mixture this particular case is relevant for quartz (SiO_2) and magnesiumoxide (MgO). These two solids do not exist in chemical equilibrium in the oxygen rich element mixture because all Mg and Si condenses into the thermodynamically more stable magnesium silicates. Because of rapid cooling in a stellar outflow, however, they become stable before olivine and pyroxene formation have consumed all of the material from which solid SiO_2 and MgO can be formed. The quartz and MgO then start to condense as separate dust components in this case[2]. This is shown in the non-equilibrium calculations of Gail & Sedlmayr [41] and Ferrarotti & Gail [31].

In the rapid cooling environment of stellar outflows one obtains a non equilibrium dust mixture which represents some transient state of the chemical system on its evolution towards chemical equilibrium which, however, never can be attained. The appropriate tool for studying this type of non-equilibrium condensation due to rather rapid cooling is the activity-temperature diagram which is shown for the particular case of condensation from the oxygen rich element mixture in Fig. 3. One readily reads off from this diagram which solids are expected to be the most stable ones which first start to condense (those with the highest activities) and which are the secondary products which do not exist in chemical equilibrium but are formed during rapid cooling. These are those for which activities cross the border $a = 1$ within a temperature difference ΔT to the first abundant condensate for which $\tau_{cool}(\Delta T)$ defined by (89) equals τ_{acc}:

$$\Delta T \approx \tfrac{1}{2}T\tau_{acc}/\tau_{exp} . \tag{94}$$

This difference ΔT amounts to about 200...300 K at $\dot{M} \approx 10^{-5}\,\mathrm{M_\odot\,yr^{-1}}$ and may be different for different cooling rates.

Model calculations for non-equilibrium condensation require to solve the set of growth equations of the type (71) for each of the possible condensates simultaneously with the equations for the consumption of gas phase material and the equations of the stellar wind. The set of equations to be solved is discussed in Gail & Sedlmayr Gail & Sedlmayr [41] and Ferrarotti & Gail [31, 32]. Solutions have been obtained so far only for the case of oxygen rich element mixtures and the case of S-Stars.

While the dust growth problem for the most abundant minerals in circumstellar shells in principle can be treated realistically since the relevant data required for the calculation have been determined in the laboratory during the last decade (see Sects. 4 and 5), realistic model calculations presently are seriously hampered by our ignorance with respect to the nucleation processes in oxygen rich environments.

---

[2] Since MgO forms a solid solution with $Fe_{1-\delta}O$ ($\delta \ll 1$) one obtains magnesiowüstite $Mg_xFe_{1-x-\delta}O$ as a dust component in oxygen rich dust shells (Ferrarotti & Gail, in preparation)

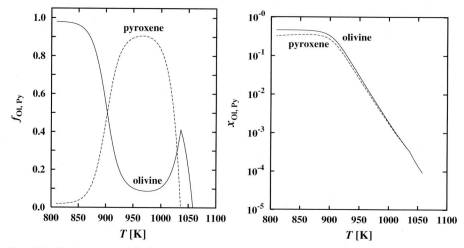

Fig. 18. Condensation of olivine and pyroxene in chemical equilibrium at total pressure $P = 10^{-10}$ bar, representative for the condensation zone in stellar winds. The condensation of solid iron is suppressed. *Left part:* Fraction of the Si condensed into olivine and pyroxene. *Right part:* Mole fractions x_{fa} and x_{fs} of the iron bearing end members fayalite and ferrosilite in olivine and pyroxene, respectively

Fe Content of Olivine and Pyroxene. Besides the well known amorphous silicates in circumstellar shells of oxygen rich stars a crystalline silicate component has been detected in the infrared spectra of many late type stars with circumstellar dust shells. The observation shows that the crystalline dust component has a low iron content (e.g. Molster et al. [78]). A low iron content of the silicates is characteristic for a dust mixture where silicate dust has equilibrated with solid iron.

Figure 18 shows the degree of condensation of olivine and pyroxene and the mole fraction of the iron rich endmember of the solid solution series if it is assumed that iron does not condense. In this case the partial pressure of Fe atoms in the gas phase is not fixed to the low vapour pressure of iron but remains high. Comparison with Fig. 7 shows that a low iron content of olivine and pyroxene formed in a stellar wind can only be expected if silicate condensation runs into completion already at temperatures $T \gtrsim 950\,\mathrm{K}$ where the vapour pressure of the iron rich component is high and the iron content of silicates is low despite of a high partial pressure of Fe in the gas phase. If the condensation process in the outflow extends to much lower temperatures one mainly obtains Fe-rich olivine and pyroxene since, due to incomplete condensation of iron, the pressure of Fe atoms in the gas phase much exceeds the vapour pressure of iron. Only if the Fe partial pressure is buffered by solid iron one gets iron poor silicates.

This favours the suggestion of Molster et al. [78] that the observed iron poor crystalline dust resides in a circumstellar disc, where pressures are higher and ample time is available for equilibration (Fig. 8 then applies to this case).

Table 7. Composition of the dust mixture in the outer regions of a protoplanetary accretion disk according to Pollack et al. [96], and fraction of the most abundant elements condensed into the different dust species. x denotes the mole fraction of the pure magnesium silicate end members of the solid solution series forming the magnesium-iron-silicates olivine and pyroxene, respectively

Dust	composition	Mg	Fe	Si	S	C	x
olivine	$Mg_{2x}Fe_{2(1-x)}SiO_4$	0.83	0.42	0.63			0.7
pyroxene	$Mg_xFe_{1-x}SiO_3$	0.17	0.09	0.27			0.7
quartz	SiO_2			0.10			
iron	Fe		0.10				
troilite	FeS		0.39		0.75		
kerogene	HCNO					0.55	

7 Dust in Protoplanetary Accretion Discs

This section considers the dust component of the material in protostellar and protoplanetary accretion discs and its evolution. We restrict our considerations to the early evolutionary phases of such discs before the onset of planet formation, when the disc material consists of a well mixed gas-dust mixture.

A precise knowledge of the composition, structure and extinction properties of the dust is important for calculating the structure and evolution of such discs as the dust extinction determines their temperature structure, the presence or absence of convection, the shielding of the discs interior from ionizing radiation from the outside and so on. Also the determination of the raw material from which later on the planetesimals and planets are formed requires an as precise as possible knowledge of the composition and structure of the dust.

The dust mixture probably is very complex; it consists of a big number of different compounds from the elements heavier than He, mainly compounds of the rock forming elements Si, Fe, Mg, Al, Ca, Na, Ni and of carbonaceous compounds, but also a number of minerals of the less abundant elements are present. For reasons of cosmic element abundance the most abundant dust components are compounds of Si, Mg, Fe, and carbon dust. They also dominate the extinction of the disc material. For the purpose of constructing models of protoplanetary discs it presently suffices to concentrate on these most abundant dust species. Other dust components are important in other respects, but have no significant influence on the disc structure and evolution.

In the following we outline the basic problems which have to be solved in order to arrive at a self-consistent description of the evolution of protoplanetary discs during their early evolution, when the raw material for formation of planetesimals and planetary bodies is fabricated.

7.1 Dust Mixture in the Outer Disc

Pollack et al. [96] carefully discussed the question which types of dust, formed from the most abundant dust forming elements (C, N, O, Mg, Si, and Fe),

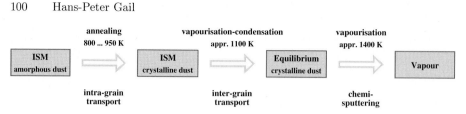

Fig. 19. Main processes responsible for the conversion of the interstellar silicate dust mixture into a chemical equilibrium dust mixture and for its final destruction

can be expected to exist in the parent molecular cloud cores, from which stars are formed, and in the cool outer parts of the resulting protostellar and protoplanetary accretion disks. They arrived at the conclusion that there exists a multicomponent mixture of several kinds of dust species, which is dominated most likely by the few species listed in Table 7.

The silicates have an amorphous lattice structure since dust in molecular clouds shows no indications for the existence of a crystalline component. The lattice structure of quartz probably also is amorphous, though from observations nothing is known about the structure of quartz. The solid iron and the troilite probably are solid solutions with Ni or NiS, respectively. The nature of the carbon dust component is not clear. The model of Pollack et al. [96] assumes kerogene, a carbon rich substance containing also significant amounts of H, N, and O. This is the carbonaceous material found in the matrix of carbonaceous chondrites.

Less abundant dust components should also be present but are not considered in the model of Pollack et al. [96] because of their unimportance for the extinction of disc matter.

7.2   Dust Metamorphosis

Viscous torques in the accretion disc induce a slow inwards migration of the disc material by which the material is transported from the cold outer disc into increasingly warmer zones. Additionally, turbulent mixing in the convectively unstable parts of the disc intermingles material from different disc regions. At sufficiently high temperature, solid diffusion and annealing processes are activated in the dust grains which tend (i) to exsolve impurities from the grain lattice and (ii) to form a regular crystalline lattice structure (Duschl, Gail, Tscharnuter [24]; Gail [35]; Hallenbeck, Nuth & Daukantas [48]; Fabian et al. [28]). The grain material develops from the dirty and amorphous composition, which is responsible for the observed extinction properties of interstellar grains, to a more clean and crystalline lattice structure. The basic chemical composition of individual grains as given in Table 7, however, is not changed by these processes. The assemblage of minerals forming the dust mixture essentially remains preserved at this stage, since there operate only *intra-grain* transport processes by solid diffusion, but no *inter-grain* transport processes via the gas phase.

This change in the lattice structure of the dust grains occurs in the region where the timescale for solid diffusion and annealing roughly equals the timescale for radial inwards migration of the disc material. The model calculations in Gail [36] and Wehrstedt & Gail [126] have shown this to occur at about 800 K for glassy silicates (cf. also Fig. 10). If the structure of the silicate dust is more similar to that of the smokes prepared in the annealing experiments of Hallenbeck et al. [48] one determines from the data in Hallenbeck, Nuth & Nelson [49] a somewhat higher annealing temperature of about 950 K (cf. Gail [36]). At temperatures above the annealing temperature the grains have a crystalline lattice structure, which at the same time increases their stability against evaporation.

At a temperature significantly higher than that required for annealing, transport processes between the surfaces of different grains via the gas phase start to operate (evaporation and re-condensation). Since the ISM dust mixture of the outer disc contains dust species, which have been formed in chemically such diverse environments as circumstellar shells of AGB-, RSG-, WC-stars, and supernovae, or by destruction and re-condensation behind shocks in the ISM, part of the mineral components of the mixture are thermodynamically unstable for the element mixture of the protoplanetary disc material. The *inter-grain* transport processes via the gas phase tend to remove the thermodynamically less stable condensed components from the ISM mineral mixture in favour of the thermodynamically most stable materials. This change in the dust composition occurs in the zone where the timescale for this kind of chemical metamorphism is comparable with the timescale of radial inwards migration of the disc material. Inside of this zone, the composition of the mineral assemblage corresponds to a chemical equilibrium state.

Figure 19 gives a schematical sketch of the dust metamorphosis in a protostellar disc and the processes involved.

7.3   Dust Mixture in the Inner Disc

The mineral mixture existing in chemical equilibrium for standard cosmic element abundances and the stability limits against evaporation/decomposition of the most abundant minerals in a chemical equilibrium mixture are shown in Fig. 4. The main components of the mixture are

- *forsterite* with composition Mg_2SiO_4
- *enstatite* with composition $MgSiO_3$
- *solid iron* which forms a solid solution with all the available Ni
- *troilite* with composition FeS which forms a solid solution with NiS.

If the Mg/Si abundance ratio in a protoplanetary disc is at the lower end but within the observed range of scattering of Mg/Si ratios in stars, also quartz (SiO_2) may be an abundant component of the equilibrium mixture (cf. Fig. 14). The frequency of such cases increases with decreasing metallicity (Ferrarotti & Gail [31]).

All condensates in the inner disc are solids; pressure and temperature conditions in the disc generally do not allow for the existence of melts.

There are two significant differences in the chemical composition between the equilibrium mineral mixture in the inner disc and the ISM dust mixture in the outer disc proposed by Pollack et al. [96]. First, since the average element mixture of the disc is oxygen rich there exists no carbon dust in the equilibrium mixture. The carbon dust is destroyed during the evolution towards the equilibrium mixture. Second, the silicate dust components olivine and pyroxene in the ISM are iron rich, while in chemical equilibrium the silicates are nearly iron free and can be assumed to be essentially pure forsterite and enstatite (cf. Fig. 8). Additionally, in most cases quartz is absent from the equilibrium mixture.

The lattice structures of the silicates in both cases also are completely different. The ISM dust has an amorphous lattice structure and a considerable fraction of impurities is build into the lattice. In chemical equilibrium, however, the substances are crystalline and the concentration of impurities solved in the solids is small.

7.4   The Three Main Dust Mixtures

The transport, diffusion, annealing, evaporation, and re-condensation processes in a protoplanetary accretion disc result in the existence of *three completely different dust mixtures in the inner, middle and outer part of the accretion disc:*

1. Amorphous dust with a strong non-equilibrium composition in the cold outer parts of the disc,
2. crystalline dust with a strong non-equilibrium composition in a certain zone of the inner parts of the disc, with an admixture of equilibrium dust from the inner zone,
3. crystalline dust with chemical equilibrium composition in the innermost parts of the disc down to the evaporation limits of the solids.

Carbon dust is only part of the first two mixtures. The equilibrium mixture is carbon dust free.

The basic chemical compositions of grains of the different kinds of dust from the first mixture are essentially the same. The isotopic mixture of the elements and the abundances of impurities in the individual grains may, however, be very different since the grains preserved the vastly different isotopic and elemental compositions of their formation sites. In meteorites they correspond to the presolar grains. The grains of the second mixture are already processed. They probably lost all or nearly all of their noble gas contents by outgassing at high temperatures. Also the isotopic anomalies of their impurities may partially be erased by internal diffusion and exchange processes with the gas phase. In meteorites, this dust component will hardly be recognized as grains of interstellar origin. The third component is home-made dust of the accretion disc. The isotopic compositions of individual grains correspond to the average isotopic mixture of the disc material after complete mixing.

The opacities of these dust mixtures are strongly different, the opacity of the equilibrium mixture being lower than that of the ISM mixture by a factor of more than ten. The reason is that the carbon in the ISM mixture accounts for nearly one half of the total opacity, that the opacity of crystalline material usually is

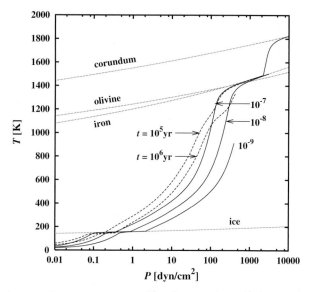

Fig. 20. The temperature-pressure stratification at the midplane in models of thin Keplerian α-discs in the one-zone approximation. *Dashed lines:* Two snapshots of the radial disc structure of a time dependent model at 10^5 and 10^6 years (Wehrstedt & Gail [126]). *Full lines:* Stationary models with the indicated accretion rates in units $M_\odot \, yr^{-1}$ (Gail [36]). *Dotted lines:* Stability limits of the indicated condensates

much lower than that of amorphous material, and that iron poor silicates have a lower opacity than iron rich silicates. A realistic modeling of the structure and evolution of protoplanetary accretion discs needs to include in the model computations the gradual change in dust composition and lattice structure

- from the ISM mixture of grains of interstellar origin, encountered in the cool outer parts of the disc,
- through the intermediate crystalline non-equilibrium mixture
- into the chemical equilibrium mixture existing in the warm inner parts of the disc.

7.5   Disc Models

In the following we discuss some details of the evolution of the dust in protoplanetary discs and present some results which are calculated for thin Keplerian disc models in the one-zone approximation, where one has averaged over the vertical structure of the disc. The viscous transport is calculated in the α-approximation. The details of how the disc models used in the present contribution are constructed are described in Gail [36] and Wehrstedt & Gail [126]. Figure 20 shows as example the variation of the pressure and temperature in the midplane of protoplanetary accretion discs.

7.6   The Advection-Diffusion-Reaction Equation

The basic equation describing the metamorphosis of the dust mixture in proto-planetary accretion discs is the combined equation for the radial advection of disc material by the disc accretion process, for the radial and vertical inwards and outwards mixing of matter by turbulent convection and by large scale circulation currents, and for the chemical reactions between dust grains and gas phase species, between the species within the gas phase, and for processes like annealing which occur within grains.

In the following the chemically or physically different dust species are distinguished by a lower index j. For each of the substances a set of fixed grain radii a_i is defined. The number density of grains having radii falling into the radius interval $[a_i, a_{i+1}]$ is divided by the number density of hydrogen nuclei. The resulting quantity, the number of dust grains of species j with radii between a_i and a_{i+1} per hydrogen nucleus, is denoted by $c_{j,i}$. All grains are assumed to have spherical shape.

The equation for time evolution of the concentration $c_{j,i}$ of dust species j having radii between a_i and a_{i+1} by advection, diffusion, and evaporation of dust in the one-zone approximation is (Gail [36])

$$\frac{\partial c_{j,i}}{\partial t} + v_{r,j,i} \frac{\partial c_{j,i}}{\partial r} = \frac{1}{nr} \frac{\partial}{\partial r} \, rnD_{j,i} \frac{\partial c_{j,i}}{\partial r} + R_{j,i} \qquad (95)$$

$v_{r,j,i}$ is the radial drift velocity of particles of species j and size a_i. $D_{j,i}$ is the diffusion coefficient of the grains induced e.g. by turbulent mixing. The terms depending on $v_{r,j,i}$ and $D_{j,i}$ describe the transport and mixing of the different components of the mixture of gases and dust in the protoplanetary disc. The term $R_{j,i}$ describes the gains and losses. For grain growth and evaporation the gains and losses for each species (j, i) are given by (Gail [36])

$$R_{j,i} = \begin{cases} \left(\dfrac{c_{j,i+1}}{\Delta a_{i+1}} - \dfrac{c_{j,i}}{\Delta a_i} \right) \left| \dfrac{\mathrm{d}\,a_{j,i}}{\mathrm{d}\,t} \right| & \text{if} \quad \dfrac{\mathrm{d}\,a_{j,i}}{\mathrm{d}\,t} < 0 \\[2ex] \left(\dfrac{c_{j,i-1}}{\Delta a_{i-1}} - \dfrac{c_{j,i}}{\Delta a_i} \right) \dfrac{\mathrm{d}\,a_{j,i}}{\mathrm{d}\,t} & \text{if} \quad \dfrac{\mathrm{d}\,a_{j,i}}{\mathrm{d}\,t} > 0 \end{cases} . \qquad (96)$$

where $\mathrm{d}a/\mathrm{d}t$ is given by an equation of the type (65) and $\Delta a_i = a_{i+1} - a_i$.

For each of the species from the gas phase there also holds an equation of the type (95). The rate terms in this case refer to chemical reactions with other gas phase species. If they are involved in reactions with grains there is an additional rate term corresponding to consumption from or injection into the gas phase by grain growth or destruction processes, respectively:

$$R_m = \sum_j \sum_r \nu_{j,r,m} (J_j^{\mathrm{ev}} - J_j^{\mathrm{gr}}) \, \mathcal{A}_j . \qquad (97)$$

The summation is over all dust species j, which are involved in reactions with molecule m and over all relevant reactions r. $\nu_{j,r,m}$ is the number of molecules

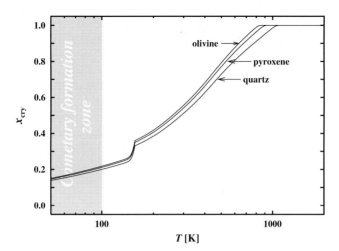

Fig. 21. Variation of the average degree of crystallization x_{cry} of the grains of interstellar origin due to annealing and radial mixing of crystallized grains as function of the midplane temperature T in a protoplanetary accretion disc. Stationary disc model with accretion rate $\dot{M} = 10^{-7}\,M_\odot\,yr^{-1}$. The feature at about $150\,K$ is due to the temperature plateau in the region of water ice evaporation (cf. Fig. 20). The grey shaded area roughly indicates the zone where cometary nuclei are expected to be formed

m injected into or removed from the gas phase in reaction r with dust species j,

$$\mathcal{A}_j = 4\pi \sum_i a_i^2 c_{j,i} \tag{98}$$

is the total surface area of all particles of dust species j, and J_j^{ev}, J_j^{gr} are the evaporation or growth current densities for grains of species j as defined in Sect. 4.2.

Equations (95) for all dust species j and grain radii i and for the relevant gas phase species have to be solved simultaneously with the equations for the disc structure and evolution. First results for stationary disc models are described in Gail [36] and for time dependent models in Wehrstedt and Gail [126].

7.7 Annealing

Annealing of the amorphous silicate grains of interstellar origin in protostellar discs and its implications for the disc structure is discussed in Duschl et al. [24], Gail [35, 36] and Wehrstedt and Gail [126].

In the model calculations [36, 126] the annealing process is modeled by solving (77) for the growth around each growth centre and defining the average degree of crystallization by (79). It is shown how this simple model can be used to determine the distribution function of the degrees of crystallization for an ensemble of dust grains. One defines a grid $0 \leq x_i \leq 1$ $(i = 1, \ldots, I)$ of totally I discrete degrees of crystallization for the dust grains. The gain and loss term in

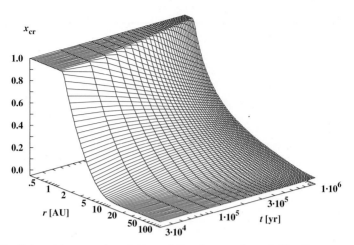

Fig. 22. Radial and temporal variation of the degree of crystallization x_{cry} of olivine grains due to annealing and radial mixing of crystallized grains for a time dependent disc model (Wehrstedt & Gail [126]). The model calculation shows that crystalline dust is mixed into fairly distant outer regions of the disc during its evolution. At the same time the inner zone of completely crystallized dust shrinks as the disk is depleted and generally becomes cooler

the advection-diffusion-reaction equation (95) for the concentration c_i of grains with a degree of crystallization between x_i and x_{i+1} is shown in [36] to be given by

$$R_i = \left(\frac{c_{i-1}}{x_i - x_{i-1}} - \frac{c_i}{x_{i+1} - x_i} \right) \frac{\mathrm{d}\, V_{\mathrm{cr}}^{\frac{1}{3}}}{\mathrm{d}\, t} \tag{99}$$

where V_{cr} is defined by (77). A solution of the system of differential equations (95) with the appropriate boundary conditions (all grains amorphous in the cold disc, all grains are crystalline in the warm inner part of the disc) then yields the probability distribution c_i for the different degrees of crystallization from which the average degree of crystallization x_{cr} can be determined.

Calculations have been performed for stationary disc models by Gail [36], for time dependent models by Wehrstedt and Gail [126], and for simple semianalytic disc models and assuming only two states (either crystalline or amorphous) by Bockelée-Morvan et al. [8]. Some results are shown in Fig. 21 for a stationary disc model and in Fig. 22 for a time dependent model which confirm the expectations based on Fig. 10: Annealing temperatures of about 800 K for olivine and pyroxene dust grains, and at about 900 K for quartz grains. The low average annealing temperature found in the calculations is important in so far, as dust destruction processes by evaporation occur at a much higher temperature. Both processes, crystallisation and evaporation, occur in different zones.

An important result of the model calculations is that crystalline dust formed at about 800 K is transported by diffusion into cool outer parts of the disc, even into the region beyond 10 AU (the grey shaded area in Fig. 21), where it may

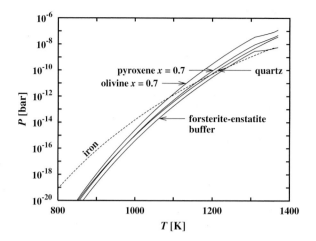

Fig. 23. Partial pressure of SiO in chemical equilibrium between the indicated solid silicon compounds and the hydrogen rich gas phase. The total pressure is $P = 10^{-4}$ bar. The dashed line shows the vapour pressure of iron

be incorporated into the planetesimals formed in the cold outer disc. Probably this mechanism is responsible for the considerable fraction of crystalline silicates observed to exist in cometary nuclei (e.g. Wooden et al. [129, 130]).

Calculations of the degree of crystallization of dust in protostellar discs based on the results of Hallenbeck et. al [48, 49] have not yet performed. A rough estimate (Gail [36]) is that crystallization occurs at about 950 K for their amorphous silicate material. Also in this case crystallization occurs at a significantly lower temperature than the evaporation of the silicate dust.

7.8   Evaporation of the Interstellar Dust Component

Accretion transports the dust material of interstellar origin into warm inner disc regions where the dust starts to evaporate. It is important to note that the pristine silicate dust components olivine, orthopyroxene and quartz are not stable in an environment with Solar System element abundances:

1.) Quartz does not exist in chemical equilibrium in such an environment because the vapour pressure of SiO molecules is much higher for quartz than for olivine and pyroxene. In an equilibrium state the silicon is condensed into olivine and pyroxene and not into quartz.

2.) The interstellar olivine and orthopyroxene are assumed to have a rather high Fe/(Fe+Mg) ratio (cf. Table 7), while in chemical equilibrium the iron content is low (cf. Fig. 8). Since the upper stability limit of the iron-magnesium silicates considerably decreases with increasing iron content, the iron rich interstellar silicates are unstable with respect to evaporation and re-condensation of their material into the iron poor silicates forsterite and enstatite and into solid iron.

The details of the evaporation mechanism of theses dust components in the inner disc depend on the degree of mixing in the disc:

1. If silicate material with equilibrium composition (forsterite + enstatite) from the inner disc region is mixed into the zone where the ISM-silicates start to

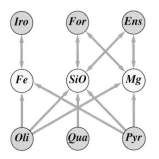

Fig. 24. Exchange of material between the different dust components (denoted by obvious abbreviations) and the vapour spezies (Fe, Mg, SiO) if the partial pressure of SiO is buffered by the presence of forsterite and enstatite. The exchange with the H_2O reservoire is not shown

evaporate, the gas phase pressure of SiO (and Mg) in the evaporation zone of the ISM-silicates is *buffered* by the low vapour pressure of the forsterite-enstatite mixture.

2. If, on the other hand, no outwards mixing of material from the inner zone occurs, the partial pressure of SiO (and Mg) in the gas phase is determined by the evaporation of the ISM-silicates itself.

The mode of conversion to an equilibrium mixture in the second case is different from the case where equilibrium grains are already present and act as buffer for the vapour pressure. Since mixing seems to be important in protoplanetary discs, only the first case needs to be considered.

Due to annealing the silicates have a crystalline lattice structure at the instant when they start to evaporate. This allows to use thermodynamic data of crystalline silicates for calculating their evaporation, which are well defined contrary to the thermodynamic data of amorphous materials. Figure 23 shows the partial pressures of SiO molecules in a state of chemical equilibrium between gas and solids with respect to the following reactions:

$$Mg_{2x}Fe_{2(1-x)}SiO_4 + 3H_2 \longrightarrow 2xMg + 2(1-x)Fe + SiO + 3H_2O$$
$$Mg_xFe_{(1-x)}SiO_3 + 2H_2 \longrightarrow xMg + (1-x)Fe + SiO + 2H_2O$$
$$SiO_2 + H_2 \longrightarrow SiO + H_2O$$

They are calculated for a typical pressure of $P = 10^{-4}$ bar in the zone where silicate destruction occurs (cf. Fig. 20), and a composition of olivine and pyroxene of $x = 0.7$ (cf. Tab. 7). Figure 23 shows that the vapour pressures of the iron rich silicates and quartz exceeds the vapour pressure of the forsterite-enstatite buffer by at least a factor of three. If evaporation and growth is not limited by diffusion in the gas phase (case A in Fig. 9) all material evaporated off from the grains of interstellar origin will precipitate on forsterite, enstatite, and on solid iron with only negligible re-condensation on the interstellar grains. The exchange of material between instellar dust and the disc made dust is schematically depicted in Fig. 24

Since only evaporation needs to be considered in the rate term (96) of equations (95) for calculating the local abundance and size spectrum of the olivine, orthopyroxene, and quartz grains of interstellar origin, one has for instance for

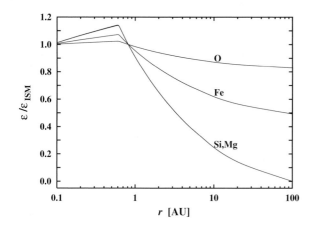

Fig. 25. Radial variation of the abundance ϵ of the elements Fe, Mg, Si, and O not bound in the ISM dust components, normalized to their ISM abundances ϵ_{ISM}. Disc model with an accretion rate of $\dot{M} = 1 \times 10^{-7}\,M_{\odot}\,\mathrm{yr}^{-1}$

olivine (Eq. 67)

$$R_{\mathrm{ol},i} = \left(\frac{c_{j,i-1}}{\Delta a_{i-1}} - \frac{c_{j,i}}{\Delta a_i} \right) \alpha_{\mathrm{ol}} v_{\mathrm{th,SiO}} \frac{p^{\mathrm{ol}}_{\mathrm{SiO,eq}}}{kT}, \qquad (100)$$

and similar terms for the other silicates. Terms of a similar structure appear as source terms in the equations for the evaporation products Mg, Fe, SiO, H_2O. Olivine, for instance, contributes the following source terms to the equations (95) for SiO, Mg, Fe and H_2O

$$
\begin{aligned}
R_{\mathrm{SiO}} &= \alpha_{\mathrm{ol}} v_{\mathrm{th,SiO}} \frac{p^{\mathrm{ol}}_{\mathrm{SiO,eq}}}{kT} \mathcal{A}_{\mathrm{ol}} \quad, & R_{\mathrm{Mg}} &= 2x R_{\mathrm{SiO}} \\
R_{\mathrm{Fe}} &= 2(1-x) R_{\mathrm{SiO}} \quad, & R_{H_2O} &= 3 R_{\mathrm{SiO}}
\end{aligned}
\qquad (101)
$$

and similar terms for the other silicates. From the solution of the equations (95) one obtains $\mathcal{A}_j$ for the dust species and from this the local production rates for Mg, Fe, SiO, and H_2O by dust evaporation which allows to calculate the local abundance of the elements Mg, Fe, Si, and O not bound in the interstellar dust grains.

Figure 25 shows the radial variation of the local abundances of the main dust forming elements not bound in the ISM dust components calculated for a stationary disc model. A conspicuous feature of the abundance distributions is the enrichment of these elements over their interstellar values in the region of dust evaporation. This results from the diffusive transport of vapours from the evaporation zone into cooler regions, opposing the effect of inwards transport by accretion, and has already been predicted by the analytical studies of e.g. Morfill and Völk [82] and Morfill et al. [83].

7.9 Iron Dust

The characteristic timescales for transport and mixing in protoplanetary discs are so big that dust growth or evaporation occurs under near-equilibrium conditions (Duschl et al. [24]). The left part of Fig. 26 shows the radial variation of

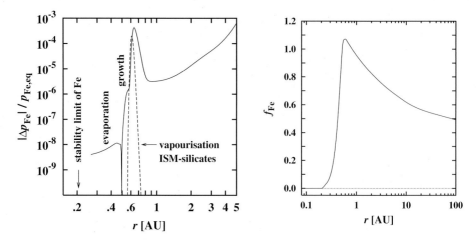

Fig. 26. *Left:* Relative deviation of the Fe partial pressure in the gas phase from the vapour pressure of solid Fe. In the region $r > 0.5\,\mathrm{AU}$ labeled by 'growth', the gas phase concentration of Fe atoms exceeds that of the equilibrium vapour pressure, in the region labeled by 'evaporation' it is less than the vapour pressure. The dashed line shows the evaporation rate of the interstellar Mg-Fe-silicates in arbitrary units. *Right:* Radial variation of the fraction f_{Fe} of Fe condensed into iron grains, normalized to the ISM abundance of Fe. Model with an accretion rate $\dot{M} = 1 \times 10^{-7}\,\mathrm{M_\odot\,yr^{-1}}$

the relative deviation of the gas phase partial pressure and the vapour pressure of Fe, calculated from (95) for Fe atoms in the gas phase by including the evaporation of ISM olivine and orthopyroxene as source terms, and evaporation and condensation of solid iron as source and sink terms:

$$R_{\mathrm{Fe}} = v_{\mathrm{th,SiO}} \left(2x_{\mathrm{ol}}\alpha_{\mathrm{ol}} \frac{p^{\mathrm{ol}}_{\mathrm{SiO,eq}}}{kT} \mathcal{A}_{\mathrm{ol}} + x_{\mathrm{py}}\alpha_{\mathrm{py}} \frac{p^{\mathrm{py}}_{\mathrm{SiO,eq}}}{kT} \mathcal{A}_{\mathrm{py}} \right)$$
$$+ \alpha_{\mathrm{ir}} v_{\mathrm{th,Fe}} \frac{p^{\mathrm{ir}}_{\mathrm{Fe,eq}} - p_{\mathrm{Fe}}}{kT} \mathcal{A}_{\mathrm{ir}} . \tag{102}$$

Obviously, the deviation from the equilibrium vapour pressure is very small. A local peak occurs where the ISM-silicates evaporate. At large r the relative deviation increases since the vapour pressure extremely rapidly decreases as the temperature becomes low while the gas phase abundance does not decrease as rapid because diffusion transports very small quantities of vapour even to large distances.

Since the deviation of the gas phase abundance from the equilibrium vapour pressure is very small, the fraction of iron condensed into solid iron can be calculated accurately as the degree of condensation as in chemical equilibrium. The result, neglecting for the possibility of FeS formation, is shown in the right part of Fig. 26. Also for iron one observes a local enrichment of iron due to vapour transport into cooler regions.

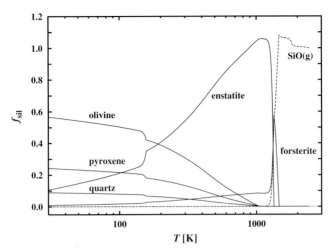

Fig. 27. Radial variation of the fraction of Si condensed into the individual Si-bearing condensates (full lines) and the fraction of Si bound in SiO molecules (dashed line). The abundances are normalized to the interstellar Si abundance

7.10   Conversion of Silicates into the Equilibrium Mixture

Figure 27 shows the fraction of the Si bound into olivine, orthopyroxene, and quartz, as calculated from the solutions of (95) for these dust species, as function of the midplane temperature in a stationary disc model. The last ISM grains evaporate at about 1 100 K. At this temperature they are already completely crystallized (cf. Fig. 21) and their vapour pressure can and has to be calculated for crystalline materials. Further, by interdiffusion all spatial compositional inhomogeneities with respect to the cations Fe^{2+} and Mg^{2+} within silicate grains are erased for $T \gtrsim 700$ K for olivine and $T \gtrsim 900$ K for pyroxene (cf. Fig. 13). If the grains arrive in the evaporation zone (cf. Fig. 13) the silicate grains can be assumed internally to be chemically homogeneous.

The characteristic timescales for a change of temperature and pressure experienced by a dust grain in the disc are so long (cf. Fig. 10) that their growth and evaporation occurs under near equilibrium conditions. The deviations of the gas phase partial pressures of the vapour compounds of the silicates, i.e. SiO, Mg and Fe[3], then are very small and forsterite and enstatite essentialy are in chemical equilibrium with the gas phase. The fraction of Si bound in forsterite and enstatite in this case also is shown in Fig. 27. One also readily sees that the region of evaporation of the equilibrium dust components forsterite and enstatite is well separated from the region where the interstellar silicates evaporate. These processes operate in different regions of the disc.

Figure 28 shows in a cumulative representation the distribution of the Si between the different silicate dust species at different radii, calculated for a

---

[3] H_2O in any case is so abundant that abundance variations with varying degree of condensation of the silicates can be neglected

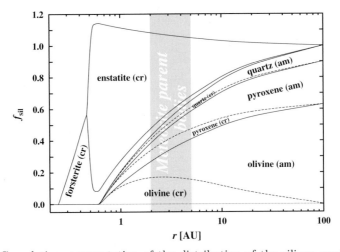

Fig. 28. Cumulative representation of the distribution of the silicon condensed into dust between the different silicate dust components (interstellar and equilibrium ones) within the protoplanetary disc, and the fraction of amorphous (am) and crystalline (cr) interstellar silicates. The equilibrium dust components forsterite and enstatite are always crystalline. The total Si abundance is normalized to the interstellar Si abundance. An enhancement of the Si abundance in the region of evaporation of the interstellar silicates is due to diffusive mixing effects. The grey shaded area roughly indicates the region where the parent bodies of meteorites are located. Stationary disc model with an accretion rate $\dot{M} = 1 \times 10^{-7}\,\mathrm{M}_\odot\,\mathrm{yr}^{-1}$

stationary accretion disc model. This is the mixture of silicate dust species from which the planetesimals are formed. Note the local enrichment of Si by diffusive mixing effects in the region where the interstellar silicates evaporate.

Table 8 shows the composition of the silicate mixture at 3 AU and 20 AU. These radii are roughly representative for the regions where the parent bodies of meteorites and the cometary nuclei, respectively, have been formed in our Solar System. The results presently are obtained on the basis of a rather crude modeling of the accretion disc (one-zone approximation, stationary, α-disc) and cannot be considered as completely realistic, but they probably outline the trends for the composition of the silicate mineral mixture in an accretion disc:

- A high fraction of equilibrated silicate dust from the warm inner region and a rather high degree of crystallization in the zone where the parent bodies of meteorites are formed.
- Mostly interstellar dust with an admixture of up to about 20% annealed interstellar dust and equilibrated dust from the warm inner disc region in the zone where cometary nuclei are formed.

These trends are in accord with what is observed for dust from comets (e.g. Wooden et al. [129, 130]). For meteoritic matrix material the predicted fraction of enstatite in the mixture seem to be somewhat high compared to forsterite,

Table 8. Composition of the silicate mixture at 3 and 20 AU in a stationary protoplanetary disc model with $\dot{M} = 1 \times 10^{-7}\,M_\odot\,yr^{-1}$. The numbers are fractions of the silicon contained in the different dust species. 'cr' and 'am' denote crystalline or amorphous dust, respectively

| r | olivine | | forsterite | pyroxene | | enstatite | quartz | |
AU	cr	am	cr	cr	am	cr	cr	am
3	0.159	0.158	0.037	0.067	0.070	0.458	0.023	0.027
20	0.073	0.469	0.010	0.031	0.202	0.129	0.011	0.076

but the general trends fits with the observed composition of the matrix material of primitive meteorites (eg. Scott et al. [111], Brearley et al. [12], Buseck & Hua [15]). The somewhat high pyroxene content in the present model calculation results from the assumption of complete chemical equilibrium between forsterite and enstatite, which possibly cannot be attained because of the slow forsterite-enstatite conversion (cf. Fig. 13).

7.11   Processing of Other Dust Components

At temperatures below $\approx 720\,K$ part of the solid iron in the protoplanetary accretion disc reacts with H_2 to form FeS. The kinetics of this process was discussed by e.g. Fegley & Prinn[30] and Fegley [29]. It has been studied in the laboratory and discussed in detail by Lauretta et al. [67] and by Tachibana & Tsuchiyama [117]. Figure 13 shows the reaction timescale for conversion of a $0.1\,\mu m$ iron grain into FeS at a pressure of $10^{-3}\,bar$ and Solar System abundances, calculated from the linear reaction kinetics rate coefficient of Lauretta et al. [67]. For small grains (size $\lesssim 10\,\mu m$) the reaction is fast enough that the abundance of FeS can be assumed to be as in chemical equilibrium. No model calculation results for the radial distribution of the thickness of FeS rims on Fe grains in an accretion disc are presently available.

The kinetics of formation of magnetite is discussed in Fegley [29], Hong & Fegley [52] and of other materials in Fegley [29].

According to equilibrium calculations there exist several dust components bearing Al and Ca which are stable in different regions of the p-T-plane (cf. Fig. 4). If condensates from different zones of the disc are intermingled by mixing and transport processes, dust components moved from their stability region into a zone where they are unstable start to convert into the locally stable species. No experimental data for the processes by which the different species are converted into each other (e.g. corundum to gehlenite, corundum to spinel, gehlenite to diopside ...) presently seem to be available.

8   Concluding Remarks

In this contribution we have discussed some problems related to the formation and evolution of mineral mixtures in circumstellar dust shells and protoplanetary accretion discs. We have seen that numerous experimental investigations have been undertaken during the last decade to study a lot of processes which are important for the condensation, evaporation and for the chemical and physical processing of cosmic dust. Many phenomenological coefficients entering the basic equations describing such processes now are determined thanks to these experimental efforts. This enables more realistic model calculations for the abundant magnesium-iron-silicates and for the iron dust component than that which were possible in the past, where such coefficients usually had simply to be guessed.

On the theoretical side much progress has been achieved during the last years in modeling the structure and evolution of protoplanetary accretion discs in 1, 2 and even 3 spatial dimensions. Combining such calculations with realistic modelings of the composition and evolution of the dust component in discs will certainly dramatically improve our understanding of the early history of our Solar System and generally of the early stages of the evolution of planetary systems.

For circumstellar dust shells there is an urgent need for combining time dependent hydrodynamical calculations of stellar winds with multicomponent dust formation models in order to obtain realistic models for oxygen rich AGB stellar winds.

There remain, however, important deficiencies in our knowledge. For the important group of Al-Ca-compounds no such type of laboratory measurements of dust evaporation or growth and of dust processing are available as for the magnesium-iron-silicates. This probably results from the much higher temperatures required for such experiments for the more refractive Ca-Al-compound than for the silicates. Technical improvements in the high temperature experimental equipment will hopefully improve this situation in the near future.

The main shortcoming in the whole field, however, is our only rudimentary knowledge (or better, our ignorance) of the basic processes responsible for dust nucleation. More theoretical and experimental efforts are required to solve this fundamental problem.

Acknowledgements

This work has been performed as part of the projects of the special research programmes SFB 359 *"Reactive flows, diffusion and transport"* and SFB 439 *"Galaxies in the Young Universe"* which are supported by the Deutsche Forschungsgemeinschaft (DFG).

References

[1] C.C. Allen, R.V. Morris, H.V. Lauer Jr., D.S. McKay: Icarus **104**, 291 (1993)

[2] E. Anders, N. Grevesse N.: Geochimica et Cosmochimica Acta **53**, 197 (1989)

[3] P.W. Atkins: Physical Chemistry, 5th edn. (Oxford University Press, Oxford 1994)

[4] I. Barin: Thermochemical Data of Pure Substances, Vol. I + II, 3rd edn. (VCH Verlagsgesellschaft Weinheim 1995)

[5] T.J. Bernatowicz, R. Cowsik, P.C. Gibbons, K. Lodders, K., B. Fegley Jr., S. Amari, R.S. Lewis: Astrophysical J. **472**, 760 (1996)

[6] M. Blander, J.L. Katz: Geochimica et Cosmochimica Acta **31**, 1025 (1967)

[7] T. Blöcker: Astrophysics & Space Sci. **275**, 1 (2001)

[8] D. Bockelée-Morvan, D. Gautier, F. Hersant, J.-M. Huré, J.-M., F. Robert: Astronomy & Astrophysics **384**, 1107 (2002)

[9] A.I. Boothroyd, I.-J. Sackmann: Astrophysical J. **510**, 232 (1999)

[10] J. Bouwman, G. Meeus, A. de Koter, S. Hony, C. Dominik, L.B.F.M. Waters: Astronomy & Astrophysics **375**, 950 (2001)

[11] J.B. Brady: 'Diffusion Data for Silicate Minerals, Glasses and Liquids'. In: *Mineral Physics & Crystallography, A Handbook of Physical Constants*, ed. by Th.J. Ahrens (American Geophysical Union, 1995) pp. 269–290.

[12] A.J. Brearley, E.R.D. Scott, K. Keil, R.N. Clayton, T.K. Mayeda, W.V. Boynton, D.H. Hill: Geochimica et Cosmochimica Acta **53**, 2081 (1989)

[13] J.R. Brucato, L. Colangeli, V. Mennella, P. Palumbo, E. Bussoletti: Astronomy & Astrophysics **348**, 1012 (1999)

[14] D.K. Buening, P.R. Buseck: J. Geophys. Res. **78** 6852 (1973)

[15] P.R. Buseck, X. Hua: Ann. Rev. Earth Planet. Sci. **21**, 255 (1993)

[16] S. Chakraborty, J.R. Farver, R.A. Yund, D.C. Rubie: Phys. Chem. Minerals **21**, 489 (1994)

[17] S. Chakraborty: J. Geophysical Research **102**, 12317 (1997)

[18] C. Chang, A.B.C. Patzer, E. Sedlmayr, D. Sülzle: European Physical J. D **2**, 57 (1998)

[19] Ch. Chang, A.B.C. Patzer, E. Sedlmayr, T. Steinke, D. Sülzle: Chemical Physics Letters **324**, 108 (2000)

[20] M. W. Chase: NIST-JANAF Thermochemical Tables. 4th edn. Journal of Physical and Chemical Reference Data, Monograph No. 9 (1998)

[21] I. Cherchneff, J.R. Barker, A.G.G.M. Tielens: Astrophysical J. **401**, 269 (1992)

[22] I. Cherchneff, Y.H. Le Teuff, P.M. Williams, A.G.G.M. Tielens: Astronomy & Astrophysics **357**, 572 (2000)

[23] D.D. Clayton, E.A.-N. Deneault, B.S. Meyer: Astrophysical J. **562**, 480 (2001)

[24] W.J. Duschl, H.-P. Gail, W.M. Tscharnuter: Astronomy & Astrophysics **312**, 624 (1996)

[25] B.T. Draine: Astrophysics & Space Sci. **65**, 313 (1979)

[26] D.S. Ebel, L. Grossman: Geochimica et Cosmochimica Acta **64**, 339 (2000)

[27] B. Edvardsson, J. Andersen, B. Gustafsson et al.: Astronomy & Astrophysics **275**, 101 (1993)

[28] D. Fabian, C. Jäger, Th. Henning, J. Dorschner, H. Mutschke: Astronomy & Astrophysics **364**, 282 (2000)

[29] B. Fegley Jr.: Space Sci. Rev. **92**, 177 (2000)

[30] B. Fegley Jr., R.G. Prinn : 'Solar nebula chemistry: implications for volatiles in the solar system'. In: *The Formation and Evolution of Planetary Systems* ed. by H.A. Weaver, L. Danly (Cambridge University Press, Cambridge 1989) pp. 171–205

[31] A.S. Ferrarotti, H.-P. Gail: Astronomy & Astrophysics **371**, 133 (2001)

[32] A.S. Ferrarotti, H.-P. Gail: Astronomy & Astrophysics **382**, 256 (2002)

[33] R. Freer: Contrib. Mineral. Petrol **76** 440 (1981)

[34] M. Frenklach, E.D. Feigelson: Astrophysical J. **341**, 372 (1989)

[35] H.-P. Gail: Astronomy & Astrophysics **332**, 1099 (1998)

[36] H.-P. Gail: Astronomy & Astrophysics **387**, 192 (2001)

[37] H.-P. Gail, E. Sedlmayr: Astronomy & Astrophysics **206**, 153 (1988)

[38] H.-P. Gail, E. Sedlmayr: Astronomy & Astrophysics **177**, 186 (1987)

[39] H.-P. Gail, E. Sedlmayr:, 'Dust formation in M stars'. In: *The Molecular Astrophysics of Stars and Galaxies* ed. by T.W. Hartquist, D.A. Williams (Oxford University Press, Oxford 1998) pp. 285–312

[40] H.-P. Gail, E. Sedlmayr: Faraday Discussion **109**, 303 (1998)

[41] H.-P. Gail, E. Sedlmayr: Astronomy & Astrophysics **347**, 594 (1999)

[42] D.R. Gies, D.L. Lambert: Astrophysical J. **387**, 673 (1992)

[43] R.C. Gilman: Astrophysical J. **155**, L185 (1969)

[44] A.E. Glassgold: Ann. Rev. Astron. & Astroph. **34**, 241 (1996)

[45] N. Grevesse, A. Noels: 'Cosmic abundances of the Elements'. In: *Origin and Evolution of the Elements*, ed. by N. Prantzos, E. Vangioni-Flam, M. Cassé (Cambridge University Press, Cambridge 1993), pp. 15–25

[46] M.A.T. Groenewegen, L.B. van den Hoek, T. de Jong: Astronomy & Astrophysics **293**, 381 (1995)

[47] L. Grossman: Geochimica et Cosmochimica Acta **36**, 597 (1972)

[48] S.L. Hallenbeck, J.A. Nuth III, P.L. Daukantas: Icarus **131**, 198 (1998)

[49] S.L. Hallenbeck, J.A. Nuth III, R.N. Nelson: Astrophysical J. **535**, 247 (2000)

[50] A. Hashimoto: Nature **347**, 53 (1990)

[51] A. Hashimoto: Meteoritics & Planetary Science **33**, A65 (1998)

[52] Y. Hong, B. Fegley Jr.: Meteoritic & Planetary Science **33**, 1101 (1998)

[53] N. Imae, A. Tsuchiyama, M. Kitamura: Earth Planet. Sci. Lett. **118**, 21 (1993)

[54] H. Inaba, S. Tachibana, H. Nagahara, K. Ozawa: Lunar and Planetary Science Conference **XXXII**, 1837 (2001)

[55] K.S. Jeong, Ch. Chang, E. Sedlmayr, D. Sülzle: J. Physics B **33**, 3417

[56] J.F. Kerridge: Icarus **106**, 135 (1993)

[57] R. Keller: 'Polyaromatic hydrocarbons and the condensation of carbon in stellar winds'. In: *Polycyclic Aromatic Hydrocarbons and Astrophysics* ed. by A. Legér, L. d'Hendecourt, N. Boccara (Reidel, Dordrecht 1987) pp. 387–397

[58] A. Klügel: Contrib. Mineral. Petrol. **141** 1 (2001)

[59] A. Kouchi, T. Yamamoto, T. Kozasa, T. Kuroda, J.M. Greenberg: Astronomy & Astrophysics **290**, 1009 (1994)

[60] T. Kozasa, H. Sogawa: 'Formation of crystalline silicate around oxygen-rich AGB stars'. In: *Asymptotic Giant Branch Stars* ed. by T. Le Bertre, A. Lèbre, C. Waelkens (Astronomical Society of the Pacific 1999) pp. 239–244

[61] A. Krot, B. Fegley Jr., K. Lodders, H. Palme: 'Meteoritical and astrophysical constraints on the oxidation state of the solar nebula'. In: *Protostars and Planets IV* ed. by V. Mannings, A.P. Boss, S.S. Russel (University of Arizona Press, Tucson 2000) pp. 1019–1054

[62] O. Kubaschewski, C.B. Alcock: Metallurgical Chemistry, 5th edn. (Pergamon Press, Oxford 1983)

[63] Landolt-Börnstein: *Zahlenwerte und Funktionen* Vol. 5b, ed. by K. Schäfer (Springer Verlag, Heidelberg 1968)

[64] J.W. Larimer: Geochimica et Cosmochimica Acta **31**, 1215 (1967)

[65] J. Lattanzio, M. Forestini: In: *IAU Symposium 191: Asymptotic Giant Branch Stars* ed. by T. Le Bertre, A. Lèbre, C. Waelkens (Astronomal Society of the Pacific, 1999) pp. 31–40

[66] J.M. Lattimer, D.N. Schramm, L. Grossman: Astrophysical J. **219**, 230 (1978)

[67] D.S. Lauretta, D.T. Kremser, B. Fegley Jr.: Icarus **122** 288 (1996)

[68] P. Lenzuni, H.-P. Gail, Th. Henning: Astrophysical J. **447**, 848 (1995)

[69] R.D. Lide: CRC Handbook of Chemistry and Physics, 76th ed. (CRC Press, Boca Raton, 1995)

[70] K. Lodders, B. Fegley Jr.: Meteoritics **30**, 661 (1995)

[71] K. Lodders, B. Fegley Jr.: 'Condensation Chemistry of Carbon Stars'. In: *Astrophysical Implications of the Laboratory Study of Presolar Meterials* ed. by T.J. Bernatowicz, E.K. Zinner (America Institute of Physics, New York 1997) pp. 391–423

[72] K. Lodders, B. Fegley Jr.: 'Condensation Chemistry of Circumstellar Grains'. In: *Asymptotic Giant Branch Stars* ed. by T. Le Bertre, A. Lèbre, C. Waelkens (Astronomical Society of the Pacific 1999) pp. 279–289

[73] R.A. Mendybaev, J.R. Beckett, L. Grossman, E. Stolper: Lunar and Planetary Science Conference **XXIX**, 1871 (1998)

[74] G. Meynet, A. Maeder, G. Schaller, D. Schaerer, C. Charbonnel: Astronomy & Astrophysics Suppl. **103**, 97 (1994)

[75] D.J. Misener: 'Cationic diffusion in olivine to 1400°C and 35 kbar'. In: *Geochemical Transport and Kinetics*, ed. by A.W. Hoffman, B.J. Giletti, H.S. Yoder, R.A. Yund (Carnegie Inst. of Washington, Washington D.C. 1974) pp. 117–129

[76] Molster F.J., 2000, PhD Thesis, University of Amsterdam

[77] F.J. Molster, L.B.F.M. Waters, A.G.G.M. Tielens, M.J. Barlow: Astronomy Astrophysics **382**, 184 (2002)

[78] F.J. Molster, L.B.F.M. Waters, A.G.G.M. Tielens: Astronomy & Astrophysics **382**, 222 (2002)

[79] F.J. Molster, L.B.F.M. Waters, A.G.G.M. Tielens, C. Koike, H. Chihara: Astronomy & Astrophysics **382**, 241 (2002)

[80] M. Morioka: Geochimica et Cosmochimica Acta **44**, 759 (1980)

[81] M. Morioka: Geochimica et Cosmochimica Acta **45**, 1573 (1981)

[82] G.E. Morfill, H.-J. Völk: 1984, Astrophysical J. **287**, 371 (1984)

[83] G.E. Morfill, W.M. Tscharnuter, H.-J. Völk: 'Dynamical and chemical evolution of the protoplanetary nebula'. In: *Protostars & Planets II*, ed. by D.C. Black, M.S. Matthews (University of Arizona Press, Tucson 1985) pp. 493–533

[84] H. Nagahara, I. Kushiro, B.O. Mysen, H. Mori: Nature **331**, 516 (1988)

[85] H. Nagahara, I. Kushiro, B.O. Mysen: Geochimica et Cosmochimica Acta **58**, 1951 (1994)

[86] H. Nagahara, K. Ozawa: Meteoritcs **29**, 508 (1994)

[87] H. Nagahara, K. Ozawa: Geochimica et Cosmochimica Acta **60**, 1445 (1996)

[88] H. Nagasawa, T. Suzuki, M. Ito, M. Morioka: Phys. Chem. Minerals **28**, 706 (2001)

[89] R.H. Nichols Jr., G.J. Wasserburg: Lunar & Planetary Science Conference XXVI, 1047 (1995)

[90] R.H. Nichols Jr., R.T. Grimley, G.J. Wasserburg: Meteoritics & Planetary Science **33**, A115 (1998)

[91] L.R. Nittler, C.M.O. Alexander, X. Gao, R.M. Walker, E. Zinner: Astrophysical J. **483**, 475 (1997)

[92] K. Ozawa, H. Nagahara: Geochimica et Cosmochimica Acta **64**, 939 (2000)

[93] A.B.C. Patzer, C. Chang, E. Sedlmayr, D. Sülzle: European Physical J. D **6**, 57 (1999)

[94] A.B.C. Patzer, A. Gauger, E. Sedlmayr: Astronomy & Astrophysics **337**, 847 (1998)

[95] A.R. Philpotts: Pinciples of Igneous and Metamorphic Petrology (Prentice Hall, Englewood Cliffs 1990)

[96] J.B. Pollack, D. Hollenbach, S. Beckwith, D.P. Simonelli, T. Roush, W. Fong: Astrophysical J. **421**, 615 (1994)

[97] H.R. Pruppacher, J.D. Klett: Microphysics of Clouds and Precipitation (Reidel, Dordrecht 1978)

[98] A. Putnis: An Introduction to Mineral Sciences (Cambridge Universty Press, Cambridge 2001)

[99] F.J.M. Rietmeijer, J.A. Nuth III: Astrophysical J. **527**, 395 (1999)

[100] F.J.M. Rietmeijer, J.A. Nuth III, J.M. Karner: Phys. Chem. Chem. Phys. **1**, 1511 (1999)

[101] F.J.M. Rietmeijer, J.M. Karner: J. Chem. Phys. **110**, 4554 (1999)

[102] E.E. Salpeter: Astrophysical J. **193**, 579 (1974)

[103] S.K. Saxena: Thermodynamics of Rock-Forming Crystalline Solutions (Springer, Berlin 1973)

[104] S.K. Saxena, G. Eriksson: 'Chemistry of the Formation of the Terrestrial Planets'. In: *Chemistry and Physics of Terrestrial Planets*, ed. by S.K. Saxena (Springer, New York 1986), pp. 30–105

[105] S.K. Saxena, N. Chatterjee, Y. Fei, G. Shen: Thermodynamic Data on Oxides and Silicates (Springer, Heidelberg 1993)

[106] G. Schaller, D. Schaerer, G. Meynet, A. Maeder: Astronomy & Astrophysics Suppl. **96**, 269 (1992)

[107] H. Schmalzried, A. Navrotsky: Festkörperthermodynamik (Verlag Chemie, Weinheim, 1975)

[108] H. Schmalzried: Chemical Kinetics of Solids (Wiley-VCH, Weinheim 1995)

[109] C.S. Schwandt, R.T. Cygan, H.R. Westrich: Contrib. Mineral. Petrol. **130**, 390 (1998)

[110] J.M. Scalo, J.E. Ross: Astronomy & Astrophysics **48**, 219 (1976)

[111] E.R.D. Scott, D.J. Barber, C.M. Alexander, R. Hutchinson, J.A. Peck: 'Primitive material surviving in chondrites: matrix'. In: *Meteorites and the early solar system* ed. by J.F. Kerridge, M.S. Matthews (University of Arizona Press. Tucso 1988) pp. 718–745

[112] E. Sedlmayr, D. Krüger: 'Formation of Dust Particles in Cool Stellar Outflows'. In: *Astrophysical Implications of the Laboratory Studies of Presolar Material* ed. by. T.J. Bernatowicz, E.K. Zinner (American Institute of Physics 1997) pp. 425–450

[113] C.M. Sharp, W.F. Huebner: Astrophysical J. Suppl. **72**, 417 (1990)

[114] C.M. Sharp, G.J. Wasserburg: Geochimica et Cosmochimica Acta **59** 1633 (1995)

[115] W.R. Smith, R.W. Missen: Chemical Reaction Equilibrium Anaysis: Theory and Algorithms. (Wiley, New York 1982)

[116] H. Sogawa, T. Kozasa: Astrophysical J. **516**, L33 (1999)

[117] S. Tachibana, A. Tsuchiyama: Geochimica et Cosmochimica Acta **62**, 2005 (1998)

[118] S. Tachibana, H. Nagahara, K. Ozawa: Lunar and Planetary Science Conf. **XXXII**, 1767 (2001)

[119] S. Tachibana, A. Tsuchiyama, H. Nagahara: Lunar and Planetary Science Conf. **XXIX**, 1539 (1998)

[120] S. Tachibana, A. Tsuchiyama, H. Nagahara: Lunar and Planetary Science Conf. **XXXI**, 1588 (2000)

[121] S. Tachibana, A. Tsuchiyama, H. Nagahara: Geochimica et Cosmochimica Acta **66**, 713 (2002)

[122] S.P. Thompson, C.C. Tang: 2001, Astronomy & Astrophysics **368**, 721 (2001)

[123] A.G.G.M. Tielens: 'The destruction of interstellar dust'. In: *Formation and Evolution of Solids in Space*, ed. by J.M. Greenberg, A. Li (Kluwer, 1999), pp. 331–375

[124] A. Tsuchiyama, S. Tachibana, T. Takahashi: Geochimica et Cosmochimica Acta **63**, 2451 (1999)

[125] J. Wang, A.M. Davis, R.N. Clayton, A. Hashimoto: Geochimica et Cosmochimica Acta **63**, 953 (1999)

[126] M. Wehrstedt, H.-P. Gail: Astronomy & Astrophysics **385**, 181 (2002)

[127] S.J. Weidenschilling, J.N. Cuzzi: 'Formation of Planetesimals in the Solar Nebula'. In: *Protostars and Planets III*, ed. by E.H. Levy, J.I. Lunine (University of Arizona Press, Tucson 1993) pp. 1031–1060

[128] B.J. Wood, O.J. Kleppa: Geochimica et Cosmochimica Acta **45**, 534 (1981)

[129] D.H. Wooden, D.E. Harker, C.E. Woodward, H.M. Butner, C. Koike, F.C. Witteborn, C.W. McMurtry: Astrophysical J. **517**, 1034 (1999)

[130] D.H. Wooden, H.M. Butner, D.E. Harker, C.E. Woodward: Icarus **143**, 126 (2000)

[131] T. Yamamoto, T. Chigai, S. Watanabe, T. Kozasa: Astronomy & Astrophysics **380**, 373 (2001)

The Mineralogy of Interstellar and Circumstellar Dust

Frank J. Molster[1] and Laurens B.F.M. Waters[2,3]

[1] Research support division, Research and Scientific Support Department of ESA, ESTEC, Keplerlaan 1, 2201 AZ Noordwijk, The Netherlands
[2] Astronomical Institute "Anton Pannekoek", University of Amsterdam, Kruislaan 403, 1098 SJ Amsterdam, The Netherlands
[3] Instituut voor Sterrenkunde, Katholieke Universiteit Leuven, Celestijnenlaan 200B, B-3001 Heverlee, Belgium

Abstract. The study of dust in space was for a long time hampered by the lack of resolution and wavelength coverage in the infrared. The launch of the Infrared Space Observatory (ISO) changed this dramatically. Its unprecedented wavelength range (2.4 - 200 μm) together with its relatively high spectral resolution ($\lambda/\Delta\lambda = 2000-150$) made this instrument ideal to study dust in space. Many new dust species have been found, in particular of oxygen-rich species. The quality of the data allows a detailed mineralogy of individual species. The ISO database can be used to carry out an inventory of the occurrence of dust species in various circumstellar and interstellar environments. The picture that emerges is that of a very rich circumstellar dust mineralogy, while the interstellar medium shows only a limited amount of species. We present an overview of this inventory, as well as of the mineralogy of the dust species found. The implications for our understanding of dust processing in different astrophysical environments are also discussed.

1 Introduction

Interstellar space is not empty, but filled with gas - mostly hydrogen - and small solid particles, referred to as *dust*. The presence of small interstellar dust particles was first deduced in the early part of the 20th century from the reddening of the colours of stars with a similar spectral type. Starlight was found to be polarized, which can be explained if some of the light is scattered by non-spherical aligned dust particles in the line of sight. The properties of reflection nebulae can be understood if they contain small solid particles. Studies of interstellar extinction and its wavelength dependence have resulted in constraints on the gas to dust ratio in the interstellar medium (ISM), the size distribution of interstellar grains, and to some extent their chemical composition, see Li & Greenberg [1] for an excellent review.

The ubiquitous presence of dust became even more evident with the advent of infrared astronomy around 1960, and in particular with the far-IR all-sky survey carried out with the IRAS satellite in 1983. The infrared wavelength range is where the bulk of the thermal emission from dust grains, heated by optical and ultraviolet starlight, is emitted. Some galaxies contain so much dust that they emit more than 90 per cent of their energy at infrared wavelengths! The study of dust impacts on our understanding of the physics and chemistry of the ISM, the star formation process, and on the last phases in the life of stars.

Thermal emission from dust has been detected from a wide variety of regions in our and in external galaxies, ranging between the densest molecular clouds to the diffuse low-density clouds at high galactic latitude, and even from intergalactic clouds. IR excess emission from stars is often a strong indication for the presence of *circumstellar* dust. Indeed, stardust is found around stars of a wide variety of mass, luminosity and evolutionary state. This dust can be related to their infancy, i.e. the star- and planet formation process, or it can be due to episodes of mass loss that stars experience at various epochs during their life. There is a strong relation between the dust produced by stars during their life, dust in the ISM and dust in molecular clouds and star forming regions. Stars are believed to be the main dust factories in galaxies. The *life cycle* of dust, from their production by stars, their life in the ISM and their incorporation into new generations of stars and planets is one of the central themes of astromineralogy.

Perhaps the most powerful tool in the field of observational astromineralogy is infrared spectroscopy of the vibrational resonances in the solid. The IR spectral region contains most of the relevant resonances of abundant species. These resonances provide information about chemical composition, size, shape, lattice structure, abundance and temperature of the grains. The resonances can be studied in absorption against bright background sources (stars) or in emission.

Infrared spectroscopy of dust in space started around 1970 with the detection of broad emission bands in the mid-infrared wavelength region using ground-based telescopes. These bands were attributed to amorphous silicates and polycyclic aromatic hydrocarbons (PAHs). Large contributions to infrared spectroscopy were made by the NASA Kuiper Airborne Observatory (KAO) and by the Low Resolution Spectrometer (LRS) on board IRAS, which operated between 8 and 23 μm. A landslide of new results in the field of astromineralogy has resulted from observations made with the Short Wavelength Spectrometer (SWS [2]) and Long Wavelength Spectrometer (LWS [3]) on board of the Infrared Space Observatory (ISO [4]). These two instruments provided an uninterrupted spectral coverage between 2 and 200 μm with a resolution $\lambda/\Delta\lambda$ between 150 and 2000. We will mostly use the ISO data to illustrate the presence of minerals in space.

This chapter describes our current understanding of the composition of interstellar and circumstellar dust as derived from infrared spectroscopy. We concentrate on refractory materials and restrict the discussion to solids, leaving out large molecules such as PAHs and ices.

2   Observations and Identification

Apart from the dust in our own solar system, which can be collected and analyzed in situ or brought back to the laboratories on earth, all other dust grains can only be studied via their influence on electro-magnetic radiation. In principle there are 3 ways in which dust interacts with light; it can absorb, scatter and emit radiation; in general it will do all three at the same time. Since the amount of absorption, scattering and/or emission depends on the grain shape, size, composition and the wavelength of the light, it allows us to get quite accurate information about the properties of the dust present outside our solar system. The main characteristic dust bands lie in the infrared region, where most of the vibrational and translational bands are found. In the remainder of this section we will therefore discuss the infrared properties of cosmic dust grains.

2.1 The Identification of Solids in Space

The identification of solids using astronomical spectra of interstellar and circumstellar dust requires laboratory data of astrophysically relevant samples whose properties are determined at the relevant temperatures. Ideally, "real" samples of stardust or interstellar dust should be used for this purpose, but such materials are either not available or extremely rare, and at any rate difficult to handle. Materials that could be used in this context are lunar samples, meteorites and Interplanetary Dust Particles (IDPs). However, it is by no means evident that these materials are indeed representative. An exception could be the IDPs of cometary origin, such as used by Bradley et al. [5] for amorphous silicates and by Keller et al. [6] for FeS.

Alternatively, laboratory samples of natural (i.e. found on Earth) and of synthetic minerals are a useful starting point to compare to the astronomical spectra. Of course, the simplest and most abundant minerals found in the solar system are the prime candidates to use. However, it should be kept in mind that the local conditions in space can be very different from those that prevailed during the formation of the solar system, resulting in the formation of minerals that are rare in the solar system.

There are a couple of methods to obtain the spectral properties of materials, examples are absorption-, reflection-, emission- and raman-spectroscopy. The most frequently used methods for astronomical purposes are reflection and absorption (also referred to as transmission) spectroscopy. The absorption spectra of very small particles ($< 1\mu$m) turn out to be very helpful for identification purposes and to obtain relative strength ratios of the features. Apparently the size and shape of the particles after grinding down is a good analogue for, but not necessarily similar to, the material that has been found around stars. For all the identification purposes in the rest of this section we have used absorption spectra. However, for modeling purposes, where not only the absorption but also the scattering is important, it is necessary to know the complete optical constants of the materials. Since the shape distribution of the laboratory samples is often not well characterized, absorption spectra are less helpful to determine the optical properties. Reflection spectra, although less convenient for identification purposes, are preferred to derive the optical constants.

In Fig. 1 we show the infrared spectrum of MWC922, together with the (calculated) emissivity of forsterite at 90 K, roughly the temperature of the forsterite around MWC922 [7]. It is clear that the grain shape is of great importance in the

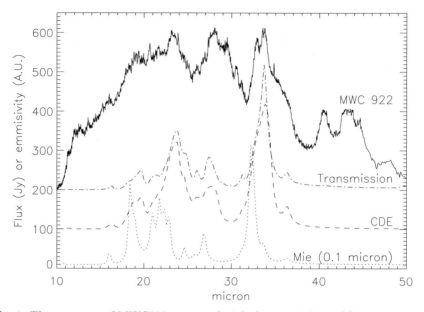

Fig. 1. The spectrum of MWC922 compared with the emissivities of forsterite at 90 K derived directly from transmission spectra [dashed dotted line 8], and calculated for two different grain shape distributions; Mie calculations of 0.1 μm spheres (dotted line) and a continuous distribution of ellipsoids (CDE; dashed line). The optical constants used for these calculations are derived from reflection measurements [9, 10].

position and width of the features. For forsterite, the Mie calculations (0.1 μm spheres) clearly differ from the positions measured in transmission for the same material. The continuous distribution of ellipsoids (CDE) gives already a much better fit, although the features seem to be a little too broad. The best match to the astronomical data is given by the absorption spectra. This indicates that the grain shapes of ground down grains used for transmission are probably a better analogue for the actual shape distribution of astronomical grains than spherical or ellipsoidal grains.

The detailed interpretation of the presently available astronomical spectra requires (full) radiative transfer calculations using optical constants measured in the laboratory and assumptions about the source geometry, and the composition, abundance, size, and shape of the grains. The laboratory spectra should have a large wavelength coverage (preferably from the UV to the mm wavelengths) and with relatively high spectral resolution ($\frac{\lambda}{\Delta\lambda} \geq 200$ for the infrared) and should be of chemically and structurally well characterized materials with an appropriate grain size (and shape). Unfortunately, only in recent years such thorough studies of cosmic dust analogues have become available, while many older studies are difficult to use due to only partial wavelength coverage and/or poor sample characterization. This can lead to some confusion in the literature as to the assignment of some bands to certain minerals. Nevertheless, even in

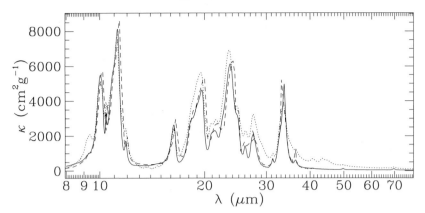

Fig. 2. The mass absorption coefficients of forsterite derived from different laboratory measurements by Koike et al. [8, solid line], Jäger et al. [13, dotted line], Koike et al. [12, dashed line]. Note that the Jäger et al. data have been multiplied by a factor 2 to match the other measurements. Figure taken from [27].

those cases where samples have been properly characterized, disturbing differences still remain (see Fig. 2). Useful laboratory measurements of well qualified materials include amorphous [11] and crystalline silicates [12, 8, 13, 14], oxides [15], metallic iron [16], metal sulfides [17], carbon bearing minerals like amorphous carbon [18], SiC [19, 20], TiC [21] and carbonates [22], sometimes also measured at different temperatures [23, 24, 25, 26].

As shown in Fig. 1, there are still problems to convert the optical properties in a realistic way to absorption and scattering efficiencies. These properties can only be exactly calculated for a restricted set of regular shapes (spheres, ellipsoids, needles etc.), while in nature dust particles do not have these regular shapes. Although CDE nicely reproduces the spectra in the case of forsterite, there are still quite some problems with this approach. First of all, CDE is not suitable for all materials (see e.g. [6]) and secondly it assumes that all particles are significantly smaller than the wavelength considered. Several methods exist to calculate the optical properties of irregular grains (like the discrete dipole approximation (DDA) [28, 29] and T(ransition)-matrix method [30, 31]), however these methods often require long calculation times to calculate even a single grain, not to mention whole grain populations, and is not applicable to all sets of optical constants (especially not when very strong resonances are involved). But in general even with these restricted set of shapes, one can get a reasonable estimate of the kind, amount and location of the circumstellar dust [e.g. 32, 33, 34].

Besides grain size and grain shape effects, also the temperature influences the shape and position of the features. Laboratory experiments show that optical constants are temperature dependent. Lowering the temperature will result in narrowing and shifting of most features [see e.g. 23, 24, 25, 26]. As the bond length decreases, the energy of the associated phonon mode increases, and there-

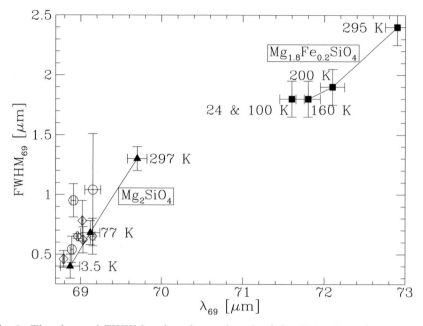

Fig. 3. The observed FWHM and peak wavelength of the 69.0 micron feature in the spectra of the dust around stars (open diamonds for the sources with a disk and open circles for the sources without a disk) and in the laboratory at different temperatures (filled triangles - forsterite [Fo_{100}; 25], and filled squares - olivine [Fo_{90}; 24]). The temperatures are indicated at each point, and within the resolution the 24 K and 100 K for Fo_{90} are similar. Note that the measurements were not corrected for the instrumental FWHM (≈ 0.29, for the ISO observations, and 0.25 and 1.0μm for the laboratory observations of respectively Fo_{100} and Fo_{90}). Figure taken from [7].

fore the infrared peak shifts to shorter wavelengths (see Fig. 3). This phenomenon can be used to estimate the temperature of the dust species [35]. Despite all this, optical constants have often only been measured at room temperature.

The precise mineralogical composition is also very important. Measurements of several samples in the same solid solution series but with slightly different compositions allow us to determine the composition of the minerals quite accurately. The spectral resolution of ISO allows us for example to see the difference between Fe-rich, Fe-poor and Fe-free silicates of the same solid solution series (see e.g. Fig. 3). The purity and characterization of the laboratory samples is therefore very important for an accurate description of the dust found around other stars.

Alternatively, it is possible to derive empirical dust properties from astronomical observations covering a wide wavelength range. For example, in this way optical constants for amorphous ("astronomical") silicates have been constructed [36, 37]. However, since we do not have a sample of the material, their exact composition has remained unclear. Nevertheless, this approach has proven very practical in many studies of interstellar dust. Of course, it should not be forgotten that for a long time the spectral resolution of the astronomical observations lacked the quality needed for detailed investigations of the dust properties. This changed with the launch of the Infrared Space Observatory (ISO). The ISO spectra do allow for detailed investigations of the circumstellar and interstellar dust composition.

3   Observational Astromineralogical Results

In Table 1 we have listed the dust species which were found spectroscopically in astrophysical environments. Note that for most dust species this is based on the identification of only one (strong) feature. In the case of the amorphous and crystalline silicates, $MgAl_2O_4$, FeS and the nano-diamonds more than one band could be attributed to these species. Most of the solar system materials have been analyzed by other means than spectroscopy (e.g. chemically, TEM, etc.), and the identification of the different dust species is therefore very robust.

Table 1. Overview of the presence of the different dust species in astronomical environments. $\sqrt{}$: indicates an insecure or dubious detection.
ISM = interstellar medium; SS= solar system material, IDPs & meteortites; (p)AGB = (post) asymptotic giant branch star; PNe = planetary nebulae; PS = proto stars; TT = T-Tauri stars; Her = Herbig Ae/Be stars; O = O-rich; C = C-rich; Mass stars = massive stars (e.g. Luminous Blue Variables, Wolf Rayet stars etc.)

	ISM	Young stars			SS	AGB		pAGB		PNe		Mass.
		PS	TT	Her		O	C	O	C	O	C	stars
O-rich dust												
Amorphous silicates	√	√	√	√	√	√	√	√		√	√	√
Crystalline silicates			√	√	√	√	√	√	√	√	√	√
Carbonates		√:			√					√		
Al_2O_3					√	√						
$MgAl_2O_4$					√	√						
SiO_2				√	√			√:	√:	√:		
[Mg,Fe]O		√:		√:	√	√						√:
C-rich dust												
Carbonaceous dust	√				√		√		√		√	
SiC					√		√		√		√	
TiC					√				√		√	
(nano-)diamonds				√	√				√			
Other dust species												
MgS	√:						√		√		√	
FeS			√	√							√	
Metallic Fe					√	√						

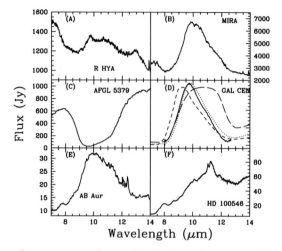

Fig. 4. Examples of 10 μm amorphous silicate band shapes in different astronomical environments. (A) R Hya, a low mass loss AGB star with a 10 μm band dominated by simple oxides and a minor amorphous silicate contribution, (B) Mira, an AGB star showing a strong amorphous silicate band very similar in shape to that seen in the ISM, (C) the OH/IR star AFGL 5379, an AGB star with very high mass loss rate showing a strong amorphous silicate absorption, (D) the galactic centre 10 μm silicate band shape (solid line) which is observed in absorption, but shown in emission here to illustrate the band shape, and compared to laboratory samples of 0.1 and 2 μm amorphous olivines (dotted and long-dashed lines respectively) and small pyroxene grains (short dashed line), (E) the Herbig Ae/Be star AB Aur, and (F) the Herbig Ae/Be star HD 100546, whose silicate band shows prominent contributions from polycyclic aromatic hydrocarbons and from crystalline silicates.

3.1   Dust Produced in O-rich Environments

Amorphous Silicates: Amorphous silicates are the most abundant grain species in interstellar space, and are responsible for the broad bands at about 9.7 (see Fig. 4) and 18 μm. These are seen in absorption against bright background sources (such as the galactic centre or proto-stellar cores), and in emission or self-absorption in many oxygen-rich cool red giants and supergiants. The 9.7 μm band is due to the Si-O stretch resonance, while the 18 μm band is caused by the Si-O-Si bending mode in the SiO_4 tetrahedron.

Both resonances unfortunately show only a minor dependence on chemical composition of the material, as several laboratory studies have shown [e.g. 11]. Therefore, it is difficult to derive strong constraints on e.g. the Fe/Mg ratio or on inclusions such as Ca and Al. Based on analysis of the temperature of dust condensing in the outflows of late-type stars, Jones & Merrill [38] suggested that the amorphous silicates should have a substantial opacity in the near-IR, requiring the addition of metals like iron. This can be either as cations into the lattice or as metal grains incorporated into a silicate grain. This is often referred to as 'dirty silicates', meaning amorphous (or glassy) silicates with unknown composition and/or inclusions. Draine & Lee [36] introduced 'astronomical silicates' to describe *empirically* the absorption and emission properties of interstellar silicates from observations, without detailed composition determination. A similar approach was used by Pegourie [39], Ossenkopf [37] and Suh [40], all using infrared spectra of M-giants. While these empirically derived optical constants are practical in fitting spectra, they do not provide more insight into the chemical composition of amorphous silicates.

More recent studies try to improve on this situation by using optical constants from well characterized materials to fit the astronomical spectra of amorphous silicates [e.g. 41, 11, 42, 43, 34]. The first and the latter two studies conclude that amorphous silicates in evolved cool giants are well represented by an olivine stoichiometric ratio ([Mg,Fe]$_2$SiO$_4$), but that an additional source of opacity is needed in the 3-8 μm wavelength range. Kemper et al. [34] show for OH 127.8+0.0 and Harwit et al. [44] for VY CMa that a good fit can be achieved with small metallic Fe particles. We note that Harwit et al. [44] used spherical grains and needed much more Fe than was available compared to cosmic abundances, while Kemper et al. [34] used a continuous distribution of ellipsoids and remained within the atomic mass budget of Fe. This would imply the Fe particles are not spherical.

The 10 and 20 μm spectral regions in oxygen-rich AGB stars show a considerable spread in the peak position and width of the 10 and 20 μm bands (see Fig. 4). Part of these variations can be attributed to the presence of other dust components, such as simple oxides (see Sect. 3.1) but also a change in silicate composition has been observed. While amorphous silicates in evolved stars [e.g. 45] may have an olivine stoichiometric ratio, amorphous silicates around young stars appear however to be more similar to pyroxene [[Mg,Fe]SiO$_3$; 42]. Such a change may be due to the impact of supernova shock waves, which will alter the Si/O ratio by sputtering [45].

Possibly the best analogue for the amorphous silicates in the ISM and around young stars are the GEMS (Glasses with embedded metals and sulfides) found in IDPs that are believed to come from comets (see Bradley, this volume). Their properties show many similarities with the amorphous silicates in the ISM [5].

Crystalline Silicates: Crystalline silicates are the highly ordered counterparts of the amorphous silicates. For a long time the crystalline silicates were only known to be present o earth, in solar system comets [46, 47], IDPs [48, 49], and in the dust disk of β-Pictoris [50, 51]. A crystalline olivine feature was also reported in the polarized 10 μm spectrum of AFGL2591 [52]. In all these cases the crystalline silicates were found by infrared spectro(polari)metry around 10 micron, except for the IDPs that were found with the aid of transmission electron microscopy in a laboratory, and later confirmed by infrared spectroscopy [e.g. 53, 49].

In hindsight, we can understand why crystalline silicates were only discovered to be ubiquitous after ISO was launched. Before ISO the main observational MIR/FIR window was around 10 μm. In this wavelength range the crystalline

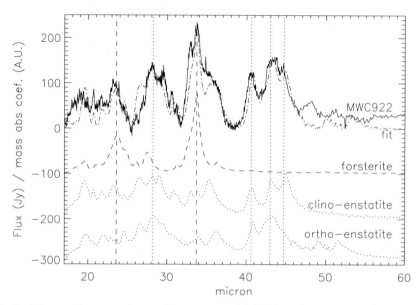

Fig. 5. The continuum subtracted spectrum of MWC922 [solid line; 57], compared with the calculated emission spectrum of forsterite (at 90 K; dashed line), and clino and ortho-enstatite (at 100 K; dotted line). The temperatures have been derived by Molster et al. [7], when they were fitting the continuum subtracted spectrum with only forsterite and enstatite (50% ortho and 50% clino enstatite). The fit is shown as the dashed-dotted line. The vertical lines denote the diagnostic features of forsterite (dashed lines) and enstatite (dotted lines).

silicates are either overwhelmed by emission from the much more abundant and warmer amorphous silicates, and/or simply too cold to show any detectable feature. The (relatively) low temperature of the crystalline silicates suppresses the intrinsically strong crystalline silicate features around 10 micron. Thanks to the extended wavelength range (up to 200 μm) of the spectrographs on board ISO, the composition of the cold dust could be studied in detail for the first time. At present, crystalline silicates have been found around young stars [54], comets [55], and evolved stars [56], but not convincingly so in the interstellar medium. It should be noted that crystalline silicates were not only found in the outflows oxygen-rich evolved stars, but also around stars that show clear evidence of a carbon-rich chemistry. The presence of the crystalline silicates in these environments has likely to do with previous mass loss episodes. We will discuss the properties of the crystalline silicates in what follows.

The sharp infrared features [see e.g. 13, for an assignment of the bands of forsterite] of the crystalline silicates allow a quite accurate identification of the crystalline materials. The two most abundant crystalline silicates that have been found are forsterite (Mg_2SiO_4) and enstatite ($MgSiO_3$, which can have a mono-clinic and orthorhombic crystallographic structure, called respectively clino and ortho-enstatite). A fit to the continuum subtracted spectrum of MWC922 with

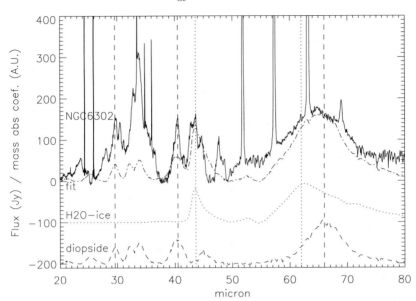

Fig. 6. The continuum subtracted spectrum of NGC6302 [solid line; 57], compared with the calculated (using optical constants derived from laboratory experiments) emission spectrum of diopside (at 70 K; dashed line), and crystalline water ice (at 40 K; dotted line). The temperature of diopside is chosen the same as the temperature found for the other pyroxenes [7], and water ice has been chosen to fit both the 40 and 60 micron complex. The combined result is shown as the dashed-dotted line. The vertical lines indicate the wavelengths of diagnostic features of diopside (dashed lines) and crystalline water-ice (dotted lines).

only these two species can be found in Fig. 5. The infrared spectra of ortho and clino-enstatite are very similar, only at wavelengths beyond of 40 μm clear differences in the absorption spectra become visible (see Fig. 5). A comparison of the strength of the individual features of ortho and clino-enstatite shows that around most stars the abundance is about equal. For some stars, with very high mass loss rates, ortho-enstatite may be more abundant than clino-enstatite [7].

Besides the above mentioned Mg-silicates, there is also evidence for Ca-pyroxenes such as diopside [58]. In Fig. 6, we show the calculated emission spectra of diopside and crystalline water ice. Both have a rather broad feature near the peak of the 60 micron band. However, the observed 60 μm band is broader than the individual H_2O ice and diopside bands, while a sum of both materials gives a satisfactory fit the 60 micron region (note that the addition of the carbonate dolomite improves the fit even further, see Fig. 8). Unfortunately, the low abundance of diopside in combination with blending of the short wavelength diopside bands with those of forsterite and enstatite make it hard to unambiguously identify diopside based on only the shorter wavelength bands. This implies that we can only clearly identify this material in systems which have very cool dust (T<100 K), such as OH/IR stars and planetary nebulae.

We now turn to the chemical composition of the crystalline silicates, focusing on the olivines and pyroxenes. Their chemical composition can be rather accurately determined because the wavelength and strength of the bands are very sensitive to differences in the Fe/Mg ratio. Laboratory studies show that a simple relation exists between the position of the crystalline silicate bands and the Fe/Mg ratio in the lattice. The peak position shift in the frequency space due to the inclusion of Fe goes roughly linear with the percentage of [FeO] in the silicate [13].

$$100 * x/\Delta\nu \; = \; -1.8 \quad \text{for olivines } (Mg_{(2-2x)}Fe_{2x}SiO_4) \text{ and} \qquad (1)$$

$$100 * x/\Delta\nu \; = \; -1.5 \quad \text{for pyroxenes } (Mg_{(1-x)}Fe_xSiO_3), \qquad (2)$$

with $0 \le x \le 1$ and $\Delta\nu = \nu_x - \nu_0$ where ν_x and ν_0 are respectively the wavenumber of the feature for composition x and $x = 0$. This implies that the shift in the wavelength domain is proportional to λ^2 and therefore clearest at the longest wavelengths. Therefore it is difficult to determine the exact Mg/Fe ratio from observations limited to only the 10 micron region.

Focusing on the 69 micron band, Fig. 3 shows the evidence for the high Mg/Fe ratio of the crystalline olivines in the outflows of evolved stars. In fact the data is even consistent with the absence of iron in the matrix (i.e. forsterite). It is also clear that the forsterite crystals are rather cold. A similar result holds for the pyroxenes based on the 40.5 micron feature. The crystalline pyroxenes found in the dusty winds of evolved stars invariably show evidence for very Mg-rich crystals (enstatite).

The determination of the exact composition of the crystalline silicates around young stars is a little more complicated. There is only one star for which a 69 micron feature has been found and even that one is rather noisy [59]. The characteristic features of crystalline silicates are found at wavelengths below 45 μm, especially in the 10 micron complex region. The absence of the 69 micron feature in the spectra of most young stars may have several causes. The abundance of crystalline silicates may be too low, or the temperature of the crystalline silicates may be too high (a lower temperature will result in a stronger and narrower feature, and a lower temperature will also relatively enhance the 69 micron feature with respect to other IR-features). It is also possible that there is some Fe in the matrix, which will also weaken the 69 micron feature [12, 13]. Finally, the low 60-70 μm flux of many young stars inhibits the detection of the weak 69 micron band. However, the short wavelength crystalline silicate bands in young stars (and in Hale-Bopp) do point to a low Fe content, which may be as low as in the evolved stars.

Due to the large number of bands available for analysis, an independent determination of the temperature of the different species of crystalline silicates can be made. It is interesting to note that in evolved stars the crystalline silicates are usually cooler than the amorphous ones [7]. This indicates that these two types of grains are not in thermal contact, and either they are spatially distinct, or they have significantly different optical properties, which leads to different equilibrium temperatures. The temperature difference can be explained in a straightforward way if we consider differences in the Fe/Mg ratios of co-spatial amorphous and crystalline silicates. As has been shown, the crystalline silicates are very Fe-poor. In contrast, it has been argued that amorphous silicates contain Fe, either in the matrix or as a metal inclusion, in order to reach the observed temperatures of these grains in circumstellar dust shells [38, 37, 34]. Adding a modest amount of iron already increases the opacity in the near-IR significantly [11]. Radiative transfer modeling shows that, if one assumes that the amorphous silicates have Fe/Mg $\approx$ 1 and that the crystalline silicates have no Fe, the temperature differences are explained by the difference in near-IR opacity. Both the amorphous and crystalline silicates can then be co-spatial [e.g. 32, 33].

Radiative transfer calculations can be used to constrain the abundance of crystalline silicates in different environments. The abundance of crystalline silicates in the winds of evolved stars is typically of the order of 10% or less. From simple modeling of the infrared spectra it has been deduced that enstatite is roughly 3 times as abundant as the forsterite in these outflows [7]. It should be noted that the derived abundances depend on the laboratory spectra used in the analysis. Using different sets of lab data sometimes changes the abundances by a factor two! It should be noted that the above mentioned values for the abundances are only derived for stars with a relatively high mass loss rate. Crystalline silicates have not been found around low mass-loss-rate stars [56, 60]. This might be a temperature effect. Indeed, Kemper et al. [61] show that the temperature difference between the crystalline and amorphous silicates, will cause the emission from the crystalline silicates to be overwhelmed by the warm amorphous silicates in the low mass-loss-rate stars. In some peculiar objects, likely as a result of binary interaction, the crystalline silicate abundance can be very high, up to 75 % of the small grains [62].

The abundance of crystalline silicates in Hale-Bopp has been estimated by several authors [63, 64, 65]. The derived abundance of crystalline silicates ranges between more than 90% [63] and about 7% [65] of the total dust mass. These studies use different Fe contents for the silicates and different grain size and shape distributions. Since the presence of Fe also reduces the strength of the features of crystalline olivine significantly [see e.g. 13], the abundance of the crystalline silicates in this comet remains uncertain. For other comets only spectra of the 10 micron region are available. This limits the determination of the abundance very much because it is very sensitive to the temperature of the different dust components, which is very difficult to obtain from a small wavelength range (8-13 μm).

Up to now we do not have strong evidence for the presence of crystalline silicates in the ISM. From the absorption profiles of protostars an upper limit of about 2% for the abundance of crystalline forsterite has been derived [42]. This is automatically the upper limit for the abundance of forsterite in the ISM. Cesarsky et al. [66] suggest that several emission bands in the ISO spectra of the Orion nebula may be due to crystalline silicates. We note however that the peak position, shape and width of the bands seen in Orion differ from those of crystalline silicates in other environments. In addition, PAH emission bands may complicate the situation. If the bands seen by Cesarsky et al. [66] are indeed crystalline silicates, these would have to be of a different composition from those seen elsewhere. We conclude that so far no convincing evidence has been found for the presence of crystalline silicates in the ISM.

The ISO data show that the abundance of crystalline silicates in Young Stellar Objects (YSOs) is higher than the upper limits that have been established for the interstellar medium. A detailed analysis of the ISO spectrum of the young Herbig Ae/Be star HD 100546 [59, 65] shows that in the 100-200 K temperature range, crystalline silicates are the *dominant* small grain species, i.e. even more abundant that small amorphous silicates. This is quite exceptional however, and most other young stars show much more modest abundances [67]. The higher abundance of crystalline silicates in the environments of young stars compared to the interstellar medium points to an in situ formation mechanism, probably by annealing in the proto-planetary disk.

The ISO observations suggest that a relation may exist between the abundance of crystalline silicates and the *geometry* of the circumstellar dust shell. Fig. 7 shows the correlation between the strength of the 33.6 micron feature (attributed to forsterite) and the IRAS 60μm/mm flux ratio. The first value gives an indication of the abundance ratio between amorphous and crystalline silicates in the dust, while the second ratio is a rough indication of the average grain size. It is interesting to note that those sources which seem to have a large abundance of crystalline silicates, also show evidence for the presence of a disk-like structure [68] and for grain coagulation. It should be added that the opposite is not true, if stars do have a disk (and large grains) it does not automatically imply that they have a high fraction of crystalline silicates. Whether the crystallization of amorphous silicates is related to grain coagulation, or that we are simply dealing with two different processes, which both require long timescales has not yet been established (see also Sect. 4).

Carbonates: Many studies have searched for carbonates ([Ca,Mg,Fe,...]CO_3) in the solar system and interstellar space (see Kemper et al. [69] for an overview). The strongest mid infrared (MIR) band of carbonates is around 6.8 μm and is due to the C=O stretching mode. Other relatively strong MIR features are found near 11 and 14 micron and are caused by respectively the out of plane and in-plane bending modes of C=O. The features found beyond 25 μm are due to a translation of the metal cations and represent motions perpendicular and parallel to the plane of the carbonate anions [70]. So far, searches for the MIR features have not been successful: while many objects show spectral bands in the 6.8 and 11-14 μm region, these bands have all been attributed to other species.

Barlow [71] discovered an emission band in NGC 6302 ranging from 88–98 μm. A similar band has been found in NGC 6537 [57]. Recently, Kemper et al. [22] showed that these bands are evidence for the presence of carbonates in these two planetary nebulae. This claim was strengthened by the presence of an emission band at 65 μm (see Fig. 8). The 92 μm band is attributed to calcite (CaCO$_3$), which gives a very good match between laboratory data and

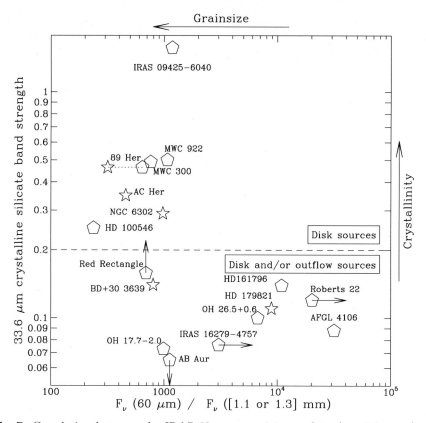

Fig. 7. Correlation between the IRAS 60 μm over 1.1 mm (stars) or 1.3 mm (pentagons) flux ratio and crystalline silicate band strength, measured from the crystalline forsterite band at 33.6 μm ($[F_{33.6\ \mu m\ peak} - F_{cont}]/F_{cont}$). The selected sources have dust colour temperatures above $\approx$ 100 K, so that the Planck function peaks at wavelengths shorter than 60 μm. The emptiness of the upper right corner of this diagram is an indication that the presence of highly crystalline dust is correlated with disks and grain-growth. All stars above the line ($(F_{33.6\ \mu m\ peak} - F_{cont})/F_{cont} = 0.2$ have relatively small 60 μm over mm-flux ratios and have indications for the presence of a disk (e.g. by direct imaging and/or the spectral energy distribution). The stars below this line are predominantly normal outflow sources, without any evidence for a disk, although exceptions exist (e.g. Roberts 22). NGC 6302 and BD+30 3639 have been shifted by a factor 2 along the x-axis since it is estimated that only half of the mm continuum flux is due to dust emission while the other half is due to free-free emission. Figure taken from [68].

ISO observations. The 65 μm band seen in the ISO spectra is a blend and has contributions from diopside and crystalline H_2O ice. The analysis of Kemper et al. [22] showed that the inclusion of dolomite ($CaMg(CO_3)_2$) substantially improves the fit in the 60–70 μm region. The low temperature and the low abundance of the carbonates prevents any detection of the carbonate features

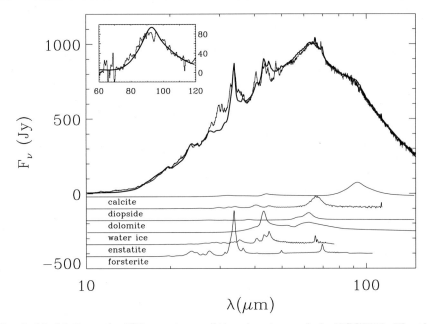

Fig. 8. Model fit to the ISO spectrum of the planetary nebula NGC6302. The thin line represents the observed spectrum. The thick line represents an optically thin dust model fit, consisting of amorphous olivine, metallic iron, forsterite, enstatite, H_2O-ice, diopside, calcite and dolomite. All contributions, except the featureless contributions from metallic iron and the amorphous olivine, are included in the bottom part of the figure. The inset shows the observed spectrum (thin line) and the model fit (thick line) from 60-120 μm, where the contributions of all species except calcite, are subtracted from both the observations as well as the model, therefore showing only the observed calcite feature. Figure taken from [22].

in the 5–15 μm region. Calcite and dolomite are also the two most abundant carbonates in the solar system.

In the spectrum of NGC1333-IRAS4, a young protostar, there is some evidence for a band at 90 μm, which is rather similar the $CaCO_3$ band [72]. Since the material in the vicinity of this protostar presumably has only experienced typical interstellar medium conditions it might imply that, if the identification can be confirmed, carbonates are also present in the ISM.

Al_2O_3: The presence of corundum (Al_2O_3) has been inferred from mid-infrared IRAS-LRS spectra of oxygen-rich red giants, that show a broad 11 μm feature [e.g. 75]. This band is often found to co-exist with the 9.7 μm amorphous silicate Si-O stretch, and tends to be seen in stars with weak dust emission. Cami showed that ISO spectra of low mass loss red giants show a broad 11 μm band consistent with amorphous alumina-oxide (see Fig. 9 [74]).

A narrow dust feature at 13 μm was discovered in the IRAS-LRS spectra of bright M Mira variables [76], which has been attributed to γ-Al_2O_3 [77]. How-

G Her

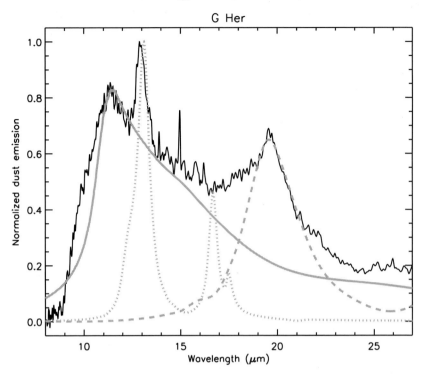

Fig. 9. The continuum subtracted spectrum of the AGB star G Her, with possible identifications of the emission bands. The grey solid line represents amorphous Al_2O_3, the dotted line spinel, and the dashed line $Mg_{0.1}Fe_{0.9}O$. see also [73]. Figure taken from [74]

ever, more detailed studies showed that γ-Al_2O_3 does not provide a satisfactory match to the band [78]; attempts to improve the match using α-Al_2O_3 were successful, if a continuous distribution of ellipsoids is assumed as particle shape. However a second band of α-Al_2O_3 near 21 μm has not been detected, suggesting that α-Al_2O_3 cannot be the carrier of the narrow 13 μm band.

The discovery of presolar Al_2O_3 in meteorites [79, 80, 81, 82], whose isotopic composition points to oxygen-rich red giants as their origin, supports the spectroscopic identification of Al_2O_3 in the mid-IR spectra of red giants. Dust nucleation models for oxygen-rich outflows from cool giants based on thermodynamic considerations predict Al_2O_3 to be one of the first condensates [83]. However, detailed quantum mechanical calculations show that homogeneous nucleation of molecular Al_2O_3 from the gas phase is unlikely in these environments [84]. The Al_2O_3 grains, which bear isotopic evidence of a formation around an AGB stars and found in meteorites, are likely formed via grain surface reaction in an AGB wind [85].

Al_2O_3 has not been detected in the diffuse interstellar medium. This may prove difficult given the strong amorphous silicate resonance.

MgAl$_2$O$_4$: Posch et al. [73] suggest (annealed) spinel (MgAl$_2$O$_4$) as the carrier of the 13 micron feature detected in spectra of oxygen-rich red giants. Speck et al. [86] raised doubts about this identification and proposed SiO$_2$ as the carrier. However, SiO$_2$ should have a very strong feature around 9 μm, which is not observed. Fabian et al. [87] seem to settle the dispute with the discovery of two other bands (around 16.8 and 32 μm) which can also be attributed to spinel.

Spinel presolar grains have been found in meteorites [88, 89], and, as with Al$_2$O$_3$, is predicted to be a condensate in the winds of oxygen-rich stars. The 13 μm band has not been detected in the diffuse ISM, suggesting it is not produced in large quantities by red giants, or that it does not survive long in the ISM.

[Mg,Fe]O: Mg$_x$Fe$_{1-x}$O shows a fairly broad resonance which shifts from about 16.5 to 19 μm as x varies from 1 to 0 [15]. In addition, the band shape and strength depend strongly on grain shape. These properties make it difficult to identify this simple oxide spectroscopically. Cami [74] and Posch et al. [90] propose that the broad 19.5 μm band seen in low mass loss red giants is due to spherical Mg$_x$Fe$_{1-x}$O grains with $x \approx 0.1$, see also Fig. 9. Molster et al. [32] show that the 18 μm amorphous silicate band cannot account for all emission in the 20-28 μm spectrum of AFGL 4106. An additional broad dust band near 23 μm is inferred and a continuous distribution of ellipsoids (CDE) of FeO particles has been proposed as its carrier [32]. Similarly, Bouwman et al. [91] suggest *non-spherical* FeO grains to account for additional emission in the 23 μm region of Herbig Ae/Be stars. In contrast Demyk et al. [42] find that *spherical* inclusions of FeO improve the fit to the dust absorption profile around the protostars RAFGL7009S and IRAS19110+1045 to the 18 micron band. Recent findings suggest that the 23 μm band in the spectra of young stars is not due to FeO with a CDE distribution, but more likely to FeS and that the assumption of a CDE shape distribution seem to fail for Fe-oxides [6]. The reason for the failure of this assumption is likely related to the fact that Fe-oxides have a cubic crystal symmetry. Therefore the crystal will not grow in a preferred direction and the shape of the particle will be more spherical like. So, we conclude that there is some evidence for the presence of spherical Mg$_x$Fe$_{1-x}$O grains (with $x \approx 0.1$) in dust shells surrounding evolved stars, but a firm theory to explain the existence of only this stoichiometric ratio is still lacking.

SiO$_2$: Silica shows bands at approximately 8.6 and 20.5 μm. An analysis of ISO spectra of young stars suggests that a small fraction of the silicate dust in the disks surrounding Herbig Ae/Be stars is SiO$_2$ [92]. This material could be formed by annealing of amorphous silicates. Natta et al. [93] analyse the 10 μm silicate band in T Tauri stars, and find evidence for a $\approx$ 9 μm contribution by SiO$_2$. The analysis of Bouwman et al. [92] shows that the temperature of the silica is roughly equal to that of the crystalline silicates. Since silica has a very low opacity in the UV to NIR this phenomenon is best explained if the crystalline silicates and the silica are in thermal contact.

Molster et al. [27] found a relative sharp band at 20.5 μm in many oxygen-rich evolved stars. They tentatively attributed this band to silica. However, it should be noted that the width of this feature is considerably narrower than in the laboratory spectra of silica. This could be a temperature effect, since the laboratory spectra were taken at room temperature, while the silica around these evolved stars is likely much cooler. It would be useful to measure laboratory spectra of crystalline SiO_2 at low temperature. For glassy SiO_2 there is no large influence of the temperature on the width of the feature [23].

Metallic Fe: The presence of metallic Fe cannot be demonstrated on the basis of the detection of an infrared resonance, because there are no infrared active vibrational modes known in this material. Nevertheless, evidence for its presence in oxygen-rich dust shells surrounding evolved stars has been found on the basis of the high near-IR opacity that metallic Fe particles provide [44, 34]. Radiative transfer models of dust shells around M supergiants and AGB stars using only laboratory opacities of silicates fail to reproduce the 3-8 μm spectral region. The inclusion of metallic Fe particles results in a satisfactory fit. The shape of these particles is very likely non-spherical. To produce enough opacity a spherical shape requires more iron than is available on cosmic abundance grounds. It is not clear from these models in what form the metallic Fe particles are included in the dust: they may be individual particles or they may be inclusions in amorphous silicates as is found in GEMS.

3.2   Dust Produced in C-rich Environments

Carbonaceous Dust Amorphous carbon does not have a clear spectroscopic signature in the infrared. Its presence is mostly based on the need for an extra source of opacity in the spectra of carbon stars. Graphite would also have a very similar and flat spectrum. However, a narrow feature at 11.52 μm, which is predicted for graphite, has not been found in carbon-rich stars. This places an upper limit of 3% on the solid carbon being in the form of small graphite particles [95]. Also the slope of the SED at the long wavelength side points to amorphous carbon instead of graphite, which would produce a steeper decline.

Graphite is often suggested as the carrier of the 2175 Å extinction feature. Small (< 150Å) graphite grains can reproduce the feature and part of the UV extinction. However, it should be noted that up to now, apart from this feature no other signature of graphite has been found in the ISM, also no source has yet been identified that could account for the production of graphite. According to Draine [96] only 15% of the carbon in the ISM has to be in graphite grains to explain the 2175 Å feature. It is unclear if the present day observations can already put such an upper limit on its abundance. Part of the popularity of graphite arises from the fact that graphite is one of the few carbonaceous materials that is well characterized and for which good laboratory optical constants are available [97].

Graphite grains have been found in meteorites [98], however they are not alike the grains, required to explain the 2175 Å feature. The inferred ISM graphite

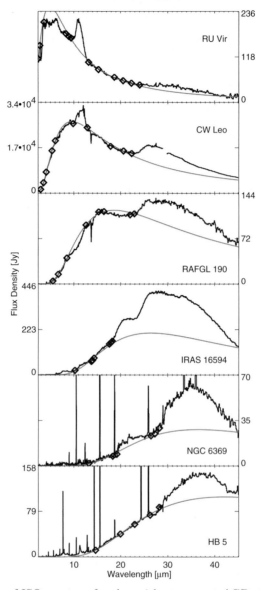

Fig. 10. Overview of ISO spectra of carbon-rich stars, post-AGB stars and planetary nebulae [94]. The 11.3 μm band is prominently in emission in RU Vir and CW Leo, and is in absorption in RAFGL 190. The 30 μm MgS band is present in all objects, but shifts its position from about 26 to 35 μm going from AGB to planetary nebula. The emission lines in the spectra of NGC 6369 and HB 5 are due to ionized gas in the AGB remnant.

grains are very small (radii $\leq 150\text{Å}$) well crystallized and abundant, while the graphite grains found in meteorites are much larger, with radii in the range of 1 to 5 μm [99], and often contain scales of poorly graphitized carbon. It is therefore likely that meteoritic graphite grains and interstellar "graphite" particles responsible for the 2175 Å feature are unrelated.

In many carbon-rich stars there is emission found between 3–4 and 11–17 μm, which is indicative for vibrationally excited C-H bonds. A closer look reveals emission from H-atoms bound to sp^2 and sp^3 sites [100, 101]. The combination of these two bonds together with the emission plateau from 6–9 μm is found in hydrogenated amorphous carbon (HAC), which is also known as a-C:H or diamond-like amorphous carbon. On earth this material has been very well studied (see Robertson for a review [102]), but the exact structure of the HACs in circumstellar environments is still not well known. Hony et al. show that HAC grains in radiative equilibrium do not explain the features found in HD56126. However, these authors suggest that transiently heated small HAC grains with a H/(H+C) ratio of 0.35 may explain the observed features [103].

An infrared absorption feature at 3.4 μm, characteristic of dust in the diffuse ISM, is normally attributed to C-H stretching vibrations of aliphatic hydrocarbons [104, 105].

Nano-diamonds: It has long been assumed that nano-diamonds are presolar and formed in interstellar shocks [106], in the atmospheres of carbon stars, and/or in supernovae. The discovery of an anomalous Xe-HL component associated with the nano-diamonds in primitive meteorites gave strong support for a supernova origin of these grains [107, 108, 109]. However, their bulk ^{13}C/^{12}C composition is solar and only 1 in 10^6 nano-diamonds contains a Xe atom. Since it is not (yet) possible to measure the isotopic composition of a single nano-diamond, the origin of (most of) these grains is not clear. The lack of very strong isotopic anomalies, as for instance seen in SiC or Al$_2$O$_3$, has precluded a firm identification with a particular stellar birth site for these grains (see also the contributions in this book).

A study by Guillois et al. [110] showed that, while pure diamonds have no infrared-active single phonons, hydrogenated nano-diamonds do show spectral structure in the 3-4 μm wavelength region. They propose that an emission band seen in a small number of stars at 3.53 μm (and two weaker ones at 3.43 and 3.41 μm) can be attributed to C-H bonds in hydrogenated nano-diamonds. So far, the 3.53 μm band has been identified in only five objects: the young stars HD 97048 [111], Elias I [112] and MWC297 [113], and in the (peculiar) post-AGB stars HR 4049 [114] and HD52961 [115]. These stars, while in vastly different evolutionary phases, all have a circumstellar disk with dust temperatures exceeding 1000 K, and have a hydrogen-rich atmosphere. Laboratory studies suggest that high temperatures and pressures are required to form nano-diamonds, consistent with the conditions that prevail in the sites where they have been identified.

Van Kerckhoven et al.[116] suggest an *in situ* formation of the nano-diamonds in the five objects with 3.53 μm emission, which of course opens the possibility that (part of) the nano-diamonds in the solar system were formed by the same process. This notion has gained support by the recent finding that cometary IDPs have a much lower abundance of nano-diamonds than the primitive meteorites do [117], suggesting a gradient in the abundance of nano-diamonds in the protosolar cloud. Such a gradient could naturally result from the production in the inner solar system of nano-diamonds and their inclusion in proto-planetary bodies.

FeS: Although FeS is a solid which is predicted to condense in the winds of carbon-rich red giants, until recently it has eluded detection outside the solar system. In contrast, FeS is abundant in meteorites and in IDPs. In fact, FeS carries the bulk of sulphur in the solar system. Recently, FeS has been detected in the ISO spectra of two carbon-rich planetary nebulae [103], by means of emission in the 23 μm band. Hony et al. also find evidence for weaker bands of FeS (troilite) at 33, 38 and 44 μm, albeit much weaker than expected on the basis of laboratory spectra. Laboratory spectra of different kinds of FeS (such as Pyrite (FeS$_2$), Troilite (FeS) and Pyrrhotite (Fe$_{1-x}$S)) all show a broad emission band at approximately 23 μm. At longer wavelengths, these materials show large differences in their spectra. Hony et al. suggest that in space different kinds of FeS coexist, weakening the spectral signature of individual species at the longer wavelengths, but keeping the prominent FeS band near 23 μm.

FeS has also been identified in the ISO spectra of some young intermediate-mass Herbig Ae/Be stars [6], on the basis of a 23 μm excess. This 23 μm excess matches the 23 μm band seen in laboratory spectra of FeS taken from IDPs [6].

MgS: A broad emission band near 30 μm was first noted in the spectra of three carbon-rich evolved stars [118]. Later studies showed that the 30 μm band is present in objects ranging from mass-losing carbon stars to post-AGB stars and planetary nebulae (e.g. [119, 120, 121, 122, 123, 94], and Figs. 10, 11). MgS has been tentatively detected in KAO spectra of the galactic centre [124] and it may be present in the peculiar massive star η Car (P. Morris, private communication).

MgS has been suggested for the identification of this feature [125]. Based on the optical constants derived by Begemann et al. [17] a single resonance at 26 μm is predicted for spherical grains. This feature shifts and broadens significantly if grain shape effects are taken into account. Transmission spectra taken by Nuth et al. [126] show a strong and broad around 36 μm. The strength of the observed emission has cast some doubt on this identification (some objects emit about 30 per cent of their light in the 30 μm band). Begemann et al. [17] convincingly demonstrate that MgS is the most likely candidate to cause the 30 μm band in IRC+10216.

ISO spectroscopy has shown that the 30 μm band shows considerable variations in strength, and width. In addition, the peak position varies from 26 to about 35 μm [127, 121, 123]. This led to the suggestion that the band is caused by two dust components whose relative strengths vary between sources [123]. However, Hony et al. [94] show that the shift in peak position of the feature can be understood as a temperature effect: the large intrinsic width of the band

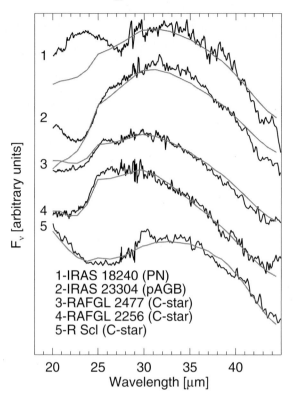

Fig. 11. Model fits to the 30 μm band using a CDE shape distribution for MgS grains and a varying MgS temperature to account for the shift in band position. Taken from [94].

causes the peak to shift to longer wavelengths as it is weighed with black bodies with decreasing temperature (see Fig. 11). Both the width and the peak position change systematically as the temperature of the underlying continuum changes, corresponding to an evolution from the asymptotic giant branch to the planetary nebula phase [94]. A weaker band near 26 μm remains visible in some AGB stars after the strong MgS contribution is removed. This component is tentatively identified with spherical homogeneous MgS particles [94].

MgS is expected to condense in a carbon-rich environment [128]. However, sulfur is found to be only marginally depleted from the gas-phase in the diffuse ISM [129] which suggests that MgS cannot be a major source of sulfur in the diffuse ISM. MgS is unlikely to survive in the oxygen-rich environment which the ISM is believed to be. In meteorites no presolar MgS grains have been found so far.

The 21 μm Feature: A prominent emission band near 21 μm was first noted by Kwok et al. [130] in the IRAS-LRS spectra of a small group of carbon-rich post-AGB stars. These objects became known in the literature as the "21 μm

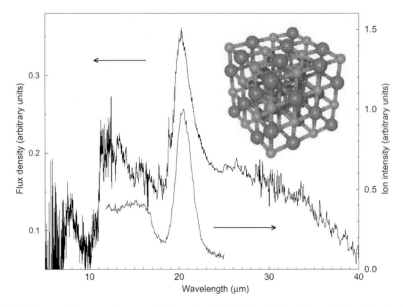

Fig. 12. The emission spectrum of the post-AGB object SAO 96709, taken by the ISO satellite (upper curve, left axis) and the wavelength spectra of the TiC nano-crystal recorded in the laboratory (lower curve, right axis). Also shown is a pictorial representation of a typical (4x4x4 atom) TiC nano-crystal. The small spheres are carbon atoms, and the large spheres are Titanium atoms. Figure taken from [21].

sources"; some 15 have been found to date. ISO-SWS spectra show that the band is actually peaking at 20.1 μm [131], however it is still referred to as the 21 μm band. The band shape hardly varies between sources. Evidence for the presence of the 21 μm feature has also been found in two carbon-rich planetary nebulae [132]. This shows that the carrier can survive the harsh conditions that prevail in such objects. So far the 21 μm band has not been seen in carbon-rich AGB stars, which indicates that its carrier is either not produced until the very end of the AGB, or that its excitation requires optical and/or UV photons.

Various identifications for the carrier of the band have been suggested (e.g. SiS_2 [133, 134]; nano-diamonds [135]), none of these however provided a convincing fit. Recently, it has been suggested that TiC nano-clusters could be the carrier of the band [21]. This identification is inspired by the remarkable similarity between the laboratory spectra of TiC nano-particles (both in peak position and in width and shape of the band, see also Fig. 12), and by the presence of TiC inclusions in presolar amorphous carbon grains [136], that can be traced to carbon-rich AGB stars. A detailed analysis of the strength of the 21 μm band in the post-AGB star HD 56126 led Hony et al. (in prep.) to the conclusion that the TiC identification may not be tenable unless the nano-particles absorb optical and near-UV light from the star very efficiently. Speck & Hofmeister[137] suggest SiC nanoparticles as a carrier. While this identification would relax the abundance problems encountered with TiC nano-particles, it also predicts a prominent 11.2 μm band which is not observed.

SiC: Silicon Carbide (SiC) was suggested to be part of the ISM by Friedemann [138]. It was first identified in the mid-IR spectra of carbon-rich AGB stars [139], following the detection of an 11.2 μm emission band in these objects [140]. Early spectral surveys of the SiC band were carried out by e.g. Forrest et al. [141] and Cohen[142]. There is considerable variation in the wavelength of the peak, ranging between 11.2 and 11.7 μm, [e.g. 143, 144, 145], while the full width at half maximum of the band is about 1.3–2.0 μm. The band is commonly observed in mass-losing AGB stars as well as in post-AGB stars and some planetary nebulae (e.g. [146, 142, 147, 94], and Fig. 10).

SiC is predicted to be a high-temperature condensate in the outflows of carbon stars [e.g. 148]. Kozasa et al. [149] suggest that SiC grains form by homogeneous nucleation, and that these grains develop an amorphous carbon mantle as they move out in the envelope. For high mass loss rates the amorphous carbon may dominate, suppressing the SiC spectral signature. More recent observations of SiC in extreme carbon stars however revealed a self-absorbed SiC band [147, 150, 94], showing that even at these high mass loss rates SiC remains detectable.

Presolar SiC grains have been identified in meteorites [151]. The isotopic ratios of noble gases extracted from bulk, as well as isotopic ratio ^{12}C/^{13}C found in individual grains strongly points to carbon stars as their main production site, while novae and supernovae may be other sources of interstellar SiC grains [e.g. 152]. All presolar SiC grains have a cubic lattice structure (so-called β type SiC). In contrast, the SiC seen in carbon stars has long been attributed to hexagonal or α-SiC, but recently doubts have arisen whether or not the shape of the 11.3 μm resonance of SiC allows a discrimination between both types [153]. Interestingly, some presolar SiC grains have diameters in excess of 10 μm [109, and references therein] which is difficult to understand in the context of grain formation in AGB winds.

SiC has not been detected spectroscopically in the diffuse ISM [154]. However, its detection may be difficult given the proximity of the strong amorphous silicate band at 9.7 μm. It is also possible that the bulk of SiC grains may not survive in the oxidizing environment which prevails in the ISM, or that the contribution of carbon stars to the dust production in the galaxy has been overestimated.

3.3   Questionable Identifications

FeSi: Ferrarotti et al. [155] propose an identification of an emission band in the ISO-LWS spectrum of AFGL 4106 at 47.5 μm with FeSi. Their chemical condensation calculations show that FeSi will only be formed in the outflows of stars with a C/O ratio near unity [155]. However, this band has been seen in many oxygen-rich (C/O < 1) dust shells [71, 27] but to our knowledge it is not prominent in S-type (C/O $\approx$ 1) AGB stars. Finally, this mineral is expected to show an additional band at 32 μm, which has not been found so far. We conclude that the identification of FeSi is not yet very convincing.

Hydrous Silicates: Layer lattice silicates have been found in solar system material collected on earth, like the Murchison meteorite and in IDPs [53, 156, 49]. The spectra of comets are less clear. The presence of hydrous silicates in space has been suggested based on the shape of the 10 μm silicate feature in some comets [53, 157, 46], but a definite answer could not been given yet. Still, hydrous silicates seem to be a well established component in IDPs and in other bodies found in the inner solar system, however they have not been detected convincingly in interstellar space. Malfait et al. [158] find a prominent, broad emission band near 100 μm in the ISO-LWS spectrum of the young intermediate mass star HD 142527, which they attribute to the hydrous silicate montmorrilonite based on laboratory data [159, 160]. This identification is supported by the presence of very strong crystalline H_2O ice bands at 43 and 60 μm in HD 142527. However, a large abundance of 15 per cent of hydrous silicates is required to fit the ISO spectrum [158]. It is not clear what could cause such a high abundance of hydrous silicates in this object.

4   Life-Cycle of Dust

The rich mineralogy of refractory dust species that has been described in the previous sections reflects the equally rich diversity in physical and chemical conditions that prevail in space. Observations show that the *circumstellar* mineralogy is by far the richest, when compared to the interstellar one. Indeed, a disappointingly small number of interstellar dust species have been identified unambiguously by means of infrared spectroscopy. This may seem surprising since it is generally accepted that stars are the dominant dust production sites in galaxies. While observational selection effects may play a role (interstellar dust has low temperatures and is usually detected in absorption, circumstellar dust is detected in emission and absorption), the paucity of detected dust species in interstellar space is probably real and suggests that the composition of interstellar dust is not merely a sum of stellar dust production rates of the different circumstellar dust species: dust is (strongly) processed in the interstellar medium, and seems to reach a fairly homogeneous composition with only a few dominant species.

Most dust forms either in the slowly expanding wind of low mass evolved stars (M < 8M$_\odot$) and the massive red supergiants, or in the expelled layers after a supernova (SN) explosion. After being expelled into the harsh ISM, it is exposed to many destructive processes [161]. On the other hand dust also forms in these environments, and because of the totally different formation conditions, different dust species than the ones which entered the ISM will form. Eventually, the dust will end up in a star formation region. Here the dust will either be accreted onto the proto-star, 'survive' as a planet(esimal), or be expelled back into the ISM, via a bipolar outflow.

In this section we will discuss the dust (trans-)formation and destruction processes based on astronomical observations and their comparison with laboratory measurements. For a more comprehensive review of the dust processes in outflows and disks we refer to the chapter by Gail in this book.

4.1   Dust Formation in Evolved Stars

There are two main classes of evolved stars that produce significant amounts of dust: i) the evolved low-mass stars, with their slow expanding winds during the AGB phase, and ii) the high-mass stars that go through a red supergiant phase during which they have a stellar wind with properties somewhat similar to those of the low-mass AGB stars. There are also some rare, unstable massive stars, called Luminous Blue Variables, that produce some dust. Below we discuss the dust formation in AGB stars; a similar description holds for the red supergiants.

The numerous AGB stars in our galaxy make it easy to study the dust formation in their slowly expanding winds. Based on the C/O ratio in their atmosphere, one can distinguish three types of objects: The oxygen-rich (C/O < 1) M-type stars, the carbon-rich (C/O > 1) C-stars and the S-type stars that have C/O $\approx$ 1. The type of dust that forms in the outer atmosphere of cool evolved stars depends very much on the C/O ratio of the gas from which the dust condenses. When this ratio is larger than unity, all oxygen will be trapped in the very stable CO molecule and carbon containing dust will form. On the other hand when this ratio is smaller than unity (which is the case when a star is born) all carbon atoms will be trapped in CO and the remaining oxygen will determine the dust properties, silicates and oxides will form. The type of dust that forms not only depends on the molecular and atomic abundances of the gas from which it condenses, but also on the local physical conditions that prevail (such as temperature, pressure, UV radiation, presence of seed particles, etc.). We note that the supergiants only have oxygen-rich members.

Oxygen-Rich Stars: In M giants with low mass loss rates ($\dot{M} < 10^{-7}$ M$_\odot$/yr) the dominant dust species that form are simple metal oxides [74, 90]; amorphous silicates are not prominent in the spectra. The low density in the wind likely prevents the formation of relatively complex solids like the silicates, because that requires multiple gas-solid interactions. It is not clear at present whether these simple oxides also form in the innermost regions of AGB winds with higher mass loss rates.

The most abundant dust species in the winds of oxygen-rich stars with intermediate and high mass loss rates are the silicates. The formation of silicates is not well understood. Homogeneous silicate dust formation directly from the gas phase is not possible, it requires first the formation of condensation seeds. It is logical to search for high temperature condensates that can form through homogeneous nucleation that could serve as condensation seeds for other materials. Gail & Sedlmayr [162] propose that in O-rich winds TiO$_2$ (rutile) clusters can act as condensation nuclei for silicates. Quantum mechanical calculations seem to confirm this possibility [163]. However, no evidence has been found yet for the presence of rutile in the spectra of outflows of evolved stars; probably the low gas-phase abundance of titanium prevents detection.

Another high temperature condensate that could serve as a seed for the formation of other materials is crystalline Al_2O_3, also because it has been associated with the 13 micron feature. However, the 13 micron feature is likely not due to crystalline Al_2O_3 (see Sect. 3.1). Detailed quantum mechanical calculations show that homogeneous nucleation of molecular Al_2O_3 from the gas phase is very unlikely in a circumstellar environment [84]. The detected amorphous Al_2O_3 and the meteoritic evidence for the formation of Al_2O_3 in the winds of AGB stars, likely result from reactions in a grain mantle [85]. In stars with very high mass loss rate, ZrO_2 clusters may form in large enough quantities to play a role as seed particle [164]. Again, no spectroscopic evidence of its existence around evolved stars has been found so far.

One of the main discoveries of ISO in the field of solids in space has been the detection of crystalline silicates in the outflows of AGB stars with high mass loss rates. It should be kept in mind that the abundance of crystalline silicates compared to the amorphous silicates is modest, typically 10 to 15 per cent, and only in very special cases the crystalline silicates dominate. Dust nucleation theories are now challenged to explain the co-existence and difference in properties of crystalline and amorphous silicates in AGB stars: (i) crystalline silicates are Fe-poor, amorphous ones contain Fe in some form, (ii) crystalline and amorphous silicates have different temperatures and thus are separate populations of grains. "Partially crystalline" grains do not seem to exist, which can be explained from a thermodynamic point of view: thermal crystallisation timescales are a steep function of temperature, and it requires extreme fine-tuning in the thermal history of a particle to observe it partially crystalline.

Although the crystalline lattice structure is energetically the most favorable state for silicates, the observations show that most of the silicate grains are amorphous or glassy. This requires an explanation. First of all, some threshold energy has to be overcome before an amorphous silicate will crystallize. At temperatures above 1000 K the crystallization takes place on a timescale of seconds to hours, while below 900 K it will take years to more than a Hubble timescale [165]. If silicate dust condenses much quicker than the crystallization timescale, it will be amorphous. This implies that if the silicates form or quickly cool off below 900 K, they will remain amorphous. Since the majority of the silicates that forms is amorphous, this may imply a formation below the glass temperature.

An alternative theory assumes that the silicates condense as crystals but their crystal structure will get destroyed in time. Tielens et al. [166] propose a scenario to explain the difference between crystalline and amorphous silicates by this destruction mechanism. They propose that gas phase Fe diffuses into the Mg-rich crystals around 800 K (which is for all practical purposes below the crystallization temperature) and destroys the crystal structure during its intrusion. In this scenario all grains would form crystalline and only later the bulk would turn amorphous as Fe is adsorbed. Very high spatial resolution far-IR observations of the dust around AGB stars should indicate if there is indeed a rather sudden decrease of the crystalline silicate abundance in the dust shell at a temperature of roughly 800 K.

Finally, the velocity difference between dust and gas particles can be high enough for destruction and sputtering. E.g., for silicate grains the velocity difference with a helium atom should be about 30 km/sec for sputtering, and these velocities are not unlikely [167]. Colliding crystalline silicate grains might be heated high enough to become a melt or even evaporate. In the first case, the very quick cooling timescale (depending on the grain size), might result in amorphous silicates. Although it remains difficult to explain the different opacities for amorphous and crystalline silicate grains in this scenario.

As mentioned above, the crystalline olivines and pyroxenes formed in the winds of AGB stars with high mass loss are Fe-free [7]. Interestingly, it is also the Fe-free olivines and pyroxenes that are the first silicates that are expected to condense. This will take place at temperatures where they will crystallize very quickly. Thermodynamic calculations show that the Fe-containing silicates will condense at a somewhat lower temperature, and since there is no evidence for Fe-containing crystalline silicates, they apparently remain amorphous. A possible explanation for this phenomenon is that as soon as some dust forms, radiation pressure will accelerate the dust (and through the drag-force also the gas) and the material will quickly be pushed to cooler regions. So, only the very first condensed silicates, the Fe-free silicates, might have a chance to crystallize, while the later condensed silicates (the Fe-containing silicates) are formed in an environment where the annealing timescale is much longer than the accretion timescale. We should note here that it is not clear that a ferromagnesiosilica will condense at all from the gas-phase. Gas phase condensation experiments of a Fe-Mg-SiO-H_2-O_2 vapor show only condensates of a magnesiosilica and of a ferrosilica composition but not of a ferromagnesiosilica composition (see Fig. 13).

While in most cases crystalline silicate dust is only a minor component in circumstellar dust shells, there are some evolved stars with a very high abundance of crystalline silicates. Remarkably, these always have a disk-like dust distribution [68]. Because the temperature in these disks is well below the annealing temperatures of amorphous silicates, Molster et al. [68] suggested that a low-temperature crystallization process is responsible for the higher fraction of crystalline silicates. Several processes were discussed by Molster et al. [68], but no conclusive answer could be given. Recently, partial crystallization at room temperature due to electron irradiation has been detected (Fig. 14 and [169]). Although this is a promising effect, there are still open questions. It is not sure that this process can create highly crystalline material in sufficient quantities. Investigations are still necessary to find out whether the particles can become completely crystallized or will simply become partially amorphous and partially crystalline. Note, that electron irradiation only causes local crystallization unlike thermal crystallization, which acts on the whole grain at the same time. Observations indicate that the amorphous and crystalline silicates are two different grain populations which are not in thermal contact [7] and it is unclear that this

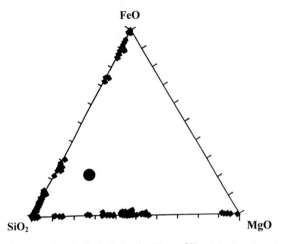

Fig. 13. Ternary diagram MgO-FeO-SiO$_2$ (oxide wt%) with the chemical compositions of gas to solid condensed grains from a Fe-Mg-SiO-H$_2$-O$_2$ vapor. The "average bulk solid" composition (the big dot) is roughly the gas phase composition which might have been somewhat less SiO$_2$-rich. Figure taken from [168].

can be achieved by this process. Finally, the compositional implications (with or without Fe) of this process are also not well studied yet.

Theoretical calculations predict that in both oxygen-rich as well as carbon-rich stars metallic Fe can condense [164]. At lower temperatures it can react with both oxygen as well as sulfur to form respectively FeO and FeS. Although there is a tentative detection of FeO in AFGL4106 [32], recent findings bring doubts to this claim and identify the applicable feature with FeS [6]. In principle one would expect that at lower temperatures FeS would react further to form FeO in oxygen-rich environments. However, it is very likely that in the outflows of these stars the temperature and density of the dust particles drop so rapidly that equilibrium reactions simply freeze out and do not take place anymore.

The presence of carbonates around two planetary nebulae [22] shows that the normal assumed dust formation processes are not always applicable. The enormous amount of small carbonate grains (50 M$_\oplus$) in NGC6302 cannot be explained by aqueous alteration, as is observed on Earth (and probably on Mars). These conditions are unlikely to exist in the circumstellar environment of evolved stars. Kemper et al. note that in both objects with carbonates there is also evidence for crystalline silicates as well as abundant crystalline H$_2$O ice [22]. The carbonate formation mechanism may involve a mobile H$_2$O-ice layer including CO$_2$ which reacts with the silicate core on which the ice was deposited, possibly via the intermediate formation of hydrous silicates. This reaction however is unlikely to occur at the temperatures a which the ice can exist (below 100 K). Also the non-detection of hydrous silicates makes this scenario unlikely. The lack of hydrous silicates also hampers a second theory, that assumes that gas phase H$_2$O will react with silicates to form hydrous silicates, which transform further to carbonates by a reaction with gas phase CO$_2$. Alternatively, the carbonates

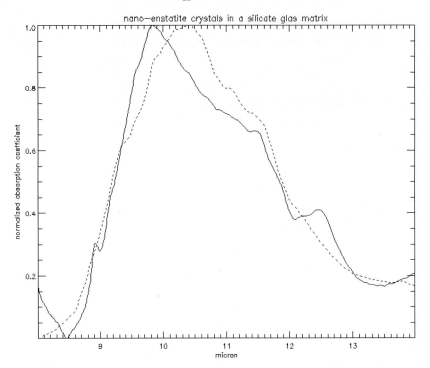

Fig. 14. The infrared absorption spectrum of an amorphous silicate smoke before (dashed line) and after (solid line) a 10 minute irradiation by the electron beam of a transmission electron microscope. Note the upcoming of the spectral features.

may form near the surface of mass-losing AGB stars from gas-phase condensation of CaO in the presence of H_2O and CO_2. Kemper et al. favour this latter mechanism. It should be noted that all methods mentioned above have not been verified yet by laboratory experiments under the conditions suitable for mass losing stars.

Recently, Rietmeijer (private comm.) suggested that the formation of carbonates proceeds via the so called weathering cycle. Due to a bombardment of energetic particles the rim of diopside will be amorphized and the lighter elements will be sputtered away [e.g. 170]. This will result in an excess of Ca. Together with CO_2 from the gas phase the rim will decompose into diopside and calcite or even dolomite. Note that diopside has been detected in these environments in contrast to the hydrous silicates.

The detection of carbonates in the circumstellar environment of evolved stars opens the possibility that *some* solar system carbonates found in primitive bodies as IDPs and the most primitive meteorites may not have formed by aqueous alteration and might even have an interstellar origin. It is clear however that the origin of the carbonates in many meteorites can most naturally be explained by aqueous alteration, as evidenced by the structure of the meteoritic matrix.

Around some oxygen-rich PNe and LBVs a small amount of PAHs have been detected together with some obvious oxygen-rich components like crystalline silicates. The origin of these carbon-rich molecules in these oxygen-rich environments is still not well understood. These stars, especially the LBVs, are not expected to have experienced a period in which the C/O ratio in their envelope was above 1. Probably the dissociation of CO (by the UV radiation field) produce some free carbon which might react with other C-atoms to form benzene rings.

After formation, dust grains in stellar outflows travel away from the star towards lower density and temperature. The grains continue to collide with gas, and, in high density winds, also grain-grain collisions (with all its consequences) can be important. The gas-grain interaction can lead to further grain growth, and to the condensation of ices (in particular H_2O ice) if conditions are favourable.

Dust Formation in Carbon Stars: In carbon stars (where the C/O ratio of the gas is higher than 1) carbon-rich dust species form. One of the first detected species was SiC [139]. There is still some debate about the exact crystal structure (see Sect. 3.2) but the SiC itself is undisputed. SiC is rather stable and forms already at relatively high temperatures, so it is therefore the expected grain to form when both Si and C are very abundant and oxygen is not.

TiC, a high temperature condensate, is often suggested to be the seed particle for dust forming processes in carbon-rich environments. Because of its low abundance it is only expected to form in significant amounts in a high density environment, which are found in the atmospheres of evolved stars at the (very) end of the AGB. This would fit with the proposal that the 21 micron feature is due to TiC, and this feature is only seen in post-AGB (directly after the huge mass loss at the end of the AGB) stars and two PN.

The chemical composition of photospheres of carbon stars and some of the physical parameters characterizing the inner stellar winds are very similar to the conditions encountered in the combustion of hydrocarbons in the laboratory. Indeed, acetylene, C_2H_2, is the dominant carbon-bearing molecule after CO in the photospheres of carbon stars [171]. The typical temperature of carbon stars (2500 K) are also found in flames rich in acetylene which are efficient sources of amorphous carbon particles in the form of soot. This natural link was used by several people to apply the well studied combustion chemistry to the modeling of the formation of amorphous carbon grains in C-stars [172]. The formation of amorphous carbon (soot) can be divided into two steps, the nucleation phase and the condensation phase. The nucleation phase starts with the formation of benzene (C_6H_6). The addition of acetylene to an aromatic radical, the subsequent extraction of hydrogen and the second addition of an acetylene molecule, resulting in the closure of another ring, leads to the growth of the PAH-molecule. When enough large aromatic molecules are formed, condensation products will (simultaneously) form. Condensation, the growth of planar species into 3-D solid (amorphous carbon) grains, is initiated by the formation of PAH dimers, where the aromatics are linked by van der Waals forces. The growth to amorphous carbon particles then proceeds through coagulation of condensation products of PAH molecules, deposition of carbon via acetylene surface reactions and surface condensation of free PAH molecules on the grain [173].

There exists a small group of carbon-rich AGB stars, which show the clear presence of amorphous and/or crystalline silicate dust in their infrared spectra. It is thought that in these cases the silicates are the remnants of a previous mass loss phase, during which the star was still oxygen-rich. The current theory assumes that these silicates are then stored (due to binary interaction) in a circum-stellar/binary disk. While the star evolved further and became carbon-rich the old (oxygen-rich) dust remained present around the evolved star [174]. Another case are the Wolf-Rayet [WC] central stars of planetary nebulae, a poorly understood class of PNe, with little or no H in their atmospheres, while the ejecta are C-rich and H-rich. ISO discovered that besides the known PAH features, also crystalline silicates are present [175, 176, 177]. Two different scenarios are proposed to explain the nature of these objects. One theory explains the chemical dichotomy by a thermal pulse either at the very end of the AGB or young PN phase [late AGB thermal pulse 178], or when the star is already on the cooling track [very late thermal pulse 179]. The H-rich and O-rich layers may be removed due to efficient mixing and subsequent nuclear burning, or by extensive mass loss, exposing processed layers to the surface. The model has both statistical and timing problems, however. Such late pulses only occur at the right moment to cause a [WC] star for a small fraction of AGB stars, and in many cases the star would already be carbon rich. The other theory solves this by storing the oxygen-rich material in a longer-lived reservoir. This is seen in objects such as the Red Rectangle, where a circumbinary disk has formed capturing earlier mass loss [180]. Accretion from a disk on the star, as is also expected to have taken place in the case of the Red Rectangle [181], may lead to the [WC] phenomenon, but this is highly speculative. The recent discovery of a disk around CPD-56°8032 [182] makes the disk storage mechanism more attractive. But this leaves the relation between disk and [WC] star open: there is no evidence for binarity in such stars. It should be emphasized that no information is yet available about the distribution of the O-rich and C-rich dust components. High spatial resolution infrared observations will reveal the location of the different dust components and might help in distinguishing between the two theories.

S stars, with their C/O ratio close to one, normally show a rather flat SED without many dust features. This has long been interpreted as evidence for the presence of carbonaceous dust. Since, the mass loss rates for those stars were based on the 60 micron flux, the assumption about the presence of carbonaceous dust resulted in rather low mass loss rates for these stars. However, recently Ferraroti & Gail [164] calculate that the 2 most likely dust components that form are solid Fe and FeSi, which both also have a rather flat spectrum, without prominent features. Furthermore, if the presence of these two materials can be confirmed, it would also bring the empirical dust mass loss rates (based on the

60 μm opacity) and the gas to dust ratio much closer to the values derived for M and C-stars.

4.2   Dust Formation in Supernovae

Unfortunately, little is known about the dust production rate in SNe. Only few infrared spectra of SN dust ejecta are available and they are not always conclusive. Arendt et al. [183] show a reasonable fit with proto-silicates to the SWS spectrum of an area centered on a fast moving knot (position N3) in the supernova remnant of Cas A. However, Douvion et al. [184] reject this identification based on a SWS spectrum taken very close to position N3, which show much similarities (but is not exactly equal). They fit their spectrum with a mixture of three dust species ($MgSiO_3$, SiO_2 and Al_2O_3) with each having two different temperatures together with some synchrotron emission. The presence of amorphous $MgSiO_3$ was predicted by Kozasa et al. [185] and fitted the ISOCAM spectra reasonably well [186]. However, we would like to note that the rather artificial temperatures (all dust species have different temperatures both for the hot and cold phases which are not directly related to opacity differences), the arbitrary dust sizes and the more or less one feature identifications, leave some room for alternative interpretations. Therefore, most of our knowledge about dust production in SNe is still based on the analysis of presolar grains, which due to their isotopic anomaly could be traced back to a supernova explosion. Selection effects make it impossible to deduce the dust production rate from these data. In addition, the large differences in physical and chemical conditions between AGB winds and SN ejecta make it difficult to assume similarities in the types of dust that may condense in SN ejecta. Nevertheless, based on these isotopic measurements in general one can say that nano-diamonds, corundum, SiC etc. do form after a supernova explosion (but none of them has been spectroscopically identified yet).

For completeness we mention here that also some novae are shown to produce dust during the eruption [e.g. 187].

4.3   The Processing of Dust in the ISM

The average lifetime of a silicate dust grain in the ISM is roughly $4 \cdot 10^8$ years, while the replenishment rate of ISM dust is about $2.5 \cdot 10^9$ years [161]. This means that interstellar dust grains are destroyed and re-formed about six times before finally entering a star forming region to be incorporated into a new generation of stars. Thus, new grains must be formed in the ISM. Below, we discuss dust formation and destruction mechanisms that may occur in the ISM.

The dust formation process in the ISM is not clear, however it seems inevitable that it takes place in the dark molecular clouds, where high densities prevail. The C/O ratio in the ISM is in general smaller than unity, which normally would result in the formation of oxygen-rich material. However, in diffuse clouds the interstellar radiation field easily dissociates the CO molecule. Thus, especially in the diffuse ISM the formation of carbon-rich dust cannot be excluded. In the dark molecular clouds a rather complex chemistry is taking place, which might also result in the formation of carbon-rich dust without a C/O ratio larger than 1, but not much is known about this process.

The most important difference between the dust formation process in the interstellar medium and the dust around stars is the temperature and density of the gas. Around stars dust forms at temperature of about 1000 K, while in the ISM dust temperatures between 10 and $\approx$ 100 K are found. These low temperatures are far below any annealing point, therefore the grains formed in the ISM will be amorphous. This is consistent with the fact that no crystalline silicates have been found in the ISM.

From depletion pattern differences between cool clouds in the Galactic disk and warm clouds in the galactic halo, it seems that the dust that forms in the ISM has an olivine stoichiometric ratio [188]. Note that this does not imply that they are olivines.

For the carbon rich dust the situation is even more complicated, based on the fact that still no conclusive identification has been made for the carbon dust component in the ISM. For many SiC particles found in meteorites it has been proven that they come from carbon stars or supernovae. They must therefore have resided in the ISM before they got incorporated in the collapsing cloud from which the solar system formed. However, up to now there is only an upper limit of the amount of Si in SiC in the ISM of 5% of that in silicates [154], which is quite low compared to the SiC abundance in carbon stars and their contribution to the ISM. Tang & Anders deduce from ^{21}Ne isotopes in SiC grain in C2 chondrites that the lifetime [189] of these grains is only 40 Myr. This is remarkably short compared to the estimated lifetime of a grain in the ISM (400 Myr). Interestingly enough this 40 Myr, is roughly equal to the oxidation time-scale for diffuse interstellar clouds. One can therefore argue that oxidation will selectively destruct the SiC grains [154] in the ISM.

Very small graphite particles have been proposed as the carrier of the 2175Å feature, but its formation site is not well known. Amorphous carbon has been proposed to form around carbon stars and interstellar shock waves might provide the energy to convert it to graphite. According to Tielens et al. [106] only 5% of the very small carbon grains (maximum size about 100Å) will be converted to diamonds, while the rest would likely become graphitic C. This process might be the source of very small graphitic grains.

Sandford et al. [190] show that the C-H stretching band shows a linear correlation with the visual extinction, except for the direction towards the galactic center. This behaviour is remarkably similar for the Si-O stretching band. This similarity suggests that the silicates, responsible for the Si-O stretch, and the aliphatic hydrocarbons, responsible for the 2950 cm^{-1} (3.4 μm) C-H stretch may be coupled, perhaps in the form of a silicate core, organic mantle grain. If true, those mantles might provide a good shielding for the destruction processes in the ISM. This will likely alter the life-cycle of the silicate dust, especially the destruction rate. How they are formed is still unclear, but the carbonaceous mantles found around IDPs (see Bradley this volume), might result from a similar kind of process and can provide a better understanding of the core mantle grains in the ISM.

The 90 μm band in NGC1333-IRAS4 attributed to $CaCO_3$, sheds (if confirmed) new light on the formation of carbonates. The material in the vicinity of this proto-star presumably has only experienced typical interstellar medium conditions and thus a gas-phase condensation mechanism for the carbonates, as suggested by Kemper et al. [22] for the planetary nebulae, seems implausible. Ceccarelli et al. [72] suggest that the low-temperature mechanism, considered by Kemper et al., may be more likely. Alternatively, the carbonates around evolved stars may have survived their residence in the ISM.

We now turn to dust destruction mechanisms. The harsh conditions that prevail in the ISM can result in the destruction or modification of interstellar dust. Interstellar shocks may shatter grains into smaller fragments, thus changing the grain size distribution by producing more small grains. The ISM is a destructive place for the crystalline silicates. We observe these grains in the outflows of evolved stars, but not along lines of sight through the ISM. The exact abundance of the crystalline silicates brought into the ISM by stars is not known yet. This depends on how much cold crystalline silicates are hidden in the warm amorphous silicate profiles. But the lack of crystalline silicate features in the ISM, makes it plausible that the crystalline silicates are destroyed or amorphized in the ISM.

In fact both processes are expected to occur in the ISM. It has been suggested that the amorphization might be due to a bombardment of the crystal structure by energetic radiation as will take place in the ISM. If the temperatures are below the crystallization temperatures there will be not enough internal energy to overcome the threshold energy necessary to fix the faults in the crystal. This is the situation in the harsh environment of the ISM, where energetic particles can destroy the matrix, and the low temperatures prevent repairs of these defects. However, Day [191] shows that MeV protons hardly have any influence on the crystalline silicates and electron irradiation might even make grains more crystalline [169].

It seems that the destructive power in the ISM mainly comes from SN shock waves. These have a different influence on the particles depending on their size and composition. Laboratory experiments of irradiation by He^+ ions, resembling supernova shock waves through the ISM, show that the small crystalline silicates are likely amorphized in the ISM [45]. Although there is a limited penetration depth, only the first micron will be completely amorphized, and larger grains might therefore survive complete amorphization, but they will suffer more from sputtering and evaporation in these shocks [161]. So, in the end it is not expected that the silicate grains, especially the crystalline ones, will survive a long stay the ISM.

The final structure of silicates exposed to He^+ irradiation shows some resemblance with the GEMS in IDPs (Bradley, this volume). In this respect it is also interesting to note that sometimes inside a GEMS a so-called "relict forsterite" grain is found, which shows evidence for heavy irradiation damage, much more than what is expected during their stay in the solar nebula. Furthermore, these irradiation experiments showed that O and Mg are preferentially sputtered away from the first 100nm of the test samples. This might explain the evolution of the amorphous silicate dust which looks more like an amorphous olivine around the evolved stars and more like an amorphous pyroxene around young stars [45]. A similar selective sputtering, but then by solar wind irradiation has been observed in solar system material [192].

4.4   Dust Processing in Star-Forming Regions

Without doubt the solar system is the best evidence available for the dramatic changes in dust composition that occur during star and planet formation. The vast literature on the mineralogy of solar system objects, ranging from planets to IDPs demonstrates that large differences exist between the properties of solids in the solar system and in interstellar space. In recent years it has become evident that these differences also exist for dust in primitive solar system objects such as comets. Clearly, understanding these differences is one of the main challenges of the field of star- and planet formation.

Astronomical observations of planet-forming disks around young stars also begin to reveal evidence for processing of dust. For instance, infrared spectroscopy and millimeter observations of disks surrounding Herbig Ae/Be stars suggest that grains in proto-planetary disks have coagulated into sizes up to cm, and probably to even substantially larger bodies that escape spectroscopic detection. Substantial changes in the composition of the gas also occur, both through freeze-out onto grains as well as through chemical reactions. Some of these reactions occur on the surface of the grains, and require irradiation with UV photons from the newly born proto-star.

Apart from an increase in average size, infrared spectroscopy has also shown that the composition of dust in proto-planetary disks has changed compared to that in the ISM. Probably the clearest example is the crystallization of the amorphous silicates around Herbig Ae/Be stars. There is no convincing evidence for the presence of crystalline silicates in the ISM. The crystalline silicates found around young stars therefore have to be formed in situ. They will be formed both by gas-phase condensation, as well as by annealing of amorphous silicates, as is also seen in our own solar system [193]. Both processes will take place close to the young star. This seems to be supported by the observations. In most of these stars there are indications for the presence of crystalline silicates, but only in the 10 micron region, indicating that the crystalline silicates are indeed hot and close to the star [67, 92]. The gas phase condensation process will produce very Fe-poor crystalline silicates, similar to what has been found around evolved stars, while the annealing process will very likely produce Fe-containing crystalline silicates. Because the amorphous silicates will in general not have the same stoichiometric ratio as the annealed crystals, some remainder material is expected to form. The presence of silica in the dust emission spectra of young stars supports this scenario. This material is very transparent and will normally

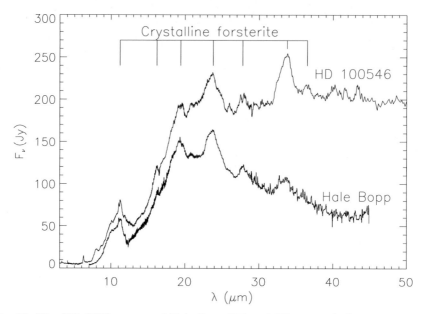

Fig. 15. The ISO SWS spectra of Hale-Bopp [55] and HD100546 [59]. The crystalline forsterite features are indicated.

be very cold relative to the other minerals. The fact that the silica features in the 10 micron region are detected, indicates that silica is much warmer than expected. It should therefore be in thermal contact with the other minerals, as is expected when it is a remainder of annealing of amorphous silicates [92].

Recently, nebular shocks have been suggested as an alternative mechanism to provide the temperatures necessary for annealing of the amorphous silicates in the proto-solar nebula at larger distances [194]. Although, this mechanism fails to explain some of the aspects of the crystalline silicates in the solar system, and will therefore not be applicable to all crystalline silicates found, it cannot be excluded as a competing mechanism. This mechanism might also play a role in the disks around evolved stars.

A spectacular result of the ISO mission has been the remarkable similarity between the spectral appearance of the solar system comet Hale-Bopp [55] and that of the Herbig Ae/Be star HD 100546 [59]. In both objects the emission bands of crystalline silicates, as usual Fe-poor, stand out (Fig 15). The origin of the high abundance of crystalline silicates in both objects is not clear. In HD 100546 the abundance of forsterite increases with distance from the star [195], which is not expected in the case of radial mixing of annealed silicates [196, 194, 197]. Bouwman et al. [195] propose a local production of small forsterite grains as a result of the collisional destruction of a large parent body.

Most considerations about the formation of crystalline silicates come from theory. However, also laboratory experiments sometimes shed new light on this subject. There are indications that the width of some features are an indication

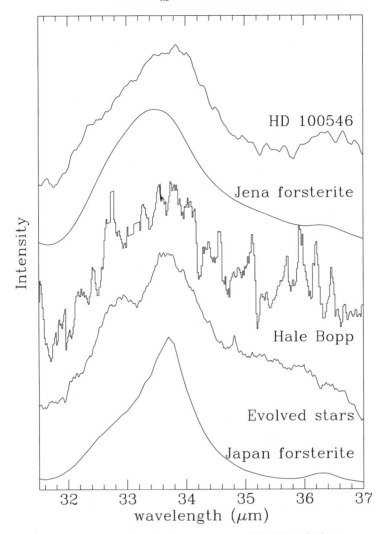

Fig. 16. The 33 micron complex of the young star HD100546 [59], the comet Hale-Bopp [55] and an average of 33 micron complexes of evolved stars with evidence for an excretion disk [27], together with 2 laboratory measurements of forsterite one meauserd in Jena [13] and one in Japan [8]. Note the difference in width of the 33.6 micron feature. [Figure taken from 7].

of the internal structure of a crystal, and therefore of their formation history. Figure 16 shows two laboratory spectra of forsterite: the forsterite sample from Jena has been made from a melt, and is likely polycrystalline, while the sample from Japan is a single crystal. Note the difference in band width between both spectra. It is interesting to compare these band shapes to those of some characteristic bands seen in the spectra of evolved and young stars. The evolved stars, with freshly made stardust, show a band shape which resembles that of the

Japan sample, i.e. corresponding to a single crystal. This may be expected, since gas-phase condensation would naturally lead to particles with a single crystal structure. On the other hand, the ISO spectrum of HD 100546 shows a resemblance to that of the polycrystalline sample from Jena. This is an indication that the forsterite in HD100546 has a polycrystalline structure. Both fractionation of a differentiated body (as proposed by Bouwman et al. [195] for HD100546), as well as annealing of amorphous silicates (as seem to occur around many other young stars) will lead to polycrystalline forsterite and is therefore in agreement with the observations. The signal to noise of the Hale-Bopp spectrum, prevents to draw conclusions about the origin of the crystalline silicates in this comet. However, it is important to stress, that there are other effects that can influence the band shape, such as blends with other dust species, and grain size and shape effects.

The presence of carbonates in evolved stars shows that there may exist alternative formation mechanisms for these materials, which has implications for the use of carbonates to reconstruct the early history of the solar system: although there is much evidence that most carbonates in the solar system are formed by aqueous alteration, this does not automatically have to be true for *all* carbonates.

While solid FeS is abundant in IDPs and primitive meteorites, for a long time FeS had not been found outside the solar system. Recently, Keller et al. [6] and Hony et al. [198] show that FeS is produced by evolved carbon-rich stars, and is also seen in proto-planetary disks surrounding young intermediate mass Herbig Ae/Be stars. In the solar system, virtually all sulfur in primitive bodies is found as FeS, suggesting an extremely efficient formation mechanism. Since sulfur is not depleted significantly from the gas-phase in the diffuse interstellar medium, this implies that the sulfur depletion occurs at some point during the star formation process, eventually leading to the formation of solid FeS. Since even in IDPs sulfur is locked up in FeS, it seems that the formation of this solid must have occurred very early in the collapse of the proto-solar cloud. Perhaps a gas-phase-solid reaction between metallic Fe and H_2S is responsible. However, this reaction only occurs at temperatures above 400 K, which seems difficult to reconcile with the low temperature conditions that prevailed in the outer regions of the proto-solar cloud.

Many theoretical studies have been performed to better understand the processing of gas and dust in proto-planetary disks. For instance, Gail [199] applies a thorough thermal equilibrium calculation to the dust mixture in a proto-stellar accretion disk. He takes into account the disk structure, opacity and the chemical compositions and abundances of the major dust species. His calculations show that substantial dust processing can occur in the inner regions of the accretion disks surrounding proto-stars.

The discovery that most nano-diamonds are formed in the accretion disks of young stars and are not presolar (Bradley this volume), opens the question about their (trans)formation. It was long accepted that the nano-diamonds in meteorites came from the ISM and should therefore be quite abundant. However, the lack of features in the ISM and the apparent increase of the abundance of nano-diamonds towards the inner solar system (Bradley, this volume) have caused some doubt to this statement. If nano-diamonds are very abundant in the ISM, it would imply that these diamonds were very easily formed via chemical vapor deposition in interstellar or circumstellar environments [i.e. 107] or via shock processing in interstellar shocks [106]. The problem is that pure nano-diamonds have no or only very weak features in the IR. They will only become detectable when they are hydrogenated [110]. And up to now only in five sources these features have been detected, two post-AGB star (HR4049 and HD52961) and three Herbig Ae/Be stars (Elias 1, HD97048 and MWC297) [110, 116]. The key question has now become are nano-diamonds only formed in special places, or are they only visible in special environments. The 5 known sites where the nano-diamonds have been found give a clue about this. All stars have a relatively hot circumstellar disk and at least four of them also an oxygen-rich gas chemistry, which is obvious for the three Herbig Ae/Be stars and more surprising for HR4049 [74]. The fifth, HD52961, has a circumbinary disk with very metal poor gas, and no obvious oxygen-rich gas chemistry has been reported. Chang et al. [200] showed that the formation of diamond crystals is facilitated by both high temperatures and the presence of oxygen. Also the formation timescales in a disk are much longer, compared to normal outflows, due to the storage of dust in a disk. Hydrogenation, necessary to detect the otherwise (almost) IR-inactive nano-diamonds, should also be possible in a carbon dominated environment. Furthermore, the size distribution of the nano-diamonds in meteorites is log-normal, which points to a growth process in a circumstellar environment and not to fractionation as is expected when the grains would have resided in the ISM [201]. Combining all this information, it seems that it is the special environment in which these diamonds are made, instead of a special environment necessary to detect them. The formation of nano-diamonds is facilitated by the presence of carbonaceous dust together with an oxygen-rich gas chemistry, long storage times and high temperatures. Interestingly enough, these were also the conditions in the inner solar system. The most likely formation mechanism is the shock-heating of carbonaceous material in the proto-stellar accretion disk. The almost complete coverage of IDPs by a carbonaceous layer (see e.g. [202] and Bradley this volume) indicates that although the overall C/O ratio is below one, a carbon based chemistry can take place in these environments.

5   Conclusions and Future Directions

Astromineralogy has gained a tremendous momentum with the launch of the Infrared Space Observatory. The spectral resolution of this satellite was perfect for the study of dust at infrared wavelengths. Where IRAS opened the infrared studies at wavelengths up to 23 μm, ISO opened the electromagnetic spectrum beyond 23 μm for systematic investigations. Some pioneering work in this wavelength region was already done by KAO, but the much improved sensitivity of ISO and its wide wavelength coverage allowed much more detailed studies.

As often in astronomy, opening a new window to the universe results in a landslide of exciting discoveries, and often in directions that were not really anticipated. ISO is no exception to this: before ISO was launched, it was expected that most discoveries related to dust(formation) would be made in carbon-rich environments. It was thought that the oxygen-rich environments would only show the broad, smooth amorphous silicates. The detection of crystalline silicates, and the evidence for carbonates in circumstellar environments came therefore as a big surprise.

A big step forward for astromineralogy is the fact, that it is now possible to determine for some dust species the exact mineralogical composition. These identifications are always done using laboratory measurements of cosmic dust analogues or of stardust samples. The importance of laboratory work for progress in the field of astromineralogy cannot be underestimated. In the first years after the launch of ISO, most effort was put into the verification and identification of all new dust features. Thanks to many laboratory experiments this has been quite successful. More than 75% of the dust features have now a (more or less secure) identification. Many identifications were for dust species which had not been detected before in space. The presence or absence of features of different dust components gives valuable information about dust (trans)formation processes. Abundance estimates are based on these features, and they provide important input for modelers. We stress that, although quite some progress has already been made, there are still many unidentified features. It is expected that the identification of these features, will give additional constrains on dust (trans)formation.

The ISO discoveries triggered many follow-up studies on dust formation and processing. These studies indicate that thermal equilibrium calculations cannot explain the diversity and abundances of the different dust species. New non-equilibrium calculations were carried out and predictions for the presence of other dust species were made. Some progress has been made in this field, but there is room for improvements. Progress can be made, especially in combination with new laboratory measurements of the infrared properties of dust species. However, the requirements for the cosmic dust analogues, are quite severe (appropriate size, no contaminations and preferably at low temperatures) and these experiments are therefore not easy to perform. Laboratory experiments are important to test the proposed physical and chemical dust processing mechanisms.

Radiative transfer models were confronted with the sudden increase of spectral detail. The infrared spectra are now accurate enough that real dust species can (and should) be incorporated. The behaviour of these real dust species is sometimes rather different from the artifical ones. The adding of more and realistic dust species leads to more complex calculations, but also to a significant increase of our knowledge about the location (and therefore evolution) of the different dust species, a good example is the space distribution of forsterite around HD100546 [195].

One of the most exciting new venues of dust studies is the possibility to quantitatively compare the dust in our own solar system to that seen in space.

Laboratory techniques are becoming increasingly powerful and provide important information about the formation history of solar system and stardust samples. New space missions (e.g. ROSETTA, STARDUST) will study planetary and cometary dust and even will retrieve samples for scrutiny in laboratories. Observationally, ISO is only the beginning of a development involving new observatories such as SIRTF, HERSCHEL and JWST, which will tremendously increase our knowledge of astromineralogy, and in particular of the mineralogy of proto-planetary systems. We will be able to study the processing of dust in newly forming planetary systems and compare directly to the history of the formation of our solar system as it is recorded in planets, asteroids and comets.

High spatial resolution observations at infrared wavelengths will allow us to study the dust formation and growth in red giants and its time dependence, thus putting tight constraints on dust nucleation models. Sensitive spectrometers will allow studies of dust in stars outside our galaxy, giving insight into dust formation in external galaxies.

The ISO archive will, however, remain a valuable source of information about astromineralogy. This is because none of the planned missions will have a similarly wide spectral coverage as ISO has. It is precisely this property of ISO, together with its spectral resolution, which made it such an efficient discovery machine. We need a new ISO!

Acknowledgements

We would like to thank Ciska Kemper, Sacha Hony, Jan Cami and Frans Rietmeijer for the supply of several figures and Ciska Kemper and Sacha Hony also for the careful reading of the manuscript. We thank Chiyoe Koike, John Bradley, and our colleagues in Amsterdam, Groningen, Leuven and Jena for the numerous discussions about dust in space and in the laboratory. This overview would not have been possible without the help of ADS and Simbad. LBFMW acknowledges financial support from an NWO "pionier" grant. Last, but not least, it is a pleasure and an honour to thank all those that have made ISO such a superb success.

References

[1] Li, A. & Greenberg, J. M., in Solid State Astrochemistry, ed. V. Pironello J. Krelowski, (Dordrect: Kluwer), in press (2002).
[2] de Graauw, T. et al., Astron. Astrophys. **315**, L49–L54 (1996).
[3] Clegg, P. E. et al., Astron. Astrophys. **315**, L38–L42 (1996).
[4] Kessler, M. F. et al., Astron. Astrophys. **315**, L27–L31 (1996).
[5] Bradley, J. P. et al., *Science* **285**, 1716–1718 (1999).
[6] Keller, L. et al., Nature **417**, 148–150 (2002).
[7] Molster, F. J., Waters, L. B. F. M., Tielens, A. G. G. M., Koike, C. & Chihara, H., Astron. Astrophys. **382**, 241–255 (2002).
[8] Koike, C., Tsuchiyama, A. & Suto, H., in Proceedings of the 32nd ISAS Lunar and Planetary Symposium, 175–178 (1999).

[9] Servoin, J. L. & Piriou, B., Phys. Status Solidi (B) **55**, 677–686 (1973).

[10] Scott, A. & Duley, W. W., Astrophys. J. Suppl. **105**, 401–405 (1996).

[11] Dorschner, J., Begemann, B., Henning, T., Jäger, C. & Mutschke, H., Astron. Astrophys. **300**, 503–520 (1995).

[12] Koike, C., Shibai, H. & Tuchiyama, A., Monthly Notices Roy. Astron. Soc. **264**, 654–658 (1993).

[13] Jäger, C. et al., Astron. Astrophys. **339**, 904–916 (1998).

[14] Chihara, H., Koike, C., Tsuchiyama, A., Tachibana, S. & Sakamoto, D., Astron. Astrophys. **391**, 267–273 (2002).

[15] Henning, T., Begemann, B., Mutschke, H. & Dorschner, J., Astron Astrophys. Suppl. **112**, 143–149 (1995).

[16] Ordal, M. A., Bell, R. J., Alexander, R. W., Newquist, L. A. & Querry, M. R., Applied Optics **27**, 1203–1209 (1988).

[17] Begemann, B., Dorschner, J., Henning, T., Mutschke, H. & Thamm, E., Astrophys. J. Letters **423**, L71–L74 (1994).

[18] Koike, C. et al., Astrophys. J. **446**, 902–906 (1995).

[19] Borghesi, A., Bussoletti, E., Colangeli, L., Orofino, V. & Guido, M., *Infrared Physics* **26**, 37–42 (1986).

[20] Pegourie, B., Astron. Astrophys. **194**, 335–339 (1988).

[21] von Helden, G. et al., Science **288**, 313–316 (2000).

[22] Kemper, F. et al., Nature **415**, 295–297 (2002).

[23] Henning, T. & Mutschke, H., Astron. Astrophys. **327**, 743–754 (1997).

[24] Mennella, V. et al., Astrophys. J. **496**, 1058–1066 (1998).

[25] Bowey, J. E. et al., Monthly Notices Roy. Astron. Soc. **325**, 886–896 (2001).

[26] Chihara, H., Koike, C. & Tsuchiyama, A., Publ. Astron. Soc. Japan **53**, 243–250 (2001).

[27] Molster, F. J., Waters, L. B. F. M. & Tielens, A. G. G. M., Astron. Astrophys. **382**, 222–240 (2002).

[28] Purcell, E. M. & Pennypacker, C. R., Astrophys. J. **186**, 705–714 (1973).

[29] Draine, B. T., Astrophys. J. **333**, 848–872 (1988).

[30] Barber, P. W. & Yeh, C., *Applied. Opt.* **14**, 2864–2872 (1975).

[31] Mishchenko, M. I., Travis, L. D. & Mackowski, D. W., J. Quant. Spectrosc. Radiat. Transfer **55**, 535–575 (1996).

[32] Molster, F. J. et al., Astron. Astrophys. **350**, 163–180 (1999).

[33] Hoogzaad, S. N. et al., Astron. Astrophys. **389**, 547–555 (2002).

[34] Kemper, F., de Koter, A., Waters, L. B. F. M., Bouwman, J. & Tielens, A. G. G. M., Astron. Astrophys. **384**, 585–593 (2002).

[35] Bowey, J. E. et al., Monthly Notices Roy. Astron. Soc. **331**, L1–L6 (2002).

[36] Draine, B. T. & Lee, H. M., Astrophys. J. **285**, 89–108 (1984).

[37] Ossenkopf, V., Henning, T. & Mathis, J. S., Astron. Astrophys. **261**, 567–578 (1992).

[38] Jones, T. W. & Merrill, K. M., Astrophys. J. **209**, 509–524 (1976).

[39] Pegourie, B. & Papoular, R., Astron. Astrophys. **142**, 451–460 (1985).

[40] Suh, K., Monthly Notices Roy. Astron. Soc. **304**, 389–405 (1999).

[41] Guertler, J. & Henning, T., Astrophys. Sp. Sc. **128**, 163–175 (1986).

[42] Demyk, K., Jones, A. P., Dartois, E., Cox, P. & D'Hendecourt, L., Astron. Astrophys. **349**, 267–275 (1999).

[43] Demyk, K., Dartois, E., Wiesemeyer, H., Jones, A. P. & d'Hendecourt, L., Astron. Astrophys. **364**, 170–178 (2000).

[44] Harwit, M. et al., Astrophys. J. **557**, 844–853 (2001).

[45] Demyk, K. et al., Astron. Astrophys. **368**, L38–L41 (2001).

[46] Hanner, M. S., Lynch, D. K. & Russell, R. W., Astrophys. J. **425**, 274–285 (1994).

[47] Hanner, M. S., in ASP Conf. Ser. 104: IAU Colloq. 150: Physics, Chemistry, and Dynamics of Interplanetary Dust, 367 (1996).

[48] MacKinnon, I. D. R. & Rietmeijer, F. J. M., *Reviews of Geophysics* **25**, 1527–1553 (1987).

[49] Bradley, J. P., Humecki, H. J. & Germani, M. S., Astrophys. J. **394**, 643–651 (1992).

[50] Knacke, R. F. et al., Astrophys. J. **418**, 440–450 (1993).

[51] Fajardo-Acosta, S. B. & Knacke, R. F., Astron. Astrophys. **295**, 767–774 (1995).

[52] Aitken, D. K., Smith, C. H., James, S. D., Roche, P. F. & Hough, J. H., Monthly Notices Roy. Astron. Soc. **230**, 629–638 (1988).

[53] Sandford, S. A. & Walker, R. M., Astrophys. J. **291**, 838–851 (1985).

[54] Waelkens, C. et al., Astron. Astrophys. **315**, L245–L248 (1996).

[55] Crovisier, J. et al., *Science* **275**, 1904–1907 (1997).

[56] Waters, L. B. F. M. et al., Astron. Astrophys. **315**, L361–L364 (1996).

[57] Molster, F. J., Waters, L. B. F. M., Tielens, A. G. G. M. & Barlow, M. J., Astron. Astrophys. **382**, 184–221 (2002).

[58] Koike, C. et al., Astron. Astrophys. **363**, 1115–1122 (2000).

[59] Malfait, K. et al., Astron. Astrophys. **332**, L25–L28 (1998).

[60] Cami, J., de Jong, T., Justtannont, K., Yamamura, I. & Waters, L. B. F. M., Astrophys. Sp. Sc. **255**, 339–340 (1997).

[61] Kemper, F., Waters, L. B. F. M., de Koter, A. & Tielens, A. G. G. M., Astron. Astrophys. **369**, 132–141 (2001).

[62] Molster, F. J. et al., Astron. Astrophys. **366**, 923–929 (2001).

[63] Wooden, D. H. et al., Astrophys. J. **517**, 1034–1058 (1999).

[64] Brucato, J. R., Colangeli, L., Mennella, V., Palumbo, P. & Bussoletti, E., Planetary and Space Science **47**, 773–779 (1999).

[65] Bouwman, J., Ph.D. thesis, University of Amsterdam, The Netherlands (2001).

[66] Cesarsky, D., Jones, A. P., Lequeux, J. & Verstraete, L., Astron. Astrophys. **358**, 708–716 (2000).

[67] Meeus, G. et al., Astron. Astrophys. **365**, 476–490 (2001).

[68] Molster, F. J. et al., Nature **401**, 563–565 (1999).

[69] Kemper, F., Molster, F. J., Jäger, C. & Waters, L. B. F. M., Astron. Astrophys. **394**, 679–690 (2002).

[70] Hellwege, K. H., Lesch, W., Plihal, M. & Schaack, G., Z. Physik **232**, 61 (1970).

[71] Barlow, M. J., Astrophys. Sp. Sc. **255**, 315–323 (1997).

[72] Ceccarelli, C. et al., submitted to Astron. Astrophys. (2003).

[73] Posch, T. et al., Astron. Astrophys. **352**, 609–618 (1999).

[74] Cami, J., Ph.D. thesis, University of Amsterdam, The Netherlands (2002).

[75] Stencel, R. E., Nuth, J. A., Little-Marenin, I. R. & Little, S. J., Astrophys. J. Letters **350**, L45–L48 (1990).

[76] Vardya, M. S., de Jong, T. & Willems, F. J., Astrophys. J. Letters **304**, L29–L32 (1986).

[77] Onaka, T., de Jong, T. & Willems, F. J., Astron. Astrophys. **218**, 169–179 (1989).

[78] Begemann, B. et al., Astrophys. J. **476**, 199–208 (1997).

[79] Huss, G. R., Hutcheon, I. D., Wasserburg, G. J. & Stone, J., in *Lunar and Planetary Science Conference*, volume 23, 29–33 (1992).

[80] Hutcheon, I. D., Huss, G. R., Fahey, A. J. & Wasserburg, G. J., Astrophys. J. Letters **425**, L97–L100 (1994).

[81] Huss, G. R., Fahey, A. J., Gallino, R. & Wasserburg, G. J., Astrophys. J. Letters **430**, L81–L84 (1994).

[82] Nittler, L. R., Alexander, C. M. O., Gao, X., Walker, R. M. & Zinner, E. K., Nature **370**, 443 (1994).

[83] Gail, H.-P. & Sedlmayr, E., in *The Molecular Astrophysics of Stars and Galaxies*, 285–312 (1998).

[84] Patzer, A. B. C., Chang, C., Sedlmayr, E. & Sülzle, D., Eur. Phys. J. D. **6**, 57–62 (1999).

[85] Jeong, K. S., Winters, J. M. & Sedlmayr, E., in *IAU Symp. 191: Asymptotic Giant Branch Stars*, volume 191, 233–238 (1999).

[86] Speck, A. K., Barlow, M. J., Sylvester, R. J. & Hofmeister, A. M., Astron Astrophys. Suppl. **146**, 437–464 (2000).

[87] Fabian, D., Posch, T., Mutschke, H., Kerschbaum, F. & Dorschner, J., Astron. Astrophys. **373**, 1125–1138 (2001).

[88] Nittler, L. R., Alexander, C. M. O., Gao, X., Walker, R. M. & Zinner, E., Astrophys. J. **483**, 475–495 (1997).

[89] Choi, B.-G., Huss, R. H., Wasserburg, G. J. & Gallino, R., Science **282**, 1284–1289 (1998).

[90] Posch, T., Kerschbaum, F., Mutschke, H., Dorschner, J. & Jäger, C., Astron. Astrophys. **393**, L7–L10 (2002).

[91] Bouwman, J., de Koter, A., van den Ancker, M. E. & Waters, L. B. F. M., Astron. Astrophys. **360**, 213–226 (2000).

[92] Bouwman, J. et al., Astron. Astrophys. **375**, 950–962 (2001).

[93] Natta, A., Meyer, M. R. & Beckwith, S. V. W., Astrophys. J. **534**, 838–845 (2000).

[94] Hony, S., Waters, L. B. F. M. & Tielens, A. G. G. M., Astron. Astrophys. **390**, 533–553 (2002).

[95] Martin, P. G. & Rogers, C., Astrophys. J. **322**, 374–392 (1987).

[96] Draine, B., in *IAU Symp. 135: Interstellar Dust*, volume 135, 313–327 (1989).

[97] Sandford, S. A., *Meteoritics and Planetary Science* **31**, 449–476 (1996).

[98] Amari, S., Anders, A., Virag, A. & Zinner, E., Nature **345**, 238–240 (1990).

[99] Zinner, E., Amari, S., Wopenka, B. & Lewis, R. S., *Meteoritics* **30**, 209–226 (1995).

[100] Buss, R. H. et al., Astrophys. J. Letters **365**, L23–L26 (1990).

[101] Kwok, S., Volk, K. & Bernath, P., Astrophys. J. Letters **554**, L87–L90 (2001).

[102] Robertson, J., Mat. Sci. Eng. **R37**, 129–281 (2002).

[103] Hony, S., Tielens, A. G. G. M., Waters, L. B. F. M. & de Koter, A., *to be submitted to* Astron. Astrophys. (2003).

[104] Sandford, S. A. et al., Astrophys. J. **371**, 607–620 (1991).

[105] Pendleton, Y. J., Sandford, S. A., Allamandola, L. J., Tielens, A. G. G. M. & Sellgren, K., Astrophys. J. **437**, 683–696 (1994).

[106] Tielens, A. G. G. M., Seab, C. G., Hollenbach, D. J. & McKee, C. F., Astrophys. J. Letters **319**, L109–L113 (1987).

[107] Lewis, R. S., Ming, T., Wacker, J. F., Anders, E. & Steel, E., Nature **326**, 160–162 (1987).

[108] Lewis, R. S., Anders, E. & Draine, B. T., Nature **339**, 117–121 (1989).

[109] Anders, E. & Zinner, E., *Meteoritics* **28**, 490–514 (1993).

[110] Guillois, O., Ledoux, G. & Reynaud, C., Astrophys. J. Letters **521**, L133–L36 (1999).

[111] Blades, J. C. & Whittet, D. C. B., Monthly Notices Roy. Astron. Soc. **191**, 701–709 (1980).

[112] Whittet, D. C. B., McFadzean, A. D. & Geballe, T. R., Monthly Notices Roy. Astron. Soc. **211**, 29P–31P (1984).

[113] Terada, H., Imanishi, M., Goto, M. & Maihara, T., Astron. Astrophys. **377**, 994–998 (2001).

[114] Geballe, T. R., Noll, K. S., Whittet, D. C. B. & Waters, L. B. F. M., Astrophys. J. Letters **340**, L29–L32 (1989).

[115] Oudmaijer, R. D., Waters, L. B. F. M., van der Veen, W. E. C. J. & Geballe, T. R., Astron. Astrophys. **299**, 69–78 (1995).

[116] Van Kerckhoven, C., Tielens, A. G. G. M. & Waelkens, C., Astron. Astrophys. **384**, 568–584 (2002).

[117] Dai, Z. R., Bradley, J. P., Joswiak, D. J., Brownlee, D. E. & Genge, M. J., in *Lunar and Planetary Institute Conference Abstracts*, volume 33, 1321 (2002).

[118] Forrest, W. J., Houck, J. R. & McCarthy, J. F., Astrophys. J. **248**, 195–200 (1981).

[119] Omont, A. et al., Astrophys. J. **454**, 819–825 (1995).

[120] Jiang, B. W., Szczerba, R. & Deguchi, S., Astron. Astrophys. **344**, 918–922 (1999).

[121] Szczerba, R., Henning, T., Volk, K., Kwok, S. & Cox, P., Astron. Astrophys. **345**, L39–L42 (1999).

[122] Hrivnak, B. J., Volk, K. & Kwok, S., Astrophys. J. **535**, 275–292 (2000).

[123] Volk, K., Kwok, S., Hrivnak, B. J. & Szczerba, R., Astrophys. J. **567**, 412–422 (2002).

[124] Chan, K. et al., Astrophys. J. **483**, 798–810 (1997).
[125] Goebel, J. H. & Moseley, S. H., Astrophys. J. Letters **290**, L35–L39 (1985).
[126] Nuth, J. A., Moseley, S. H., Silverberg, R. F., Goebel, J. H. & Moore, W. J., Astrophys. J. Letters **290**, L41–L43 (1985).
[127] Waters, L. B. F. M. et al., in *ESA SP-427: The Universe as Seen by ISO*, volume 427, 219–228 (1999).
[128] Lattimer, J. M., Schramm, D. N. & Grossman, L., Astrophys. J. **219**, 230–249 (1978).
[129] Savage, B. D. & Sembach, K. R., Anual Rev. Astron. Astrophys. **34**, 279–330 (1996).
[130] Kwok, S., Volk, K. M. & Hrivnak, B. J., Astrophys. J. Letters **345**, L51–L54 (1989).
[131] Volk, K., Kwok, S. & Hrivnak, B. J., Astrophys. J. Letters **516**, L99–L102 (1999).
[132] Hony, S., Waters, L. B. F. M. & Tielens, A. G. G. M., Astron. Astrophys. **378**, L41–L44 (2001).
[133] Goebel, J. H., Astron. Astrophys. **278**, 226–230 (1993).
[134] Begemann, B., Dorschner, J., Henning, T. & Mutschke, H., Astrophys. J. Letters **464**, L195–L198 (1996).
[135] Hill, H. G. M., Jones, A. P. & D'Hendecourt, L. B., Astron. Astrophys. **336**, L41–L44 (1998).
[136] Bernatowicz, T. J. et al., Astrophys. J. **472**, 760–782 (1996).
[137] Speck, A. K. & Hofmeister, A. M., in *Lunar and Planetary Institute Conference Abstracts*, volume 33, 1155 (2002).
[138] Friedemann, C., *Astronomische Nachrichten* **291**, 177–+ (1969).
[139] Gilra, D. P., in *IAU Symp. 52: Interstellar Dust and Related Topics*, volume 52, 517–528 (1973).
[140] Hackwell, J. A., Astron. Astrophys. **21**, 239–248 (1972).
[141] Forrest, W. J., Gillett, F. C. & Stein, W. A., Astrophys. J. **195**, 423–440 (1975).
[142] Cohen, M., Monthly Notices Roy. Astron. Soc. **206**, 137–147 (1984).
[143] Baron, Y., Papoular, R., Jourdain de Muizon, M. & Pegourie, B., Astron. Astrophys. **186**, 271–279 (1987).
[144] Willems, F. J., Astron. Astrophys. **203**, 51–70 (1988).
[145] Goebel, J. H., Cheeseman, P. & Gerbault, F., Astrophys. J. **449**, 246–257 (1995).
[146] Treffers, R. & Cohen, M., Astrophys. J. **188**, 545–552 (1974).
[147] Speck, A. K., Barlow, M. J. & Skinner, C. J., Monthly Notices Roy. Astron. Soc. **288**, 431–456 (1997).
[148] Gilman, R. C., Astrophys. J. Letters **155**, L185–L187 (1969).
[149] Kozasa, T., Dorschner, J., Henning, T. & Stognienko, R., Astron. Astrophys. **307**, 551–560 (1996).
[150] Justtanont, K., Yamamura, I., de Jong, T. & Waters, L. B. F. M., Astrophys. Sp. Sc. **251**, 25–30 (1997).
[151] Bernatowicz, T. et al., Nature **330**, 728–730 (1987).

[152] Huss, G. R. & Lewis, R. S., Geochimica et Cosmochimica Acta **59**, 115–160 (1995).

[153] Mutschke, H., Andersen, A. C., Clément, D., Henning, T. & Peiter, G., Astron. Astrophys. **345**, 187–202 (1999).

[154] Whittet, D. C. B., Duley, W. W. & Martin, P. G., Monthly Notices Roy. Astron. Soc. **244**, 427–431 (1990).

[155] Ferrarotti, A., Gail, H.-P., Degiorgi, L. & Ott, H. R., Astron. Astrophys. **357**, L13–L16 (2000).

[156] Rietmeijer, F. J. M. & MacKinnon, I. D. R., *Journal Geophysical Research Supplement* **90**, 149–155 (1985).

[157] Bregman, J. D. et al., Astron. Astrophys. **187**, 616–620 (1987).

[158] Malfait, K., Waelkens, C., Bouwman, J., de Koter, A. & Waters, L. B. F. M., Astron. Astrophys. **345**, 181–186 (1999).

[159] Koike, C., Hasegawa, H. & Hattori, T., Astrophys. Sp. Sc. **88**, 89–98 (1982).

[160] Koike, C. & Shibai, H., Monthly Notices Roy. Astron. Soc. **246**, 332–336 (1990).

[161] Jones, A. P., Tielens, A. G. G. M. & Hollenbach, D. J., Astrophys. J. **469**, 740–764 (1996).

[162] Gail, H.-P. & Sedlmayr, E., in *Chemistry and Physics of Molecules and Grains in Space. Faraday Discussions No. 109*, 303–320 (1998).

[163] Jeong, K. S., Sedlmayr, E. & Winters, J. M., in *Astronomische Gesellschaft Meeting Abstracts*, volume 17, 30 (2000).

[164] Ferrarotti, A. S. & Gail, H.-P., Astron. Astrophys. **382**, 256–281 (2002).

[165] Hallenbeck, S. L., Nuth, J. A. & Daukantas, P. L., *Icarus* **131**, 198–209 (1998).

[166] Tielens, A. G. G. M., Waters, L. B. F. M., Molster, F. J. & Justtanont, K., Astrophys. Sp. Sc. **255**, 415–426 (1997).

[167] Simis, Y. J. W., Ph.D. thesis, Leiden Observatory, The Netherlands (2001).

[168] Rietmeijer, F. J. M., Nuth, J. A. & Karner, J. M., Astrophys. J. **527**, 395–404 (1999).

[169] Carrez, P., Demyk, K., Leroux, H. & Cordier, P., *Meteoritics & Planetary Science, vol. 36, Supplement, p.A36* **36**, 36 (2001).

[170] Bradley, J. P., *Science* **265**, 925–929 (1994).

[171] Gail, H.-P. & Sedlmayr, E., Astron. Astrophys. **206**, 153–168 (1988).

[172] Cherchneff, I., Barker, J. R. & Tielens, A. G. G. M., Astrophys. J. **401**, 269–287 (1992).

[173] Cherchneff, I. & Cau, P., in *IAU Symp. 191: Asymptotic Giant Branch Stars*, volume 191, 251–259 (1999).

[174] Yamamura, I., Dominik, C., de Jong, T., Waters, L. B. F. M. & Molster, F. J., Astron. Astrophys. **363**, 629–639 (2000).

[175] Waters, L. B. F. M. et al., Astron. Astrophys. **331**, L61–L64 (1998).

[176] Cohen, M. et al., Astrophys. J. Letters **513**, L135–L138 (1999).

[177] Cohen, M., Barlow, M. J., Liu, X.-W. & Jones, A. F., Monthly Notices Roy. Astron. Soc. **332**, 879–890 (2002).

[178] Zijlstra, A. A. et al., Astron. Astrophys. **243**, L9–L12 (1991).

[179] Iben, I., Astrophys. J. **277**, 333–354 (1984).

[180] Waters, L. B. F. M. et al., Nature **391**, 868–871 (1998).

[181] Waters, L. B. F. M., Trams, N. R. & Waelkens, C., Astron. Astrophys. **262**, L37–L40 (1992).

[182] De Marco, O., Barlow, M. J. & Cohen, M., Astrophys. J. Letters **574**, L83–L86 (2002).

[183] Arendt, R. G., Dwek, E. & Moseley, S. H., Astrophys. J. **521**, 234–245 (1999).

[184] Douvion, T., Lagage, P. O. & Pantin, E., Astron. Astrophys. **369**, 589–593 (2001).

[185] Kozasa, T., Hasegawa, H. & Nomoto, K., Astron. Astrophys. **249**, 474–482 (1991).

[186] Douvion, T., Lagage, P. O. & Cesarsky, C. J., Astron. Astrophys. **352**, L111–L115 (1999).

[187] Evans, A., in *IAU Colloq. 122: Physics of Classical Nova*, 253–263 (1990).

[188] Jones, A. P., J. Geoph. Res. **105**, 10257–10268 (2000).

[189] Tang, M. & Anders, E., Astrophys. J. Letters **335**, L31–L34 (1988).

[190] Sandford, S. A., Pendleton, Y. J. & Allamandola, L. J., Astrophys. J. **440**, 697–705 (1995).

[191] Day, K. L., Monthly Notices Roy. Astron. Soc. **178**, 49P–51P (1977).

[192] Bradley, J. P., *Meteoritics* **29**, 447 (1994).

[193] Molster, F. J. & Bradley, J. P., *Meteoritics & Planetary Science, Supplement* **36**, A140–A141 (2001).

[194] Harker, D. E. & Desch, S. J., Astrophys. J. Letters **565**, L109–L112 (2002).

[195] Bouwman, J., de Koter, A., Dominik, C. & Waters, L. B. F. M., Astron. Astrophys.in press (2002).

[196] Nuth, J. A., Hill, H. G. M. & Kletetschka, G., Nature **406**, 275–276 (2000).

[197] Bockelée-Morvan, D., Gautier, D., Hersant, F., Huré, J.-M. & Robert, F., Astron. Astrophys. **384**, 1107–1118 (2002).

[198] Hony, S., Bouwman, J., Keller, L. P. & Waters, L. B. F. M., *submitted to* Astron. Astrophys. (2003).

[199] Gail, H.-P., Astron. Astrophys. **332**, 1099–1122 (1998).

[200] Chang, C. P., Flamm, D. L., Ibbotson, D. E. & Mucha, J. A., Journ. Applied Phys. **63**, 1744–1748 (1988).

[201] Daulton, T. L., Eisenhour, D. D., Bernatowicz, T. J., Lewis, R. S. & Buseck, P. R., Geochimica et Cosmochimica Acta **60**, 4853–4872 (1996).

[202] Brownlee, D. E., Joswiak, D. J., Bradley, J. P., Gezo, J. C. & Hill, H. G. M., in *Lunar and Planetary Institute Conference Abstracts*, volume 31, 1921 (2000).

The Mineralogy of Cometary Dust

Martha S. Hanner

Jet Propulsion Laboratory, California Institute of Technology, Pasadena CA 91109, USA

Abstract. Cometary dust is a mix of various silicates, with some iron sulfide; iron oxides and metallic nickel-iron are minor constitutents. Carbon in the dust is enriched relative to CI chondrites; a significant fraction of the carbon is in the form of organic refractory material. Cometary silicates consist of an unequilibrated mixture of Mg-rich crystalline and non-crystalline olivine and pyroxenes, as revealed from infrared spectroscopy and from in situ sampling of comet Halley dust. The strong similarity of all known cometary dust properties to the anhydrous chondritic aggregate class of interplanetary dust particles argues that comets are the source of these IDPs.

Comets formed in the outer regions of the solar nebula where the temperatures remained cold enough so that unaltered interstellar grains could have been incorporated into the accreting comet nuclei. The small glassy silicates in comets may indeed be interstellar grains. The origin of the crystalline silicates is not clear. If they formed in the hot inner solar nebula, then their presence in comets requires extensive mixing in the solar nebula. If they are pre-solar, then it is puzzling that their spectral signatures are not seen in the spectra of the interstellar medium or young stellar objects except in a few late-stage young stellar objects where a cloud of comets may already be present.

1   Introduction

Comets contain some of the most primitive material surviving from the early solar system. The comets we see today come from two distinct reservoirs. The Oort Cloud, at heliocentric distance 10^4 to 10^5 AU, is the source of the new and long-period comets. The Oort Cloud comets probably formed in the region of the giant planets and were dynamically scattered to their present orbits [74]. Short-period comets with low inclination orbits originate in the Kuiper Belt, at 30-50 AU [23,68]. Bodies at the large end of the Kuiper Belt population have now been detected; as of early 2001 about 400 were known [69]. It is estimated that there are 10^9 - 10^{10} objects larger than 1 km.

Thus, comets formed over a wide range in heliocentric distance, from 5 to 50 AU. These regions of the solar nebula remained cold enough that pre-existing interstellar grains, and perhaps even interstellar ices, could have survived and could have been incorporated into the forming comet nuclei. The extent to which comets do include interstellar grains, as opposed to nebular condensates, remains controversial and will be discussed in this Chapter.

While comets may have formed at low temperatures, collisions likely played a role in altering the cometary material. Reference [24] concluded that cometary-sized objects (a few km diameter or less) in the Kuiper Belt are either collisional fragments or have been deeply modified by collisions over the age of the solar system. The Oort Cloud comets likely experienced considerable collisional processing during their formation in the region of the giant planets before their ejection to the Oort Cloud [74].

In 1950, comet nuclei were first described as icy conglomerate objects a few km in size, whose activity is due primarily to the sublimation of water ice [76]. Over the past 50 years, remote sensing from Earth and the very successful space missions to comet Halley in 1986 have filled in our picture of comets. Nuclear diameters range from a few hundred meters to Hale-Bopp's $\sim$ 40 km nucleus [72]. The small fragments detected as sun-grazing comets [63] and the observed cases of nucleus fragmentation, such as comet Shoemaker-Levy 9 [64], which was torn apart during a close pass by Jupiter in 1992, indicate that the internal strength of at least some comet nuclei is low, consistent with a collisional history.

In most comets, outgassing appears to be confined to perhaps 10% of the surface, as dramatically confirmed by the Giotto images of comet P/Halley [47] and the recent DS-1 images of P/Borrelly [65]. The rest of the surface is covered with dark, non-volatile material. While water is the dominant volatile species, some comets are enriched in CO and display activity at large heliocentric distances. Many other molecules have been detected at the 0.1-5% level [15,43].

Solid particles are entrained in the outflowing gas and form a dust coma. The dust detectors on board the Halley probes detected particle masses from 10^{-16} to 10^{-3} g [57]. Small particles are dispersed by solar radiation pressure, while large particles (mm-cm sized) form dust trails along the cometary orbit [67,59]. Although small particles contribute much of the scattered light and thermal emission from the dust coma, most of the particulate mass is contained in the large particles, if the measured Halley mass distribution is typical.

Important evidence about the mineralogy of cometary dust has been acquired from the space probes that encountered comet Halley in 1986 as well as from remote sensing, primarily infrared spectroscopy. Additionally, interplanetary dust particles (IDPs), if their cometary origin can be inferred, allow detailed study of particle structure and mineralogy [6]. In the future, in situ sampling and dust sample return from active comets will augment our present knowledge. In this Chapter, our current knowledge of the mineralogy of cometary dust will be summarized and the links between cometary dust, IDPs, and interstellar dust will be discussed.

2   In Situ Sampling

The most direct means to determine the composition of cometary dust is by in situ sampling. The two Soviet Vega missions and the European Space Agency's Giotto mission sent to fly past comet Halley in 1986 each carried an impact ionization time of flight mass spectrometer to measure the elemental composition of the dust [49,50]. All together, the composition was recorded for about 5,000 particles in the mass range 10^{-16} to 10^{-11} g. Although the total mass of material measured was only a few nanograms, it was sampled over tens of thousands of km in the coma along 3 spacecraft tracks.

The sampled particles divided into 3 main types: mass spectra dominated by the major rock-forming elements, Mg, Si, Ca, Fe; mass spectra consisting primarily of the light elements H,C,O,N, the "CHON" particles; and mixed spectra containing both the rock and CHON elements. If we define mixed particles as having a ratio of carbon to rock-forming elements between 0.1 and 10, then $\sim$ 50% of the particles are mixed and $\sim$ 25% are rock and CHON respectively [27]. At some level, however, the CHON and rocky material are mixed down to the finest submicron scale in all particles [54]. The bulk abundances of the major rock-forming elements are solar (chondritic) within a factor of $\sim$ 2 [44,45]. (Conversion of the mass spectra to relative elemental abundances depends on the ion yields, which are uncertain by at least a factor of 2 because of the impossibility of calibrating the instrument at the flyby speed of $\sim$ 70 km/s). The CHON elements, especially C,H,N, are enriched relative to CI-chondrites, classifying the cometary material as more "primitive" than the carbonaceous chondrite meteorites.

The rocky material displays a wide range in Fe/Mg, but a narrow range in Si/Mg [45,55]. The range in Fe/Mg is distinctly different from that of primitive meteorites. Mg-rich (Fe-poor) silicates comprise at least 40% and maybe $\geq$ 60% of the rocky particles [44]. Iron is present in other minerals including metals (1-2%), iron oxide ($\leq$ 1%) and iron sulfides ($\sim$ 10%) [61].

Isotopic ratios are solar, within the measurement uncertainties, with the exception of $^{12}C/^{13}C$, which displays a wide range [44,46]. While low ratios (^{13}C enrichment) are uncertain due to noise of uncertain origin, definite ^{12}C enrichments, up to $^{12}C/^{13}C \sim 5000$ have been identified, indicative of presolar nucleosynthesis products.

3   Cometary Silicates

Until the next space missions enable direct analysis of the dust, infrared spectroscopy is the best method for determining the mineralogy of cometary silicates and documenting differences among comets. Small silicate particles in the optically thin coma will generate the characteristic 10 and 16-35 μm spectral features in emission. The 10 μm feature lies within the 8-13 μm atmospheric window, and ground-based spectra are available for about a dozen comets.

The first comet for which good signal/noise 10 μm spectra were obtained was comet 1P/Halley [10,12]. These spectra clearly indicated the presence of crystalline olivine from its distinctive peak at 11.2 μm (Fig. 1). It is also evident from Fig. 1 that other silicate minerals must be present to explain the full width of the feature. However, uncertainty in the correction for the telluric ozone feature near 9.5 μm made 9-10 μm spectral structure in these spectra uncertain.

Comet Hale-Bopp (C/1995 O1) exhibited the strongest silicate feature ever seen in a comet. The dust coma was observed in the infrared from 4.6 AU preperihelion to 3.9 AU postperihelion. The high signal/noise 10 μm spectra

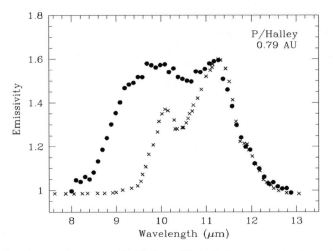

Fig. 1. Olivine feature in comet 1P/Halley. *filled circles*: spectrum from [12] divided by a 360 K blackbody; *crosses*: measured relative emissivity of ground forsterite (Mg-olivine) from [66]

have enabled the silicate mineralogy to be determined. Moreover, full 16-40 μm spectra of Hale-Bopp were acquired for the first time by the SWS instrument on board ESA's Infrared Space Observatory (ISO).

The SWS spectrum taken when the comet was at 2.8 AU from the Sun before perihelion is presented in Fig. 2 [17,18]. All of the major spectral peaks correspond to those of Mg-rich crystalline olivine [52]. Minor structure is attributed

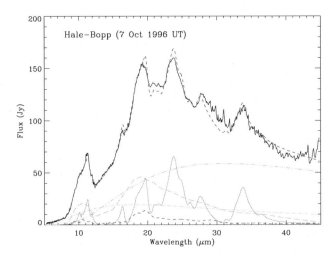

Fig. 2. ISO/SWS spectrum of comet Hale-Bopp at 2.8 AU compared with a dust model [17]. BB1: 280 K; BB2: 165 K; Cry Ol: forsterite; Cry o-Pyr: orthopyroxene; Am Pyr: amorphous pyroxene (*from* [17])

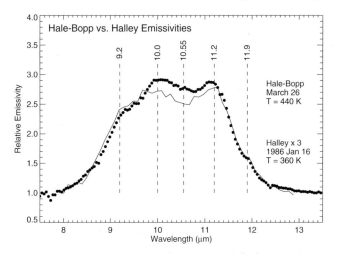

Fig. 3. The silicate feature in Hale-Bopp (*filled circles*) [36] and P/Halley (*line*) [12]. Each spectrum has been divided by a blackbody at the temperature shown and the Halley spectrum has been multiplied by 3. (*from* [36])

to crystalline pyroxene [77]. In contrast, airborne spectra of comet Halley at 1.3 AU (the only other 16-30 μm spectra of a comet) are rather different, showing the olivine feature at 28.4 μm but only very weak features at 23.8 and 19.5 μm [29,41]. This contrast is puzzling, since the 10 μm spectra of Halley did exhibit an 11.2 μm peak of normal strength (Fig. 1). Viewing restrictions on solar elongation angle prevented SWS observations when Hale-Bopp was near 1 AU, so we do not know how the 16-35 μm spectrum evolved as the grains heated.

A number of ground-based 8-13 μm spectra of Hale-Bopp were obtained [20,36,40,77]. A typical spectrum near perihelion is shown in Fig. 3. One sees maxima at 9.2, 10.0, and 11.2 μm and minor structure at 10.5 and 11.9 μm. These peaks appear consistently in spectra taken with 4 different instruments over several months under a wide range of observing conditions. The 11.2 μm olivine peak is similar to that seen in comet Halley; the 11.9 μm shoulder is also due to olivine. The 9.2 μm maximum is evidence of pyroxene grains. Figure 4 compares the Hale-Bopp spectrum with the measured emissivity of amorphous Mg-pyroxene [66]. Computed emissivities based on the refractive indices of glassy pyroxene in [22] yield a similar shape.

The broad maximum in the spectrum near 10 μm is likely due to glassy or amorphous olivine, although crystalline olivine can also contribute. The overall width of the silicate feature, its relatively flat top from 10 to 11 μm, and the structure near 10.5 μm are most likely produced by crystalline pyroxene. Crystalline pyroxenes display considerable variety in their spectral shapes. Pyroxene-rich IDPs tend to have a broad feature with peaks in the 10-11 μm region [60] and a few have a peak near 9.3 μm as well [78]. The orthorhombic orthoenstatite measured by [66] produced a narrow feature sharply peaked at 9.9 μm.

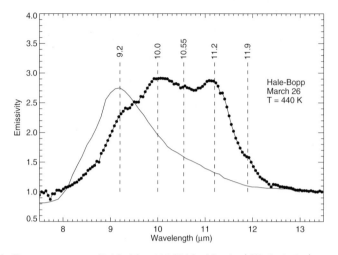

Fig. 4. Hale-Bopp spectrum divided by 440 K blackbody (*filled circles*) compared with the measured relative emissivity of amorphous Mg-pyroxene from [66] (*line*). The peak position at 9.2 μm matches the shoulder in the comet spectrum. (*from* [36])

To produce a strong emission feature, silicate particles must have radii of order 1 μm or smaller. Larger particles will display a strong feature only if they are very porous aggregates, such that the individual small constituent grains radiate essentially independently [32,34].

Several authors have modeled the Hale-Bopp silicate spectra using various mixtures of amorphous and crystalline olivine and pyroxene [11,14,28,40,77,78]. Reference [77] combined laboratory spectra of crystalline olivine and pyroxene and Mie theory calculations of the emissivity for glassy olivine and pyroxene to model both their 8-13 μm spectra and the 2.8 AU SWS spectrum. In order to interpret differences in shape between spectra near 1 AU and at 2.8 AU, the authors proposed that there was a large population of cold, Mg-rich crystalline pyroxene grains, whose thermal emission was seen only near 1 AU. These grains sharpened the 9.3 μm peak and produced structure near 10.5 μm. The authors estimated that these Mg-rich pyroxene grains constituted at least 90% of the small silicates in the coma. A subsequent paper showed that the spectra could also be fit by a selected set of porous aggregate IDP spectra, particularly the pyroxene-rich cluster subclass [78]. Temperature differences between olivine and pyroxene to explain the Hale-Bopp spectra were also suggested by [28].

A difficulty with this model is the need to explain how the small crystalline pyroxene grains avoid thermal contact with any absorbing material, in order to remain colder than the other silicates. Alternatively, emission from submicron sized silicate particles might have been masked at large heliocentric distances by organic refractory mantles that sublimated rapidly near perihelion, revealing the optically thin silicate emission. Since the pyroxene peaks are less prominent than the 11.2 μm olivine peak, they could be more easily masked.

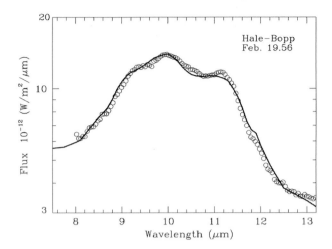

Fig. 5. Spectrum of Hale-Bopp at 1.16 AU fit to a 5-component dust model. The same emissivity template fits all of the 1997 spectra. (*from* [40])

Reference [40] used measured emissivities of amorphous and crystalline pyroxene and olivine to fit their 8-13 μm spectra. They assumed that all of the silicate components were at the same temperature and added a component of amorphous carbon grains to match the observed thermal spectral energy distribution. Figure 5 compares the model with a sample spectrum. The same emissivity template fits all of the 1997 spectra, including spectra of the jets, implying that the various silicate minerals were well-mixed. In this model, glassy pyroxene is the most abundant silicate ($\geq$ 40%) and the glassy and crystalline pyroxenes together constitute about 2/3 of the small silicate particles; the abundance of crystalline olivine is $\sim$ 20%.

Models based on their laboratory spectra of amorphous and crystalline olivine plus an amorphous carbon material (both small and large particles) were presented by [11]. When a component of amorphous enstatite was added as well [14], a good fit to the entire 8-40 μm SWS spectrum was achieved.

It is not possible to quantify uniquely the relative contributions of the various components, for several reasons. Some laboratory measurements of spectral transmission or emissivity are relative and do not give quantitative mass absorption coefficients ([52] does present Q_{ext}/a and [14] gives mass extinction coefficients). The strength of the feature depends upon the (unknown) particle size. Mie theory, in wide use for modeling scattering and emission from small spherical particles, is not applicable for modeling the crystalline olivine peaks because of the strong dependence on particle shape near resonances in the optical constants [3]. Finally, even if the emissivities were accurately known, the thermal emission depends upon the grain temperature. The temperature of a silicate particle in the comet coma is controlled by the balance between absorption of solar radiation and infrared emission and will vary with grain size, Mg/Fe ratio, and thermal contact with absorbing material.

However, we can make the general statement that pyroxenes, either glassy or crystalline, constitute the majority of small silicate particles in Hale-Bopp, while the abundance of crystalline olivine is 20% or less. This latter point, that the crystalline olivine constitutes a relatively small fraction of the cometary silicates in Hale-Bopp, is important to remember when interpreting the SWS spectrum or comparing it to other astronomical sources. The modeling illustrates the importance of analyzing both the 8-13 μm and 16-30 μm spectral regions, in order to infer the silicate mineralogy.

Hale-Bopp and P/Halley are both Oort Cloud comets. Other Oort Cloud comets with similar silicate features are Levy (C/1990 K1), Bradfield(C/1987 P1), Mueller (C/1993 A1), and Hyakutake (C/1996 B2) [37,38]. In fact, the 8-13 μm spectrum of comet Mueller is fit exactly by the same silicate model used to fit Hale-Bopp [40], although the ratio of small silicates to featureless absorbing material is somewhat lower.

The 8-13 μm spectra of four other new comets discussed in [38] are puzzling; each has a unique, and not understood, spectrum. Comet Austin (1990 V) showed a feature 20% above the continuum with a maximum near 9 μm while Okazaki-Levy-Rudenko (1989 XIX) displayed a 20% feature with a broad maximum at 11-11.5 μm.

No strong 10 μm emission feature has yet been seen in a short-period comet. Broad emission features about 20% above a black body continuum were present in spectra of Comets 4P/Faye and 19P/Borrelly [39] and 103P/Hartley 2 [17]. Filter photometry of 81P/Wild 2 reveals a feature about 25% above the continuum; no spectrum exists. Other short-period comets with 10 μm filter photometry show no feature at the 10% level. The absence of strong silicate emission in the short-period comets could be explained either by a difference in the composition between Oort Cloud and Kuiper Belt comets or by a lower abundance of submicron sized particles. Short-period comets have generally been outgassing during many orbits in the inner solar system and the smaller or more fluffy particles may have preferentially been expelled over time. (Small particles, even when producing most of the thermal emission, may constitute only a small fraction of the dust mass.) Thus, a lower abundance of isolated small grains or very fluffy aggregates of small grains is the simplest explanation for the lack of a strong silicate feature.

4   Interplanetary Dust Particles of Probable Cometary Origin

The submicron grain size, high Mg/Fe ratio, and mix of crystalline and non-crystalline olivine and pyroxenes have no counterpart in any meteoritic material, with the exception of the anhydrous chondritic aggregate interplanetary dust particles (IDPs). These are fine-grained heterogeneous aggregates having chondritic abundances of the major rock–forming elements; they comprise a major fraction of the IDPs captured in the stratosphere [6]. Typical grain sizes within the aggregates are 0.1–0.5 μm; micron-sized crystals of forsterite and enstatite are also present. These aggregate IDPs are thought to originate from comets, based on their porous structure, small grain size, high carbon content, and relatively high atmospheric entry velocities. The match between the mineral identifications in the Hale–Bopp spectra and the silicates seen in the IDPs strengthens the link between comets and this class of IDPs.

Thus, the dust composition and physical structure of chondritic aggregate IDPs can be used to augment our understanding of the nature and origin of cometary dust. These IDPs are a mixture of crystalline and non-crystalline silicate particles embedded in dark, carbon-rich matrix material. Their aggregate structure suggests that the larger particles in comets are aggregates of small grains with a range of porosities.

The dominant form of non-crystalline silicates in the chondritic aggregate IDPs are the GEMS, submicron glassy Mg-silicate grains with embedded nanometer FeNi and Fe sulfide crystals [6,7]. The GEMS show evidence of exposure to large doses of ionizing radiation, pointing to exposure in the interstellar medium prior to their incorporation into comets. They are frequently embedded in carbon-rich material having high D/H ratios, also indicative of an interstellar origin. Infrared spectra of GEMs-rich IDP samples resemble the cometary spectra, while spectra of individual GEMS reveal a single broad peak between 9.3 μm (pyroxene) and 10 μm (olivine) [9]. The FeNi inclusions provide sufficient absorption of solar radiation to heat a GEMS grain in a cometary coma to a temperature close to that of a blackbody.

5   Origins of Cometary Silicates

We have seen that cometary dust is an unequilibrated heterogeneous mixture of minerals, including both high and low temperature condensates. These various components do not necessarily share a common origin.

The temperatures in the outer solar nebula beyond 5 AU, where the comet nuclei accreted, were never higher than about 160 K, too low for significant processing of dust particles [4,5]. In the Kuiper Belt beyond 30 AU, the temperature was considerably lower. Thus, one expects that interstellar grains present in the outer solar nebula were preserved in comets. Other solid material that condensed within the solar nebula may also have accreted into the comet nuclei. Radial gradients in the temperature, composition and extent of mixing within the solar nebula should be evident today as differences in the dust properties among comets that formed at different distances.

The glassy silicate grains (GEMS) described in Sect. 4 appear to constitute the major fraction of the non-crystalline silicates in cometary IDPs and the evidence is quite strong that these are interstellar grains, based on their morphology and inferred high radiation dosage [6,7]. The GEMS must have formed at comparatively low temperatures and were never heated sufficiently to anneal. Thus, a major portion of non-crystalline cometary silicates is apparently of interstellar origin.

Crystalline silicate grains can form by direct condensation from the vapor phase at T=1200-1400 K followed by slow cooling. Crystallization can also occur by annealing of amorphous silicates at temperatures around 900 K or higher [35,53]. Enstatite whiskers, rods, and platelets found in aggregate IDPs of likely cometary origin have growth patterns, such as axial screw locations, that indicate direct vapor phase condensation from a hot gas [8]. Although rare, they show that at least some primary condensates were present in the region where comets formed. The Mg crystalline minerals enstatite and forsterite are predicted from thermodynamic models to be the first to condense in a hot gas at 1200-1400 K and only react with Fe at lower temperatures. Thus, direct condensation is a natural explanation for the preponderence of Mg-silicates in comet dust.

Direct grain condensation or annealing could have occurred in the hot inner solar nebula. Disk midplane temperatures $\geq$ 1000 K were reached inside about 1 AU, depending on the mass infall rate [4,13] During the early high mass accretion phase (mass infall rate $\geq 10^{-6}$ solar masses/yr), this hot region could have extended to 3-4 AU [1]. The main drawback to this scenario is the need to transport the crystalline grains out to the region where the comets formed at 5-50 AU. The extent of radial mixing in the solar nebula remains controversial (see also chapter by Gail in this book). Some models have predicted that small grains entrained in the outflowing gas near the mid-plane of the solar nebula would drift radially outward only $\sim$ 2-5 AU before being accreted onto larger particles [19]. Grain growth by aggregation was apparently a rapid process, in a time scale short compared with the age of the solar nebula, limiting the time for radial diffusion [58,73].

If the inner solar nebula was indeed the source of crystalline silicates, then one would expect a strong radial gradient in their abundance; that is, radial transport to 5 AU would have been more efficient than radial transport to 30 AU. Consequently, one would expect crystalline silicates to be rare or absent in the short-period Kuiper Belt comets. Crystalline olivine is difficult to substantiate spectroscopically in the short-period comets because the overall 10 μm silicate feature is weak. However, high data points at 11.2 μm are present at the 1 or 2 sigma level in spectra of P/Borrelly [39] and P/Hartley 2 [17], suggesting that crystalline olivine may be present. A definitive answer should be provided by the Stardust sample return from comet P/Wild 2.

Accretion shock heating of silicates was significant only in the inner nebula $\leq$ 1 AU [13], so it does not provide an alternative source of crystalline grains in the regions where comets formed. The crystalline silicate grains in comets would not have been formed by the short-term heating events that produced the chondrules. These mm-sized components of chondritic meteorites required heating to about 1800 K followed by rapid cooling of order 1000 K/hr [42], too rapid to allow crystal growth.

Heating is insufficient to anneal the silicates in the coma or on the nucleus. Moreover, the 11.2 μm peak is observed at the same strength relative to the 10 μm maximum in both new and long-period Oort Cloud comets over a wide range in heliocentric distances, from 4.6 AU (Hale-Bopp; [16]), where the blackbody temperature is only 130 K, to 0.79 AU (P/Halley; [12]), where the blackbody temperature is near 350 K.

It appears likely, then, that the crystalline silicates in comets were pre-existing grains in the interstellar cloud from which the solar nebula formed. Crystalline silicate grains are known to condense in the envelopes around evolved stars. Forsterite has been detected in oxygen-rich outflows of some evolved stars [70]. Yet, signatures of crystalline silicates are absent in spectra of the diffuse interstellar medium (ISM) or molecular clouds. Moreover, the spectra of most young stellar objects show no evidence of crystalline grains. Only in debris disks around young main-sequence objects such as β Pictoris [51] and in certain late-stage Herbig Ae/Be stars that are precursors of β Pictoris systems [71] does one find the spectral peaks of crystalline olivine. The ISO spectrum of HD100546 is very similar to that of comet Hale-Bopp, for example [56]. These systems are thought to have developed a population of comets that are the source of the dust [30,51,75]. Grain destruction in the ISM is an efficient process. Thus, if the comet grains formed in circumstellar outflows, one has to understand how they survived destruction in the ISM and why their spectral signatures are not seen in the ISM or young stellar objects.

It may be that the crystalline olivine and crystalline pyroxenes have different origins. The spectral peaks of crystalline pyroxene are more diverse and not necessarily as strong as the olivine peaks at 11.2 μm and 19.5 - 33.5 μm. The pyroxene spectral features could be masked more easily by absorbing mantle material or by the emission from other dust components in the spectra of young stellar objects or, if free of mantle material, the small pyroxene crystals might have been too cold for their thermal emission to be visible [78].

Thus, neither a pre-solar nor a solar nebula origin of the crystalline silicates in comets is free of problems. It is possible that *all* of the cometary silicates are of interstellar origin. Isotopic oxygen measurements of both GEMS and crystalline silicates in IDPs with the new nanoSIMS techniques will be extremely valuable in elucidating their origin; in fact, the first measurements of oxygen isotopes in crystalline silicate grains by this technique have detected non-solar isotopic ratios [6].

6   Organic Refractory Material

Already in 1982 [31], it was proposed that interstellar silicate grains possess organic refractory mantles as a result of UV photoprocessing in the diffuse interstellar medium following deposition of icy mantles in cold molecular clouds. These submicron core/mantle grains, perhaps with an additional icy mantle, were subsequently agglomerated and incorporated into comets. A mass of the organic refractory material comparable to the mass of the silicates satisfied cosmic abundances; in fact, it was argued that organic refractory mantles are a necessary repository of carbon to account for its cosmic abundance.

There is considerable evidence that comet dust includes a component of organic refractory material. As described in Sect. 2, Halley's dust contains a high abundance of CHON material, usually in combination with silicates. Significant clustering of subgroups (e.g., [H,C], [H,C,O], etc.) within the CHON classification indicated variable composition of the organic refractory component [26].

IDPs of likely cometary origin contain several percent carbon, at least some of it in an organic phase. The GEMS are embedded in an organic carbon matrix in which high D/H ratios have been detected. High D/H ratios have also been measured in complex macromolecular kerogen present in carbonaceous chondrite meteorites [48]. The deuterium enrichment and the wide range in $^{12}C/^{13}C$ recorded in the Halley dust suggest that at least some organic refractory material of interstellar origin has survived in the solar system. However, it has been pointed out that processes leading to D/H fractionation might have occurred in the cold outer solar nebula [25].

Organic refractory grains have not been detected spectroscopically in comets. An emission feature at 3.4 μm first detected in P/Halley, but also seen in other comets, was initially attributed to organic refractory grains. However, the discovery of methanol in comets and analysis of its infrared bands led to the conclusion that the 3.4 μm feature is due to gas phase methanol and other gaseous species [2,21].

The presence of organic refractory mantles on silicate grains will affect their optical properties, both absorptivity at visual wavelengths and contrast in the silicate features in the infrared [33].

7   Summary

Cometary silicates consist of an unequilibrated mixture of predominantly Mg-rich crystalline and non-crystalline olivine and pyroxenes, as revealed from infrared spectroscopy and from in situ sampling of comet Halley dust. While crystalline olivine generates strong emission peaks at 11.2 μm and (at least in comet Hale-Bopp) at 20-30 μm, it constitutes less than 20% of the small silicate particles. Iron sulfides are present at the 10% level; iron oxides and metallic nickel-iron are minor constituents. Carbon in the dust is enriched relative to CI chondrites; a significant fraction of the carbon is in the form of organic refractory material. The strong similarity of all known comet dust properties to the anhydrous chondritic aggregate class of IDPs argues that comets are the source of these IDPs.

The heterogeneous mix of cometary silicates implies that they do not necessarily have a common origin. The submicron glassy silicate particles (GEMS), ubiquitous in IDPS and presumably also in comets, are thought to have a pre-solar origin, based on their high radiation dosage. The origin (or origins) of the crystalline silicates is unclear. If formed as high temperature condensates or by annealing in the inner solar nebula, radial transport must have been more efficient during the planetesimal accretion phase than some models predict. In this case, crystalline silicates should be less abundant - or absent entirely - in the Kuiper Belt comets. If the crystalline silicates were already present in the cloud from which the solar nebula formed, then one needs to explain why their spectral signatures are not seen in interstellar dust or in young stellar objects. Isotopic measurements of individual grains in IDPs with nanoSIMS techniques may help to clarify their origin.

Although the chondritic aggregate IDPs have given us extremely interesting insight into the nature of probable cometary dust, we do not know the specific source of an individual IDP, nor the selection effects between comet ejection and Earth capture. Thus comet dust sample return and in situ analysis are very important. In the next decade, we can look forward to the Stardust sample return from the short-period comet 81P/Wild 2 in January 2006 and the encounter of ESA's Rosetta mission with short-period comet 46P/Wirtanen in 2011-2013 [62]. The goal of NASA's Stardust mission is to capture at least 1000 particles of size $\geq$ 10 μm into low-density aerogel during a slow flyby of P/Wild 2 in January 2004. The Rosetta mission carries several instruments to measure dust mass distribution, composition, and structure as it orbits with P/Wirtanen from aphelion to perihelion. In the meantime, SIRTF, NASA's infrared space telescope currently scheduled for launch in early 2003, will allow the 16-35 μm region to be studied spectroscopically for a number of comets.

Acknowledgements

This research was carried out at the Jet Propulsion Laboratory, California Institute of Technology under contract with the National Aeronautics and Space Administration.

References

[1] K.R. Bell, P.M. Cassen, J.T. Wasson, D.S. Woolum: 'The FU Orionis Phenomenon and Solar Nebula Material', In: *Protostars and Planets IV*, ed. by V. Mannings, A.P. Boss, S.S. Russell (Univ. Arizona Press, Tucson 2000) pp. 897–926

[2] D. Bockelee-Morvan, T.Y. Brooke, J. Crovisier: 'On the Origin of the 3.2- to 3.6-μm Emission Features in Comets', Icarus **116**, 18–39 (1995)

[3] C. F. Bohren and D. R. Huffman: *Absorption and Scattering of Light by Small Particles* (Wiley, New York, 1983)

[4] A.P. Boss: 'Temperatures in Protoplanetary Disks', Ann. Rev. Earth Planet. Sci. **26**, 53–80 (1998)

[5] A.P. Boss: 'Midplane Temperatures and Solar Nebula Evolution', Proc. Lunar Plan. Sci. Conf **25**, 149 (1994)

[6] J.P. Bradley: this volume

[7] J.P. Bradley: 'Chemically Anomalous, Preaccretionally Irradiated Grains in Interplanetary Dust from Comets', Science **265**, 925 (1994)

[8] J.P. Bradley, D.E. Brownlee, D.R. Veblen: 'Pyroxene Whiskers and Platelets in Interplanetary Dust: Evidence of Vapour Phase Growth', Nature **301**, 473–477 (1983)

[9] J.P. Bradley, L.P. Keller, T. P. Snow, M.S. Hanner, G.J. Flynn, J.C. Gezo, S.J. Clemett, D.E. Brownlee, J.E. Bowey: 'An Infrared Spectral Match between GEMS and Interstellar Grains', Science **285**, 1716–1718 (1999)

[10] J.H. Bregman, H. Campins, F.C. Witteborn, D.H. Wooden, D.M. Rank, L.J. Allamandola, M. Cohen, A.G.G.M. Tielens: 'Airborne and Ground-based Spectrophotometry of Comet P/Halley from 5–13 μm', A&A **187**, 616–620 (1987)

[11] J.R. Brucato, L. Colangeli, V. Mennella, P. Palumbo, E. Bussoletti: 'Silicates in Hale-Bopp: Hints from Laboratory Studies', Plan. Space Sci. **47**, 773–779 (1999)

[12] H. Campins and E. Ryan: 'The Identification of Crystalline Olivine in Cometary Silicates', ApJ **341**, 1059–1066 (1989)

[13] K.M. Chick, P. Cassen: 'Thermal Processing of Interstellar Dust Grains in the Primitive Solar Environment', ApJ **477**, 398–409 (1997)

[14] L. Colangeli, V. Mennella, J.R. Brucato, P. Palumbo, A. Rotundi: 'Characterization of Cosmic Materials in the Laboratory', Space Sci. Rev. **90**, 341–354 (1999)

[15] J. Crovisier, D. Bockelee-Morvan: 'Remote Observations of Cometary Volatiles', Space Science Rev. **90**, 19–32 (1999)

[16] J. Crovisier, T.Y. Brooke, M.S. Hanner, et al.: 'The Infrared Spectrum of Comet C/1995 O1 (Hale-Bopp) at 4.6 AU from the Sun', A&A **315**, L385–L388 (1996)

[17] J. Crovisier, T.Y. Brooke, K. Leech, et al.: 'The Thermal Infrared Spectra of Comets Hale-Bopp and 103P/Hartley 2 Observed with the Infrared Space Observatory', In: *Thermal Emission Spectroscopy and Analysis of Dust, Disks, and Regoliths* ed. by M.L. Sitko, A.L. Sprague, and D.K. Lynch ASP Conf. Series **196** 109–117 (2000)

[18] J. Crovisier, K. Leech, D. Bockelee-Morvan, T.Y. Brooke, M.S. Hanner, B. Altieri, H.U. Keller, E. Lellouch: 'The Spectrum of Comet Hale–Bopp (C/1995 O1) Observed with the Infrared Space Observatory at 2.9 Astronomical Units from the Sun', Science **275**, 1904–1907 (1997)

[19] J.N. Cuzzi, A.R. Dobrovolskis, J.M. Champney: 'Particle–Gas Dynamics in the Midplane of a Protoplanetary Nebula', Icarus **106**, 102–134 (1993)

[20] J.K. Davies, T.R. Geballe, M.S. Hanner, H.A. Weaver, J. Crovisier, D. Bockelee-Morvan: 'Thermal Infrared Spectra of Comet Hale-Bopp at Heliocentric Distances of 4 and 2.9 AU', Earth, Moon, and Planets **78**, 293–298 (1999)

[21] M.A. DiSanti, M.J. Mumma, T.R. Geballe, J.K. Davies: 'Systematic Observations of Methanol and Other Organics in Comet P/Swift-Tuttle', Icarus **116**, 1–17 (1995)

[22] J. Dorschner, B. Begemann, Th. Henning, C. Jäger, H. Mutschke: 'Steps toward Interstellar Silicate Mineralogy II. Study of Mg-Fe Silicate Glasses of Variable Composition', A&A **300**, 503–520 (1995)

[23] M. Duncan, T. Quinn, S. Tremaine: 'The Origin of Short-Period Comets', ApJ Lett. **328**, L69–L73 (1988)

[24] D.D. Durda, S.A. Stern: 'Collision Rates in the Present-Day Kuiper Belt and Centaur Regions', Icarus **145**, 220–229 (2000)

[25] B. Fegley: 'Chemical and Physical Processing of Presolar Materials in the Solar Nebula and the Implications for Preservation of Presolar Materials in Comets', Space Sci. Rev. **90**, 239–252 (1999)

[26] M.N. Fomenkova, S. Chang, L.M. Mukhin: 'Carbonaceous Components in the Comet Halley Dust', Geochim. Cosmochim. Acta **58**, 4503–4512 (1994)

[27] M. Fomenkova, J. Kerridge, K. Marti, L. McFadden: 'Compositional Trends in Rock-forming Elements of Comet Halley Dust', Science **258**, 266–269 (1992)

[28] P. Galdemard, P.O. Lagage, D. Dubreuil, R. Jouan, P. Masse, E. Pantin, D. Bockelee-Morvan: 'Mid-infrared Spectro-imaging Observations of Comet Hale-Bopp', Earth, Moon, and Planets **78**, 271-277 (1999)

[29] W. Glaccum, S.H. Moseley, H. Campins, R.F. Loewenstain: 'Airborne Spectrophotometry of P/Halley from 20 to 60 microns', A&A **187**, 635–638 (1987)

[30] C.A. Grady, M.L. Sitko, K.S. Bjorkman, M.R. Perez, D.K. Lynch, R.W. Russell, M.S. Hanner: 'The Star-Grazing Extrasolar Comets in the HD100546 System', ApJ **483**, 449–456 (1997)

[31] J.M. Greenberg: 'What are Comets Made of? A Model Based on Interstellar Dust', In: *Comets*, ed. by L.L. Wilkening (Univ. Arizona Press, Tucson 1982) pp. 131–163

[32] J.M. Greenberg, J.I. Hage: 'From Interstellar Dust to Comets: A Unification of Observational Constraints', ApJ **361**, 260–274 (1990)

[33] J.M. Greenberg, A. Li: 'What are the True Astronomical Silicates?', A&A **309**, 258–266 (1996)

[34] J.I. Hage, J.M. Greenberg: 'A Model for the Optical Properties of Porous Grains', ApJ **361**, 251–259 (1990)

[35] S.L. Hallenbeck, J.A. Nuth, P.L. Daukantas: 'Mid-Infrared Spectral Evolution of Amorphous Magnesium Silicate Smokes Annealed in Vacuum: Comparison to Cometary Spectra', Icarus **131**, 198–209 (1998)

[36] M.S. Hanner, R.D. Gehrz, D.E. Harker, T.L. Hayward, D.K. Lynch, C.C Mason, R.W. Russell, D.M. Williams, D.H. Wooden, C.E. Woodward: 'Thermal Emission from the Dust Coma of Comet Hale-Bopp and the Composition of the Silicate Grains', Earth, Moon, and Planets **79**, 247–264 (1999)

[37] M.S. Hanner, J.A. Hackwell, R.W. Russell, D.K. Lynch: 'Silicate Emission Feature in the Spectrum of Comet Mueller 1993a', Icarus **112**, 490–495 (1994)

[38] M.S. Hanner, D.K. Lynch, R.W. Russell: 'The 8–13 micron Spectra of Comets and the Composition of Silicate Grains', ApJ **425**, 274–285 (1994)

[39] M.S. Hanner, D.K. Lynch, R.W. Russell, J.A. Hackwell, R. Kellogg: 'Mid-Infrared Spectra of Comets P/Borrelly, P/Faye, and P/Schaumasse', Icarus **124**, 344–351 (1996)

[40] T.L. Hayward, M.S. Hanner, Z. Sekanina: 'Thermal Infrared Imaging and Spectroscopy of Comet Hale-Bopp (C/1995 O1)', ApJ. **538**, 428–455 (2000)

[41] T. Herter, H. Campins, G.E. Gull: 'Airborne Spectrophotometry of P/Halley from 16 to 30 Microns', A&A **187**, 629–631 (1987)

[42] R.H. Hewins: 'Experimental Studies of Chondrules' In: *Meteorites and the Early Solar System*, ed. by J.F. Kerridge and M.S. Matthews (Univ. Arizona Press, Tucson 1988) pp. 660–679

[43] W.M. Irvine, F.P. Schloerb, J. Crovisier, B. Fegley, M.J. Mumma: 'Comets: A Link between Interstellar and Nebular Chemistry', In: *Protostars and Planets IV*, ed. by V. Mannings, A.P. Boss, S.S. Russell, (Univ. Arizona Press, Tucson 2000) pp. 1159–1200

[44] E. Jessberger: 'Rocky Cometary Particulates: Their Elemental, Isotopic, and Mineralogical Ingredients', Space Science Rev. **90**, 91–97 (1999)

[45] E. Jessberger, A. Christoforidis, J. Kissel: 'Aspects of the Major Element Composition of Halley's Dust', Nature **332**, 691–695 (1988)

[46] E. Jessberger, J. Kissel: 'Chemical Properties of Cometary Dust and a Note on Carbon Isotopes' In: *Comets in the Post-Halley Era*, ed. by R.L. Newburn, M. Neugebauer, J. Rahe (Dordrecht, Kluwer 1991) pp. 1075–1092

[47] H.U. Keller, W.A. Delamare, W.F. Huebner, et al.: 'Comet P/Halley's Nucleus and its Activity', A&A **187**, 807–823 (1987)

[48] J. F. Kerridge: 'Formation and Processing of Organics in the Early Solar System', Space Sci. Rev. **90**, 275–288 (1999)

[49] J. Kissel et al.: 'Composition of Comet Halley Dust Particles from Giotto Observations', Nature **321**, 336–338 (1986)

[50] J. Kissel et al.: 'Composition of Comet Halley Dust Particles from Vega Observations', Nature **321**, 280–282 (1986)

[51] R.F. Knacke, S.B. Fajardo-Acosta, C.M. Telesco, J.A. Hackwell, D.K. Lynch, R.W. Russell: 'The Silicates in the Disk of β Pictoris', A&A **418**, 440–450 (1993)

[52] C. Koike, H. Shibai, A. Tuchiyama: 'Extinction of Olivine and Pyroxene in the Mid- and Far-Infrared', MNRAS **264**, 654–658 (1993)

[53] C. Koike, A. Tsuchiyama: 'Simulation and Alteration for Amorphous Silicates with Very Broad Bands in Infrared Spectra', MNRAS **255**, 248–254 (1992)

[54] M.E. Lawler, D.E. Brownlee: 'CHON as a Component of Dust from Comet Halley' Nature **359**, 810–812 (1992)

[55] M.E. Lawler, D.E. Brownlee, S. Temple, M.M. Wheelock: 'Iron, Magnesium, and Silicon in Dust from Comet Halley', Icarus **80**, 225–242 (1989)

[56] K. Malfait, C. Waelkens, L.B.F.M. Waters, B. Vandenbussche, E. Huygen, M.S. de Graauw: 'The Spectrum of the Young Star HD100546 Observed with the Infrared Space Observatory', A&A **332**, L25–L28 (1998)

[57] J.A.M. McDonnell, P.L. Lamy, G.S. Pankiewicz: 'Physical Properties of Cometary Dust', In: *Comets in the Post Halley Era*, ed. by R. L. Newburn, M. Neugebauer, and J. Rahe (Kluwer, Dordrecht 1991) pp. 1043–1073

[58] H. Mizuno: 'Grain Growth in the Turbulent Accretion Disk Solar Nebula" Icarus **80**, 189–201 (1989)

[59] W.T. Reach, M.V. Sykes, D. Lien, J.K. Davies: 'The Formation of Encke Meteoroids and Dust Trail', Icarus **148**, 80–94 (2000)

[60] S.A. Sandford and R.M. Walker: 'Laboratory Infrared Transmission Spectra of Individual Interplanetary Dust Particles from 2.5 to 25 microns', ApJ **291**, 838–851 (1985)

[61] H. Schulze, J. Kissel, E. Jessberger: 'Chemistry and Mineralogy of Comet Halley's Dust', In: *From Stardust to Planetesimals*, ed. by Y.J. Pendleton and A.G.G.M. Tielens ASP Conf. Series **122**, 397–414 (1997)

[62] G. Schwehm, R. Schulz: 'Rosetta Goes to Comet Wirtanen', Space Sci. Rev. **90**, 313–319 (1999)

[63] Z. Sekanina: 'Secondary Fragmentation of the Solar and Heliospheric Observatory Sungrazing Comets at Very Large Heliocentric Distances', ApJ Lett. **542**, L147–L150 (2000)

[64] Z. Sekanina, M.S. Hanner, E.K. Jessberger, M.N. Fomenkova: 'Cometary Dust', In: *Interplanetary Dust* ed. by E. Grün, B.A.S. Gustafson, S. Dermott, H. Fechtig (Springer, Heidelberg 2001)

[65] L.A. Soderblom, D.C. Boice, D.T. Britt, R.H. Brown, B.J. Buratti, M.D. Hicks, R.M. Nelson, J. Oberst, B.R. Sandel, S.A. Stern, N. Thomas, R.V. Yelle: 'Observations of Comet 19P/Borrelly from the Miniature Integrated Camera and Spectrometer (MICAS) aboard Deep Space 1 (DS1)', Bull. AAS **33**, 1087 (2001)

[66] J.R. Stephens, R.W. Russell: 'Emission and Extinction of Ground and Vapor- Condensed Silicates from 4 to 14 microns and the 10 micron Silicate Feature', ApJ **228**, 780–786 (1979)

[67] M.V. Sykes, R.G. Walker: 'Cometary Dust Trails', Icarus **95**, 180–210 (1992)

[68] C.A. Trujillo, M.E. Brown: 'The Radial Distribution of the Kuiper Belt', ApJ Lett. **554**, L95–L98 (2001)

[69] C.A. Trujillo, D.C. Jewitt, J.X. Luu: 'Properties of the Trans-Neptunian Belt: Statistics from the Canada-France-Hawaii Telescope Survey', Astron. J. **122**, 457–473 (2001)

[70] L.B.F.M. Waters, F.J. Molster, T. de Jong, et al.: 'Mineralogy of Oxygen-Rich Dust Shells', A&A **315**, L361–L364 (1996)

[71] C. Waelkens, L.B.F.M. Waters, M.S. de Graauw, et al.: 'SWS Observations of Young Main-Sequence Stars with Dusty Circumstellar Disks', A&A **315**, L245–248 (1996)

[72] H.A. Weaver, P.L. Lamy: 'Estimating the Size of Hale-Bopp's Nucleus', Earth, Moon, and Planets **79**, 17–33 (1999)

[73] S.J. Weidenschilling: 'Formation Processes and Time Scales for Meteorite Parent Bodies', In: *Meteorites and the Early Solar System* ed. by J.F. Kerridge and M.S. Matthews (Univ. Arizona Press, Tucson 1988) pp. 348–371

[74] P.R. Weissman: 'Diversity of Comets: Formation Zones and Dynamical Paths', Space Science Rev. **90**, 301–311 (1999)

[75] P.R. Weissman: 'The Vega Particulate Shell: Comets or Asteroids', Science **224**, 987 (1984)

[76] F.L. Whipple: 'A Comet Model: I. The Acceleration of Comet Encke', ApJ **111**, 375–394 (1950)

[77] D.H. Wooden, D.E. Harker, C.E. Woodward, H. Butner, C., Koike, F.C. Witteborn, C.W. McMurty: 'Silicate Mineralogy of the Dust in the Inner Coma of Comet C/1995 O1 (Hale-Bopp) Pre- and Post-Perihelion', ApJ. **517**, 1034–1058 (1999)

[78] D.H. Wooden, H.M. Butner, D.E. Harker, C.E. Woodward: 'Mg-rich Silicate Crystals in Comet Hale-Bopp: ISM Relics or Solar Nebula Condensates?' Icarus **143**, 126–137 (2000)

The *In-situ* Study of Solid Particles in the Solar System

Ingrid Mann and Elmar K. Jessberger

Institut für Planetologie, Westfälische Wilhelms-Universität, Münster
Wilhelm-Klemm-Str. 10, 48149 Münster, Germany

Abstract. *In-situ* measurements of dust from spacecraft can in principle provide information about dust properties at any given place in the solar system and under conditions that are not reproducible on Earth. Already relatively simple *in-situ* measurements have provided information about the properties of dust particles. Measurements of interplanetary dust have shown for the first time the fluffy and porous structure of interplanetary dust as well as the existence of two dust components with different properties in the inner solar system inside Earth orbit. Experiments during the missions to comet Halley have shown the cometary dust to consist of two major components to about the same amount: a component that is rich in rock-forming elements and a component that is rich in the elements H, C, N and O and is assumed to consist of refractory organic material. Although the composition of interstellar dust particles is not directly measured yet, their conditions of entry into the solar system reveal the forces that are acting on them. The forces depend on the properties of grains and allow for a comparison to astrophysical models of dust size, composition and structure. *In-situ* measurements of the mass distribution and flux rates of interstellar dust confirm models to describe large interstellar dust particle as aggregates that consist of core-mantle particles and allow to reject some of the models to describe smaller interstellar dust particles. They also show that the detected dust is physically coupled to the gas component in the local interstellar cloud that surrounds the Sun. Hence, the *in-situ* measurements of interstellar dust in the solar system provide a basis for studying not only the dust properties but also the connection to the gas phase of the interstellar medium (ISM). *In-situ* measurements with improved dust detectors are presently carried out and planned for future missions. The scientific return of the measurements can be greatly enhanced by combining detailed laboratory studies of the physics and functional principles of the detectors.

1 Introduction

The interplanetary medium of our solar system is filled with neutral and charged atoms and molecules as well as with solid dust particles. Dust particles are ejected from comets, generated by mutual collisions of solar system bodies, and emerge from atmosphereless bodies for instance due to volcanic activity, collisions, and impact erosion. The two latter mechanisms produce considerable amounts of dust from asteroids in the inner solar system and presumably from Kuiper belt objects in the outer solar system beyond the giant planets, while dust from comets is one of the major components in the inner solar system. Aside from those dust particles that originate from solar system objects, interstellar dust

particles stream through the solar system due to the motion of the sun relative
to the surrounding interstellar medium.

Dust particles are considerably larger than atoms, molecules and ions and
opposed to these particles usually show some of the properties of a solid. They
are sufficiently large to built up a solid structure which can be heterogeneous
while compared to large solar system bodies, they do not show significant internal processing. As far as their dynamics is concerned, gravitational forces affect
them in the same way as large solar system objects, but also radiation pressure
and electromagnetic forces act on them as they act on ions. Dust particles bear
the properties of their parent bodies and their study reveals different paths and
different degrees of processing of solid material from the time of the formation
of the solar system. At present, information on dust particles is gathered from
astronomical observations, from direct measurements of dust from spacecraft,
from the detection of particles entering the Earth's atmosphere and from the
collection of its fragments or even of the entire particle on Earth. Remote astronomical observations of solar light scattered by dust as well as of the thermal
emission reveal the average properties of dust particles along the line of sight
of observations and over their entire size spectrum. Among the astronomical
observations, thermal emission in many cases characterizes the material composition of dust particles, while, for instance, the polarization of the scattered light
provides information on the material, size and structure of the particles.

The direct detection of dust particles (i.e. meteoroids) is possible when they
enter the Earth's atmosphere. The decelerated meteoroidal material and surrounding atmospheric particles are ionized and produce the observable brightness, i.e. the meteor, or in case of larger infalling particles a shooting star or
a bolide. Depending on the size, structure and entry conditions, meteoroidal
material reaches the ground as meteorites where it can be collected and further studied. For particles of certain sizes entry conditions can be very smooth
and they can be collected in the stratosphere, as IDPs (Interplanetary Dust
Particles)[1]. However, as far as detection on Earth is concerned, several selection effects have to be considered: the probability for dust particles of crossing
the Earth's orbit depends on their orbital parameters; further, the particles are
focused by the Earth's gravity field and the focusing depends on the relative
velocity; similarly, heating during the entry depends on the velocity, structure,
size, and material composition.

---

[1] According to the IAU convention bodies that are considerably smaller than an asteroid and considerably larger than an atom or molecule are defined as meteoroids (cf.
Jessberger, 1981). Dust particles are micrometeoroids typically of sizes of 100 μm
and smaller. In some cases, the term "IDP" is solely used for those Interplanetary
Dust Particles (IDPs) that are collected in the stratosphere by highflying aircraft.
Their typical size ranges from 5 to 50 μm. Cosmic dust particles that are collected in
the Arctic or Antarctic Ice or from the sea floor and typically range between 100 and
1000 μm are often denoted as Micrometeorites (MMs). Dust particles in interplanetary space that make up the visible and thermal brightness of the Zodiacal light, i.e.
mainly dust in the size interval 1 - 100 μm are often referred to as Zodiacal dust.

"Full information" about dust particles would result from structural, chemical, mineralogical, molecular and isotopic analyses if combined with the orbital parameters. However, even solid interplanetary grains that are readily available for laboratory analysis hide behind their small size the wealth of information they potentially contain: A typically 15 μm particle only can be analyzed rather roughly (cf. Arndt et al. 1996, Jessberger et al. 2001) and e.g. the radiometric age of not even a single IDP has been obtained!

Figure 1 lists various kinds of dust detection and their approximate size range. It also shows the differential flux curve of dust particles at 1 AU as compiled by Grün et al. (1985) on the basis of several data sets. The observation of meteors (Ceplecha et al. 1998), the atmospheric collection of IDPs (see Bradley, this volume) and collection of micrometeorites (Maurette et al. 1991) on the ground yield data about near Earth dust and meteorites of sizes beyond several μm. In the case of meteor observations, recent headechoe measurements allow for a better estimate of the dust velocities (Pellinen-Wannberg et al. 1998, Janches et al. 2000). Thermal emission and visual scattered light measurements of the Zodiacal light give information averaged over large spatial regions mainly between 0.3 and 1.7 AU distance from the Sun and close to the ecliptic plane (Mann 1998). They describe the size interval from about 1 to 100 μm. PVDF (Tuzollino et al. 2001a; see Sect. 2.3) and impact ionization detectors (cf. Auer 2001) measure dust, predominantly of sizes below 100 μm from spacecraft. The detection of large dust particles is limited by the statistics of the low number densities. Clearly, *in-situ* measurements of cosmic dust from spacecraft cover the size range with the largest number of particles. But still the study with *in-situ* experiments is impeded by the low flux of impacting dust particles, the selection of certain sizes and locally occurring dust fluxes, the limited number of spacecraft carrying dust detectors as well as the experimental limits of the instruments. Nevertheless, they reveal the local dust dynamics resulting from the influence of the surrounding plasma and magnetic field and bear valuable information on the composition of those dust particles and dust components that do not reach the surface of the Earth.

The interest in *in-situ* dust measurements is two-fold: (a) measurements of dust fluxes and sizes as function of time and location reveal the dynamics of dust particles and the underlying physical processes, and (b) *in-situ* measurements could, in principal, provide direct information on the chemistry, structure and isotopic composition that is not biased by the entry conditions if particles hit the Earth. This article discusses the possibilities of *in-situ* experiments from the mineralogical point of view in comparison and as complement to the other measurement techniques mentioned above.

In-situ measurements of dust from spacecraft can in principle provide information about dust properties at any given place in the solar system. Namely *in-situ* measurements detect particles in different evolutionary stages from interstellar dust, primordial cometary and possibility Kuiper belt dust to highly processed dust debris from asteroids or larger bodies in the solar system. In the context of astromineralogy the basic issues that should be addressed with *in-*

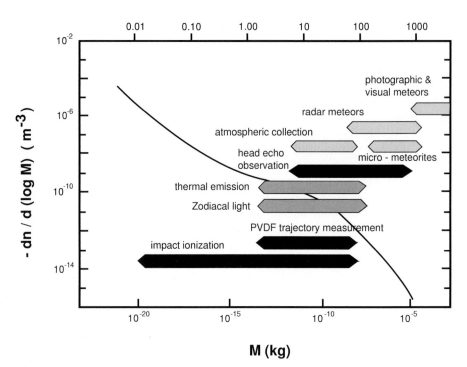

Fig. 1. Detection methods of dust and meteorites in interplanetary space. The solid line denotes the differential number density distribution as function of the mass of particles derived from various measurements at 1 AU. The upper axis denotes the size corresponding to the mass scale under the assumption of a bulk density of $3\cdot10^3$ kg/m^3. The range of detection is shown for different methods.

situ measurements and can be compared to other studies (of IDPs, astronomical observations, etc.) are properties of interstellar dust and the evolution of dust in the solar system. Although the *in-situ* analysis in space is still limited, we will show that already *in-situ* detection in terms of dust sizes and flux rates can provide valuable information for restricting the range of parameters of models that are derived from observations.

We first describe the basic functional principles of *in-situ* experiments and indirect dust measurements during previous and on-going space missions. We then will emphasize the study of dust properties based on space experiments addressing three specific topics: measurements of the interplanetary dust cloud, *in-situ* measurements at comets and *in-situ* measurements of interstellar dust entering the solar system.

2   The Basis of Dust Measurements

A process that is similar to meteor formation takes place when dust particles hit a solid target in space. Momentum is transferred onto the target and, depending on impact speed and mass of the particle, both target and projectile material are mechanically deformed; part of the material is melted and sublimates. The impact process can be distinguished according to three regimes of impact parameters: The scattering and accretion regime, the fragmentation and low vaporization regime, and the volume vaporization regime. The boundaries between these regimes depend on the relative velocity and on the mass of the particles. At relative velocities - between dust and analyzing tool - below about 1 km/s collisions may lead to the accretion or sticking of the colliding particles. At relative velocities 1 – 50 km/s collisions lead to fragmentation as well as to the ionization of surface materials. At relative velocities beyond 50 km/s the entire particle is destroyed and the produced vapor is ionized (Fig. 2). Many of the currently flying space experiments are based on the detection of the ionized impact debris (impact ionization detectors). Most spacecraft cruising in interplanetary space (Helios, Ulysses) or on fast flyby trajectories (Giotto, Vega) are in a range of relative velocities where impacting particles are partly destroyed. So-called "rendezvous missions" (for example Rosetta) where the spacecraft closely follows a solar system body in its orbit may lead to impact speeds in a range where impacting particles are accreted at a collection surface without or with few mechanical destruction.

   McDonnell (1978) gave a summary of the early measurement techniques. Early experiments detected the momentum transfer associated with a dust impact with microphones. Also the destructed target area can be determined to estimate the size of the impacting particles (see for instance Humes 1980, Simpson and Tuzzolino 1985). Moreover, the light flash that is associated with the impact can be measured. For the technical and functional principle of modern instruments we refer to a recent review by Auer (2001).

2.1   Impact Ionization Detectors

Impact ionization detectors use the fact that the particles are destroyed during the impact: the produced electrons and ions are separately detected as shown in Fig. 2. Further, time of flight measurement of the produced ions yields the elemental composition. The principal of impact ionization detectors is demonstrated with the dust analyzer onboard Cassini in Fig. 3 (Srama and Grün 1997). Figure 4 shows the first time-of-flight spectrum obtained with the Cassini experiment. The detected species stem from a complex ionisation, expansion and recombination process that takes place in the dust-plasma cloud that is formed at the high velocity impact. The interpretation in terms of the elemental compositions of the impacting dust particles is not unambigous and requires a detailed analysis of the impact ionisation process.

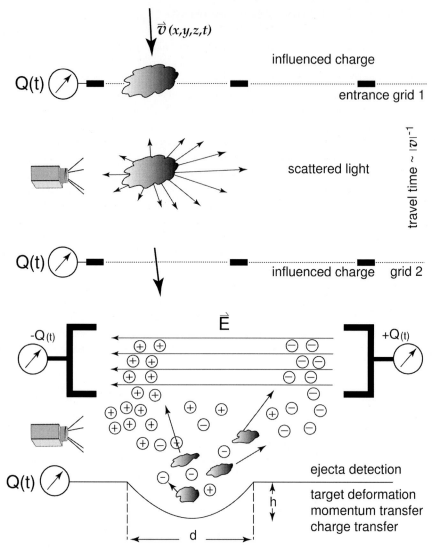

Fig. 2. Schematic view of some of the dust detection mechanisms. The electric surface charge of grains induces a charge pulse when dust particles travel through a system of conductive grids. This allows to derive the travel time of the particle. Particles may be optically detected through a scattered light signal. Upon impact on a solid target the particles transfer momentum and charge onto the target. Both, target and projectile material are mechanically deformed and solid material is ejected. Target and projectile material are heated, partly melted and ionized. Positively and negatively charged particles are separated in an electric field and can be measured. The formation of an ion cloud also causes a light flash.

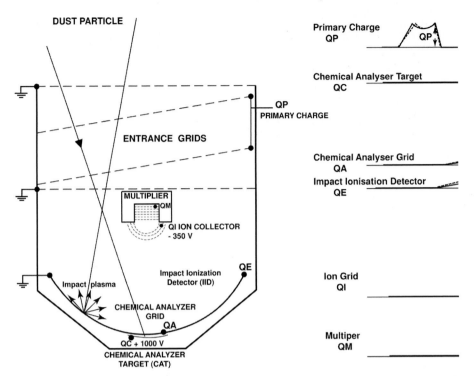

Fig. 3. The impact ionization dust analyzer aboard Cassini (Srama and Grün 1997). The left hand side of the figure show the schematics of the detector, the right hand side shows the different channels of measured signals: The particle velocity is estimated through the sequence of charges that are induced in a system of tilted entrance grids (denoted as primary charge QP on the right hand side of the figure) before they impact onto the target plate. The produced ions and electrons are detected at the target plate (QE, QA, QC) and in the ion collector (QI). Ions that are produced are accelerated due to the applied voltage and measured as a function of their flight time at a multichannelplate (QM).

2.2   Measurement of Structure and Optical Properties

A combined study of non-optical *in-situ* measurements and remote optical observations would allow for a close comparison to astronomical observations. Giese et al. (1979) had pointed out the possibility to measure the light scattered by a single cometary dust grain as it passes through a collimated light beam and suggested the analysis of cometary dust by optical scattering experiments locally from space probes. As a result, an optical probe experiment was suggested for the European fly-by mission to Comet Halley to measure the local brightness and polarization along the spacecraft trajectory (Levasseur-Regourd et al. 1981). The proposed experiments were limited by the light sources and detector units that were available for space instrumentation and indeed the instrument aboard GIOTTO used sunlight for illumination (Levasseur-Regourd et al. 1986). Devel-

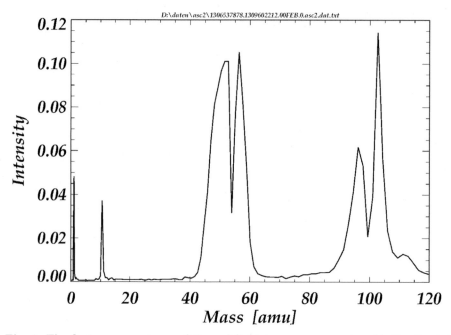

Fig. 4. The first mass spectrum of time-of-flight ion measurements with the Cassini dust analyzer corresponding to channel QM in Fig. 3.

opment of low weight and low power laser elements now allows for using them in space experiments. Laser light scattering instruments are built for the Rosetta mission and will be used to estimate the velocity vector of infalling dust particles (cf. Leese et al. 1996) This points to an important issue of dust measurements from spacecraft: How to obtain orbital information on the detected particles.

2.3   Measurement of Dust Orbits

The impact speed and direction of particles can often only be crudely estimated from the orientation of the spacecraft at the time of the impact. Tuzzolino et al. (2001a) recently attempted a more specific measurement of a dust particle velocity and impact direction from spacecraft. The Space Dust (SPADUS) instrument aboard the Earth orbiting satellite ARGOS uses Polyvinylidene Fluoride (PVDF) dust sensors combined with a velocity/trajectory system. Dust particles travel through a sandwich of sensor foils where the impact time and location is measured. This allows estimating the impact velocity and direction (Tuzzolino et al. 2001a). It permits the distinction between orbital debris and cosmic dust, as well as the determination of the orbital elements for some of the impacting particles (Tuzzolino et al. 2001b). McDonnell et al. (1999) suggested combining Aerogel collection to retrieve samples for analysis with real time detection with a position-sensitive impact sensor in order to achieve a better classification of the samples. We expect the determination of dust orbital parameters to improve for future dedicated dust experiments.

2.4   Indirect Measurements

Aside from these direct dust detections some information on dust properties is also obtained from signals gathered with other space instruments. Charged particles are produced by dust impact ionization and cause a variation of plasma parameters which is typically nonlinear and hence its interpretation is difficult. The radio noise that was detected when the Voyager 1 and 2 spacecraft crossed the Saturn rings was explained with dust grains as a possible source (Aubier et al. 1983). During the Uranus ring plane crossing of the Voyager 2 spacecraft the Planetary Radio Astronomy instrument aboard recorded a characteristic intense noise due to the impact ionization of dust particles with sizes >1 μm (Meyer-Vernet et al. 1986a). Voyager measurements at planetary rings and ICE (International Cometary Explorer) measurements during the encounter with comet Giacobini-Zinner were used to estimate dust sizes and fluxes (Meyer-Vernet et al. 1986b, Tsintikidis et al. 1994). Dust impacts in the milligram range on the Giotto spacecraft were observed to influence the time variation of the measured magnetic field; and simultaneous events were seen in the plasma analyzer and an ion mass spectrometer (Neubauer et al. 1990). Also plasma wave measurements onboard Voyager were used to estimate dust fluxes in the outer solar system (Gurnett et al. 1997).

3   *In-situ* Measurements of Interplanetary Dust

3.1   Experimental Results

Early *in-situ* experiments did not measure the elemental composition of dust, but they allowed already an estimate of parameters like density and structure. An important piece of information obtained from combining the results of *in-situ* and optical measurements is that a large fraction of dust in the interplanetary medium has a fluffy, porous structure: Measurements aboard the Helios space probes allowed the comparison of dust flux measurements and optical measurements aboard the same spacecraft. The comparison of the local dust number density and local scattered light brightness showed that previous models explaining the Zodiacal light by light scattered primarily at compact dust particles of sizes below 1 μm were not valid. Instead, Giese et al. (1978) suggested that the main contribution to the zodiacal light stems from particles greater than or equal to 10 μm in size. In order to find a consistent description of both flux measurements and optical measurement, Giese et al. (1978) proposed absorbing particles of fluffy structure as an important component of the interplanetary dust cloud. Around the same time the analysis of interplanetary dust samples collected from the stratosphere demonstrated that they are fine-grained aggregates of nonvolatile building blocks (Fraundorf et al. 1981, 1982). Laboratory measurements of dust particles of terrestrial as well as meteoritic origin were carried out to study their scattering characteristics. Single particles were electrostatically levitated and illuminated with a laser beam in order to measure the scattered light as a function of the scattering angle (Weiss-Wrana, 1983). The scattering properties of irregularly shaped dark opaque particles with very rough surface reproduce the characteristic scattering features that were derived from measurements of the Zodiacal light. As far as model calculations are concerned, Mie theory is not a good approximation for their scattering properties since particles have radii larger than the wavelength of the scattered light. In first model calculations multiple reflections of incident light on the rough surface were treated on the basis of multiple scattering theory (Mukai et al. 1982). The results were found to be in agreement with irregular particle scattering features determined through laser measurements (Weiss-Wrana, 1983, Hanner et al. 1981). Fractal dust analogues (e.g. Mukai et al. 1992) were later applied to estimate the differences in the thermal and optical properties of dust originating from comets and from asteroids (Mann et al. 1994, Wilck and Mann 1996, Kimura and Mann 1999).

Already measurements of the Helios missions indicated the existence of two dust components in the inner solar system between 0.3 and 1 AU (Fechtig 1982): they have distinctly different densities and orbits and presumably stem from different sources. Also the change of the average optical properties within the dust cloud that is derived from observations is better explained with the existence of two distinct dust components (Mann 1998). A clear distinction of such components is one of the tasks of future *in-situ* measurements.

3.2   Future Measurements of Interplanetary Dust

Dust measurements in the near solar environment were suggested for future space missions to reveal the dynamics of dust in the time variable solar magnetic field as well as the gradual sublimation of dust particles of different sizes and material composition (Mann and Grün 1995). Integrated payload packages are envisaged to measure plasma particles and dust grains simultaneously and with shared instrument constituents in order to reduce weight and costs for the experiments (cf. Tsurutani et al. 1997).

The future ESA mission Solar Orbiter (cf. Fleck et al. 2001) provides a unique opportunity for studying interplanetary dust inside the Earth orbit. The spacecraft will orbit the Sun as close as 0.21 AU (45 solar radii) and will reach heliographic latitudes as high as 38 degrees. Aside from Solar remote-sensing instruments Solar Orbiter will carry a package of *in-situ* experiments including an interplanetary dust detector (cf. Mann et al. 2001).

The solar environment is the central region of the solar system dust cloud. The dust density as well as plasma density and temperature are increasing and dust-plasma interactions are becoming particularly important. Sublimation (Mann et al. 1994) and collisional destruction (Ishimoto and Mann 1999) of dust grains, as well as the release of solar wind particles implanted into the surface of dust particles produce atoms, molecules and ions in the solar environment.

Recent measurements indicate that the solar wind carries pick up ions from a so-called inner source that is connected to these processes (Glöckler and Geiss 1998). Increasing solar radiation modifies the dust composition and material properties in the inner solar system. Dust measurements aboard Solar Orbiter could reveal this variation of dust properties and at the same time through solar wind measurements give direct information about possible interactions with the plasma. Also, the dust and meteoroid destruction by mutual collisions inward from 1 AU could shift the size distribution to smaller particles (Ishimoto and Mann 1999, Ishimoto 2000). The collision evolution of dust in the solar system has been proven by recent measurements of β–meteoroids as a product of dust collisions (Wehry and Mann 1999). Dust measurements aboard Solar Orbiter will cover the region where the collision production of β–meteoroids is expected to take place and also show the dust size distribution that is influenced by collisions.

As mentioned before future measurements may reveal the existence of different dust components in the inner solar system. Recent observations of cometary dust trails show that they eject a larger amount of (dark) meteorite fragments than previously assumed (Ishiguro et al. 2002). Dust production from comets may account for the change of average optical properties that is seen in brightness observations. Hence future measurements would help to better quantify the dust production from different sources (asteroids, long- and short-period comets) and possibly will allow to study cometary material in the inner solar system.

4   *In-situ* Studies at Comets

4.1   Experimental Results

Among the various types of investigations of cometary dust (see Hanner, this book), the missions to comet Halley around 1986 allowed for the first time the direct detection of dust around a comet. The masses and flux rates of dust were measured with the Dust Impact Detection System (DIDSY) aboard Giotto (McDonnell et al. 1986) and the Dust Counter and Mass Analyser (DUCMA) measurements from VEGA spacecraft (Simpson et al. 1986, 1987). The Halley Optical Probe Experiment (HOPE) aboard Giotto optically probed the environment of a comet inside the coma (Levasseur-Regourd et al. 1986). The missions to comet Halley also allowed for the first time *in-situ* measurements of dust composition in space. The properties of cometary dust were analyzed by the experiments PIA and PUMA 1+2 onboard GIOTTO and VEGA 1+2 (Kissel et al. 1986a,b; cf. Jessberger et al. 1988).

Measurements of dust particles from Vega spacecraft showed that the dust coma is highly variable. They suggested the existence of localized dust emission regions on the nucleus (Simpson et al. 1986, 1987) and large fluxes of low mass particles that extend to great distances beyond the coma (Simpson et al. 1986, Utterback and Kissel 1989). The data suggest the emission of large conglomerates from the nucleus, which disintegrate as they travel outward. Also DIDSY measurements showed the mass distribution of dust to vary within the coma of comet P/Halley (McDonnell et al. 1987). The HOPE instrument (Levasseur-Regourd et al. 1986) detected the light scattering at dust and the emission from gaseous species (i.e. CN, C_2, CO^+ and OH). The results from HOPE and from impact measurements (DIDSY) are consistent for distances from the nucleus in the 10^3 to 10^5 km range (Levasseur-Regourd et al. 1999) and confirm the model assumption of absorbing (and presumably porous) cometary dust particles.

The measurements of PIA and PUMA 1+2 onboard GIOTTO and VEGA 1+2 yield information about the dust composition, that can be summarized as follows (cf. Kissel et al. 1986a,b; Jessberger 1999 and references therein): The masses and densities of the detected and analysed individual dust particles range from 10^{-19} to 10^{-14} kg (approximate diameter range 0.02 to 2 μm) with densities from $0.3 \cdot 10^3$ to $3 \cdot 10^3 kg/m^3$. The abundance of small particles below 10^{-17} kg is higher than anticipated. Most of the small particles are rich in the light elements H, C, N, and O. Particles that are abundant in these light elements have a low ratio of mass to volume, i.e., low density. The detected particulates are mixtures of two end-member components, dubbed CHON (rich in the elements H, C, N, and O) and ROCK (rich in rock-forming elements as Si, Mg, Fe), respectively. The CHON component is assumed to be refractory organic material while the ROCK component is assumed to comprise silicates, metals, oxides, sulfides and others. Both end-member components do not occur as pure components but are intimately mixed down to the finest scale. CHON- and ROCK-rich particulates each comprise about 25% of the dust while 50% are MIXED. The latter group is defined by the ratio of carbon to any rock-forming element between 0.1 and 10. Several lines of evidence suggest that the refractory organic CHON-components form mantles around ROCK-rich cores.

The bulk abundances of the rock-forming elements in Halley's dust follow solar and chondritic abundances within a factor of two (Fig. 5). With the plausible assumption that the whole comet (dust and ice) has approximately solar composition (with the exception of H and N) an overall dust/ice ratio of 2 was inferred (Fig. 6). The chondritic and solar abundances shown in Fig. 5 and Fig. 6 are from Grevesse and Savaul (1998) with some corrected values from Holweger (2001) included (see Kimura et al. 2003a for a detailed discussion). The elemental abundances of cometary dust are derived from PIA/PUMA data. It should be noted that recently Allende Prieto et al. (2002) suggested the photospheric nitrogen abundances were overestimated, so that the difference between cometary and photospheric nitrogen abundances may originate from observational uncertainties.

As far as the isotopic information is concerned, no significant (i.e. greater than a factor of two) deviation from the normal composition was found for those elements for which it technically would be detectable, like Mg and S. The notable exception, however, is carbon. The dust has a variable carbon isotopic composition and in extreme cases contains isotopically light carbon with $^{12}C/^{13}C$ ratios as high as 5,000 (in one particle composed of almost pure carbon). Similar values were measured in certain presolar graphite grains extracted from carbonaceous chondrites (Amari et al. 1990). Besides the extremely high $^{12}C/^{13}C$ ratios also

$^{12}C/^{13}C$ ratios below the normal value of 89 were measured. They are, however, *strictu sensu* lower limits that may result from noise intensities of unknown origin in the mass spectra. Two conclusions can be drawn from the derived isotopic data: (a) There is no single fixed or defined or typical cometary carbon isotopic composition at least for the case of comet Halley. (b) The presence of the wide range of isotopic compositions of carbon excludes any equilibration processes affecting the carbon carrier during comet formation or later in the comet's history.

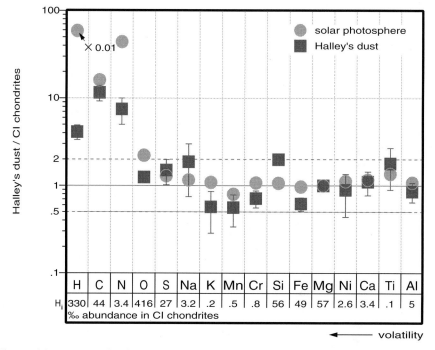

Fig. 5. Mg≡1 normalized element abundances in the dust of comet Halley and the solar photosphere, both relative to the abundances in CI-condrites. The elements are ordered according to their volatility and recent abundance values were used as explained in the text. The solar hydrogen abundance is multiplied by 0.01. Because of the lack of calibration, Halley's dust data are believed to be certain within a factor two.

Inspecting the variation of the elemental composition from particle to particle some clues as to the mineralogical composition can be drawn: Mg/Fe-ratios display a rather wide range, while the variation of the Si/Mg ratios is very limited. At least 40%, maybe more than 60% of the number of analysed particles are Mg-rich Fe-poor silicates. About 10% of particles (and therefore the next largest group) are iron- (+nickel)-sulfides while Fe-oxides play only a very minor role (<1%). Particles that are rich in refractory elements and resemble Ca,Al-rich inclusions known from chondrites (possibly the earliest inner-solar-system material formed) or carbonates or sulfates have not been unambiguously identi-

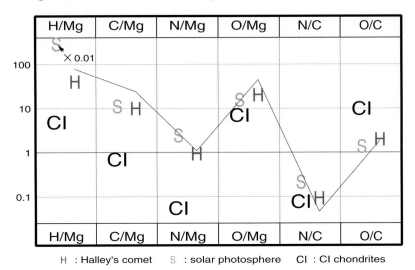

Fig. 6. Element ratios of Halley's comet (H) - dust *and* ice - compared to the same ratios in carbonaceous chondrites (CI) and the solar photosphere (S), obtained from the abundances as in Fig. 5. The solar hydrogen abundance is multiplied by 0.01. The overall similarity of Halley's comet to the solar composition as well as the lesser degree of depletion of non- condensible elements compared to CI-chondrites testify to the more primordial nature of comet Halley.

fied. The presence of non-equilibrated (high-temperature) minerals like Mg-rich silicates and Fe-sulfides, both formed above ∼600 K, yields evidence that equilibration at low temperatures is too slow a process to have affected the cometary dust particles. It also implies that these minerals after their formation never experienced elevated temperatures during their lifetime within the comet.

4.2   Future Measurements of Cometary Dust

A dust experiment of the PIA/PUMA type is currently flying aboard the NASA mission STARDUST (Brownlee et al. 1996) to carry out *in-situ* measurements (Kissel et al. 2001) and to bring back samples of interstellar and cometary dust.

A next mission that will include a whole set of *in-situ* measurements of cometary dust is the ESA mission Rosetta. Rosetta is going to be launched in 2003 and will closely follow a cometary nucleus along its trajectory up to perihelion. Its scientific payload includes several instruments designed to better understand the nature of cometary dust. Goals of the measurements cover the determination of the evolution of the dust fluxes and grain dynamics in the coma as well as studies of the dust microstructure and the dust composition. The GIADA experiment (Grain Impact Analyzer and Dust Accumulator; Leese et al. 1996; Lamy et al. 1998, Bussoletti et al. 1999) is designed to measure velocity, momentum and direction of the dust flux. The passage of grains through the instrument is detected by laser light scattering, particle momentum by piezoelectric transducers, and mass flux by means of three quartz crystal microbalances.

GIADA will be able to detect dust particles in the size range 5–1000 μm and in the velocity range 0.1–150 m/s. The MIDAS instrument aboard Rosetta is an atomic force microscope (AFM) where a micro–needle scans the surface structure of dust samples collected near the comet (Riedler et al. 1998, Romstedt et al. 1998). Aside from providing, for the first time, structural information by three-dimensional imaging of the small dust particles and thus possibly the first direct analysis of cometary mineralogy, MIDAS will also allow to better determine the size distribution of sub–μm cometary grains. While MIDAS relies on the fact that dust particles reach the spacecraft with small relative velocities and are not destroyed, experiments to study the dust elemental composition have to apply a destruction mechanism to allow for a composition analysis of the particles. The COSIMA experiment (described by Zscheeg et al. 1992) will analyze the elemental composition of collected dust particles that are ionized with an ion beam. This technique Time-of Flight Secondary-Ion-Mass-Spectrometry (TOF-SIMS) is now established in the laboratory and recently used for analysis of atmospheric IDPs (Stephan 2001).

5   *In-situ* Detection of Interstellar Dust

5.1   Experimental Results

After an early detection of interstellar dust in near-Earth-orbit (Bertaux and Blamont 1976), interstellar dust particles have been positively identified with measurements onboard Ulysses (Grün et al. 1994) and then with Galileo and Hiten. The interstellar dust is distinguished from solar system dust by its impact speed and impact direction (see Mann 1996, Kimura and Mann 2000, and references therein). Especially the long duration and the path of Ulysses (Wenzel et al. 1992) where the direction of the interstellar wind flux into the solar system is nearly perpendicular to the Ulysses orbit provided an excellent data set. Moreover, Ulysses allowed for a direct comparison of the dust direction to the direction of the interstellar neutral helium flux measured with a different instrument onboard the same spacecraft that provided further information on the interstellar wind direction (Witte et al. 1993). After flight to and flyby at Jupiter, the space probe Ulysses is orbiting the Sun on a nearly polar orbit that is inclined by 79° relative to the ecliptic with a perihelion distance of 1.3 AU and an aphelion distance of 5.4 AU. A set of 1695 dust impacts detected with the Ulysses sensor has been accumulated between October 1990 and December 1999 (Krüger et al. 2001, Grün et al. 1995, Krüger et al. 1999).

The mass of the particles detected between 1.8 and 5.4 AU ranges from 10^{-20} to 10^{-8} kg (Grün et al. 1995) which corresponds to radii a, 0.015 μm $< a < 4.1$ μm for compact spherical grains and bulk densities of $2 \cdot 10^3$ kg m^{-3}. The detection efficiency of the dust experiment varies with mass and impact speed of particles and the uncertainty of the mass measurement is a factor of 10. Still, the measured mass distribution can be approximated as $dn/dm \propto m^{-1.55}$ for masses $10^{-15.5}$ kg $\leq m \leq 10^{-11.5}$ kg (Kimura et al. 1999).

Interstellar dust of masses $m < 10^{-18}$ kg is partly prevented from entering the solar system (Mann and Kimura 2000) and deflected in the boundary region between solar wind and interstellar medium plasma (Czechowski and Mann 2001). Since 1996 Ulysses detects a reduced impact rate of smaller dust which is explained by the influence of the solar cycle variation (Landgraf 2000). The data may as well indicate that gravitational focusing influences the measured mass distribution at the large end of the mass interval (Mann and Kimura 2000). Once the different mechanisms to modify the flux of interstellar dust are quantified, the observed mass distribution may be compared to astrophysical models of interstellar dust as shown in Fig. 7. Classical models of interstellar medium dust suggested a power law size distribution as $dn/da \propto a^{-3.5}$ (Mathis et al. 1977, Draine and Lee 1984) that shows a sharp cut-off to larger masses at 10^{-16} kg ($a = 0.25$ μm). A recent model assumes composite grains of 0.003 to 3 μm in size (Mathis 1996) that are produced by coagulation growth and account for the abundances of heavy elements in the interstellar medium (ISM). Greenberg has suggested that a mantle is formed by ice accretion on silicate cores in molecular clouds (see for instance in Greenberg 1978). Li and Greenberg (1997) model simultaneously the interstellar extinction and polarisation by assuming three types of interstellar grains: Small PAH grains, very small carbonaceous particles and large core–mantle particles. They now assume the mantle to consist of organic refractory material produced by photo–processing of dirty ices in diffuse clouds where the UV irradiation is more intense.

The *in-situ* measurements of interstellar dust in the solar system could be used to estimate the mass density of dust in the local interstellar cloud (LIC) that is currently surrounding the sun. A detailed analysis that takes into account the enhancement of the dust flux as a result of the gravitational focusing of large dust particles (see Mann and Kimura 2000) shows that the measurements are in agreement with the canonical value of gas-to-dust mass ratio of 100 for the LIC (Kimura et al. 2003a) and that it is reasonable to assume that the interstellar dust in the mass range of the Ulysses measurements is coupled to the LIC gas component.

Further estimates can be made by investigating the mass distribution and its variation with distance from the Sun. The mass distribution of interstellar dust measured at $r < 3$ AU shows a gap around 10^{-17} kg which is not detected at larger distances. Although the acting forces on the dust grains do not only vary with the size but also with dust structure and material composition (cf. Mann and Kimura 2000, 2001), particles with masses $m > 10^{-17}$ kg are predominantly influenced by gravity and radiation pressure. The distance that interstellar particles approach the Sun against the repelling force of solar radiation pressure depends on the ratio β of radiation pressure and solar gravitational force $\beta = F_{\mathrm{rad}}/F_{\mathrm{grav}}$. This reduces the flux at smaller distance from the Sun. β–ratios $1.4 < \beta < 3.1$ for $m \approx 10^{-17}$ kg would explain the observational data. By comparing these values to model calculations of the radiation pressure (Fig. 8), some of the models of interstellar dust (like solid carbon or silicate particles) can

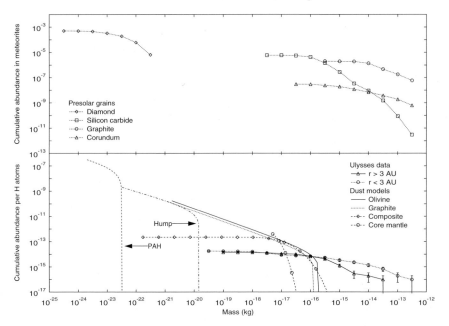

Fig. 7. The upper part of the figure shows the mass distribution of presolar grains extracted from meteorites, the lower figure shows Ulysses measurements of interstellar dust within and beyond 3 AU distance from the sun and different model assumptions for the size of dust in the interstellar medium. Both figures show the cumulative distribution.

be excluded, since they lead to radiation pressure forces that are clearly below or above the values that are needed to explain the observed interstellar dust fluxes.

The coupling of gas and dust in the LIC allows to assume that heavy elements that are depleted in the gas phase are condensed onto the dust grains. These elemental abundances combined with assumptions for a condensation sequence were used to estimate a possible composition of LIC dust that at the same time has the particles properties that are needed to explain the Ulysses data as mentioned above (Kimura et al. 2003b). The resulting composition of the core-mantle grains is that they consist of Mg-rich pyroxene and olivine with inclusions of troilite, Fe-rich kamacite, and corundum in the core and organic refractory compounds of C, N, and O in the mantle.

While the measurements that were discussed so far were not decidedly built to study interstellar dust and were planned at a time when it was uncertain whether interstellar dust particles are detectable in the solar system, some ongoing and future missions are more dedicated to the *in-situ* study of interstellar dust.

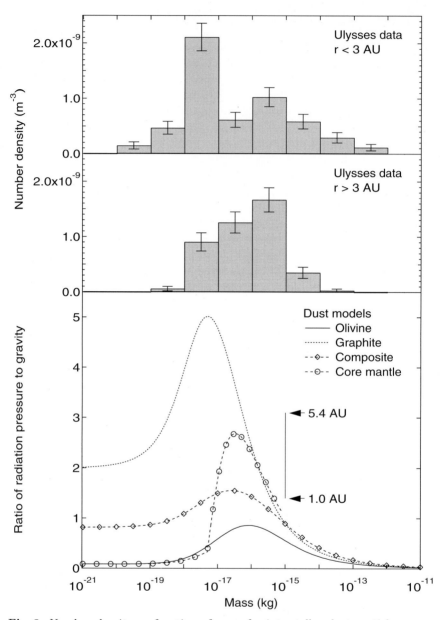

Fig. 8. Number density as function of mass for interstellar dust particles measured aboard Ulysses within and beyond 1 AU are shown in the upper panel of the figure. The lower panel shows the ratio β of radiation pressure force to gravitational force in the same mass interval calculated for different model assumptions of interstellar dust. The gap in the mass distribution of interstellar dust within 3 AU shown in the uppermost panel, may be attributed to a high radiation pressure compared to gravity for particles in this mass interval shown in the lower panel.

5.2 Future Measurements of Interstellar Dust

The Dust experiment aboard Cassini (Srama and Grün 1997) will for a portion of the cruising phase be oriented into the direction of the interstellar dust flux in order to obtain the first *in-situ* measurements of the elemental composition of interstellar dust in space. Also the dust measurements aboard STARDUST mentioned above will provide *in-situ* data of interstellar dust and a sample return. A dedicated mission to measure the composition of interstellar dust reaching the Earth orbit was proposed to ESA and NASA (Grün et al. 2000) and may in the future be realized. Future missions planned to explore the boundaries of the solar system and the interstellar medium beyond the heliopause with *in-situ* measurements from spacecraft (Liewer et al. 2000) would allow for the first *in-situ* measurements of dust in interstellar space without the restrictions that entry conditions of interstellar dust into the solar system (cf. Mann and Kimura 2000) currently impose on *in-situ* and laboratory studies of interstellar dust.

6 Discussion

In-situ experiments from spacecraft currently play an important role for solar system studies and improve our understanding of the dust evolution in the interstellar medium. Astronomical observations provide similar information, but integrated over a large set of particles providing the average properties. They are restricted to a certain size interval of particles. Moreover, most of the mineralogical observation is obtained from the information on the presence or absence of infrared features. The presence of features, however, does not only depend on the material composition, but also on the size of particles. The limits of the results that are obtained from the analysis of IDPs collected in the atmosphere, on the other hand, show that even collecting samples and bringing them back to Earth is not only costly, but also limited in the information it provides. Dedicated experiments for the *in-situ* analyses from spacecraft addressing certain clearly formulated questions are an important complementary tool. But they have to be augmented by laboratory studies of IDPs, micrometeorites and meteorites as well as by astronomical observations. Moreover, a good physical understanding of the impact process is essential for deriving reliable information from *in-situ* analyses of dust from spacecraft. Still this effort is worthwhile. Our discussion of previous *in-situ* measurements has shown that they reveal information about dust properties - sometimes very unexpected - that is not accessible to other methods of observation. As far as the study of dust as a component of the interplanetary and the interstellar medium is concerned, the dynamics of small grains reveals charging and dynamic effects that are valuable for the understanding of dust-plasma interactions. By *in-situ* analysis during spacecraft encounter, the detection of ejecta particles for one allows a study of the erosion processes. Furthermore, information about the composition of the parent body is accessible to spacecraft without the need to actually sample them on their surface. This may prove to be most important for the study of the many atmosphereless moons of the giant planets as well as of the enigmatic Kuiperbelt objects.

The discussed results from *in-situ* measurements of interplanetary, cometary and interstellar dust have shown that already previous, relatively simple experiments improved our understanding of the material properties and material evolution of dust in the solar system. Present results from *in-situ* measurements in combination with the results that were obtained with the other methods of dust studies point to the link between the different dust components. But until recently an effective proof of a direct link of cometary dust and any class of IDPs or meteorites was not existent. This situation has changed with the results from ISO IR-spectrometry that identify the mineral assemblage in comet Hale-Bopp dust as crystalline Mg-rich, Fe-poor enstatite and forsterite (minerals that are formed at high temperatures) plus lower-temperature glassy or amorphous grains (Hanner et al., this volume). Mg–rich, Fe–poor particles dominate not only Halley's dust, but were also identified as carriers of isotopic anomalies in IDPs (Stephan 2002). *In-situ* measurements of interstellar dust in the solar system show the existence of relatively large dust particles in the local interstellar medium, that most likely have a porous irregular structure. The models of interstellar dust are in accord with the formation of cometary dust from 'primordial' interstellar dust components. This is especially supported by recent studies showing the dust detected near comet Halley and the dust in the local interstellar medium to have similar element abundances (Kimura et al. 2003a). But the link between cometary dust and interstellar dust needs more detailed studies.

The strength of future *in-situ* measurements lies in their universality as far as the exploration of different "remote" regions in the solar system is concerned and in the fact that space experiments can be especially designed to address certain issues of dust analysis. By measurement of dust properties and their variation with distance from the sun, *in-situ* experiments allow for a direct study of particle processing in space: For the inner solar system we anticipate the thermal processing of asteroidal and cometary dust when approaching the sun. For spacecraft leaving the solar system, on the other hand, we expect to detect pristine dust particles in the Kuiper belt region and to detect pristine interstellar dust beyond the heliopause. As far as astromineralogy is concerned the major issues of *in-situ* experiments are: the detection of the structure of dust particles including the size distribution of their building stones, the detection of the nature and the degree of crystallization of silicates in the dust and the detection of organic components or of those elements that might be attributed to organic components in the dust particles.

7   Summary

The results derived from *in-situ* measurements of interplanetary, cometary and interstellar dust demonstrated that *in-situ* experiments are complementary to other methods of dust measurements and improved our understanding of the material properties of dust. *In-situ* studies of solar system dust provide access to different stages of the evolution of cosmic matter from interstellar dust, to pristine Kuiper belt and cometary dust, to highly processed asteroidal dust and improve our understanding of the origin and evolution of the solar system. Moreover, the measurements provide information about dust in the local interstellar medium as well as about basic physical processes that occur in cosmic environments. Future studies of cosmic dust need a combined approach from the astronomical, astrophysical, cosmochemical, and mineralogical points of view to enhance the scientific return of space measurements. Important steps in this direction are the design of dedicated space experiments combined with a supporting laboratory program to improve the detailed physical understanding of their functional principles.

Acknowledgements

This research has been supported by the German Aerospace Center DLR (Deutsches Zentrum für Luft- und Raumfahrt) under project "Mikro-Impakte: Empirische Analyse der Physik von Hochgeschwindigkeitseinschlägen" (RD-RX-50 OO0203).

References

[1] C. Allende Prieto, D.L. Lambert, and M. Asplund: *A Reappraisal of the Solar Photospheric C/O Ratio*, Astrophys. J. **573**, L137 (2002)

[2] S. Amari, E. Anders, A. Virag, and E. Zinner: *Interstellar graphite in meteorites*, Nature, **345**, 238 (1990)

[3] P. Arndt, J. Bohsung, M. Maetz, and E.K. Jessberger *The elemental abundances in interplanetary dust particles*, Meteoritics **31**, 817 (1996)

[4] M.G. Aubier, N. Meyer-Vernet, and B.M. Pedersen: *Shot noise from grain and particle impacts in Saturn's ring plane*, Geophys Res. Letters **10**, 5 (1983)

[5] S. Auer: *Instrumentation*. In: Interplanetary dust, ed. by E. Grün, B.A.S. Gustafson, S. Dermott, and H. Fechtig Astron. Astrophys. Library (Springer, Berlin and New York, 2001) pp.385–444

[6] J.L. Bertaux and J.E. Blamont: *Possible evidence for penetration of interstellar dust into the solar system*, Nature **262**, 263 (1976)

[7] J.P. Bradley, L.P. Keller, T.P. Snow, M.S. Hanner, G.J. Flynn, J.C. Gezo, S.J. Clemett, D.E. Brownlee, and J.E. Bowey: *An infrared spectral match between GEMS and interstellar grains*, Science **285**, 1716 (1999)

[8] D.E. Brownlee, D. Burnett, B. Clark, M.S. Hanner, F. Hörz, J. Kissel, R. Newburn, S. Sandford, Z. Sekanina, P. Tsou, and M. Zolensky: *STARDUST:Comet and Interstellar Dust Sample Return Mission*. In: Physics, Chemistry, and Dynamics of Interplanetary Dust, ed. by B.A.S. Gustafson and M.S. Hanner, Astronomical Society of the Pacific Conference Series (San Francisco, 1996) pp. 223-225.

[9] E. Bussoletti, L. Colangeli, J.J. Lopez Moreno, E. Epifani, V. Mennella, E. Palomba, P. Palumbo, A. Rotundi, S. Vergara, F. Girela, M. Herranz, J.M. Jeronimo, A.C. Lopez-Jimenez, A. Molina, F. Moreno, I. Olivares,

R. Rodrigo, J.F. Rodriguez-Gomez, J. Sanchez, J.A.M. Mc Donnell, M. Leese, P. Lamy, S. Perruchot, J.F. Crifo, M. Fulle, J.M. Perrin, F. Angrilli, E. Benini, L. Casini, G. Cherubini, A. Coradini, F. Giovane, E. Grün, B. Gustafson, C. Maag, and P.R. Weissmann: *The GIADA Experiment for Rosetta Mission to Comet 46P/Wirtanen: Design and Performances*, Adv. Space Res. **24**, (9) 1149 (1999)

[10] Z.K. Ceplecha, J.I. Borovicka, W.G. Elford, D.O. Revelle, R.L. Hawkes, V. Porubcan, and M. Simek: *Meteor Phenomena and Bodies*, Space Sci. Rev. **84**, (3/4) 327 (1998)

[11] A. Czechowski and I. Mann: *Dynamics of interstellar dust at the heliopause.* In: The Outer Heliosphere: The Next Frontiers, ed. by K. Scherer, H. Ficht-ner, H.J. Fahr, and E. Marsch (COSPAR Colloquiua Series, Pergamon Press, Amsterdam 2001), pp. 365-368.

[12] B.T. Draine and H.M. Lee: *Optical properties of interstellar graphite and silicate grains*, Astrophys. J. **285**, 89 (1984)

[13] H. Fechtig: *Cometary dust in the solar system.* In: Comets, ed. by H. Wilkening (University of Arizona Press, Tuscon, 1982) pp. 370-382.

[14] B. Fleck, E. Marsch, E. Antonucci, P.A. Bochsler, J.L. Bougeret, R. Har-rison, R.P. Marsden, M. Coradini, O. Pace, R. Schwenn, and J.-C. Vial: *Solar Orbiter: A high-resolution mission to the sun and inner heliosphere*, UV/EUV and Visible Space Instrumentation for Astronomy and Solar Physics, ed. by O.H. Siegmund, S. Fineschi, and M.A. Gummin, Proc. SPIE **4498**, 1 (2001)

[15] P. Fraundorf, R.I. Patel, and J.J. Freeman: *Infrared spectroscopy of inter-planetary dust in the laboratory*, Icarus **47**, 368 (1981)

[16] P. Fraundorf, R.M. Walker, and D.E. Brownlee: *Laboratory studies of in-terplanetary dust.* In: Comets, ed. by H. Wilkening, (University of Arizona Press, Tuscon, 1982) pp. 383-409.

[17] R.H. Giese, G.H. Schwehm, and R.M. Zerull: *A concept for analysis of cometary dust by light scattering experiments on future cometary probes.* In: Space research XIX, Proceedings of the Open Meetings of the Working Groups on Physical Sciences, Innsbruck 1978, (Oxford, Pergamon Press, 1979) pp. 475–478.

[18] R.H. Giese, K. Weiss, R.H. Zerull, and T. Ono: *Large fluffy particles - A possible explanation of the optical properties of interplanetary dust*, Astron. Astrophys. **65**, 265 (1978)

[19] G. Glöckler and J. Geiss: *Interstellar and Inner Source Pickup Ions Ob-served with SWICS on ULYSSES*, Space Sci. Rev. **86**, (1/4) 127 (1998)

[20] J.M. Greenberg: *Interstellar Dust.* In: Cosmic Dust, ed. by J.A.M. McDon-nell (Wiley-Interscience, Chichester, Sussex, England and New York, 1978) pp. 187-294.

[21] N. Grevesse and A.J. Sauval: *Standard Solar Composition*, Space Sci. Rev. **85**, (1/2) 161 (1998)

[22] E. Grün, H.A. Zook, H. Fechtig, R.H. Giese: *Collisional balance of the meteoritic complex*, Icarus **62**, 244 (1985)

[23] E. Grün, B. Gustafson, I. Mann, M. Baguhl, G.E. Morfill, P. Staubach, A. Taylor, and H.A. Zook: *Interstellar Dust in the Heliosphere*, Astron. Astrophys. **286**, 915 (1994)

[24] E. Grün, M. Landgraf, M. Horanyi, J. Kissel, H. Krüger, R. Srama, H. Svedhem, and P. Withnell: *Techniques for galactic dust measurements in the heliosphere*, J. Geophys. Res. **105**, 10403 (2000)

[25] E. Grün, M. Baguhl, N. Divine, H. Fechtig, M.S. Hanner, J. Kissel, B.A. Lindblad, D. Linkert, G. Linkert, I. Mann, J.A.M. McDonnell, G.E. Morfill, C. Polanskey, R. Riemann, G. Schwehm, N. Siddique, P. Staubach, and H.A. Zook: *Two Years of Ulysses Data*, Planet. Space Sci. **43**, 971 (1995)

[26] E. Grün, P. Staubach, M. Baguhl, D.P. Hamilton, H.A. Zook, S. Dermott, B.A. Gustafson, H. Fechtig, J. Kissel, D. Linkert, G. Linkert, R. Srama, M. S. Hanner, C. Polanskey, M. Horányi, B.A. Lindblad, I. Mann, J.A.M. McDonnell, G.E. Morfill, and G. Schwehm: *South-north and radial traverses through the interplanetary dust cloud*, Icarus **129**, 270 (1997)

[27] E. Grün, M.Landgraf, M.Horanyi, J.Kissel, H. Krüger, R. Srama, H. Svedhem, and P. Withnell: *Techniques for galactic dust measurements in the heliosphere*, J. Geophys. Res. **105**, 10403 (2000)

[28] D.A. Gurnett, J.A. Ansher, W.S. Kurth, and L.J. Granroth: *Micron-sized dust particles detected in the outer solar system by Voyager 1 and 2 plasma wave instruments*, Geophys. Res. Letters **24**, (24) 3125 (1997)

[29] M.S. Hanner, R.H. Giese, K. Weiss, and R. Zerull: *On the definition of albedo and application to irregular particles*, Astron. Astrophys. **104**, 42 (1981)

[30] H. Holweger: *Photospheric Abundances: Problems, Updates, Implications*, Joint SOHO/ACE Workshop 2001, ed. by R.F. Wimmer-Schweingruber (American Institute of Physics Press, New York) AIP Conference Proceedings **598**, 23 (2001)

[31] D.H. Humes: *Results of Pioneer 10 and 11 meteoroid experiments - Interplanetary and near-Saturn* J. Geophys. Res. **85**, 5841 (1980)

[32] H. Ishimoto: *Modeling the number density distribution of interplanetary dust on the ecliptic plane within 5 AU of the Sun*, Astron. Astrophys. **362**, 1158 (2000)

[33] H. Ishimoto and I. Mann: *Modeling the particle mass distribution within 1 AU of the Sun*, Planet. Space Sci., **47**, (1-2) 225 (1999)

[34] M. Ishiguro, J. Watanabe, F. Usui, T. Tanigawa, D. Kinoshita, J. Suzuki, R. Nakamura, M. Ueno, and T. Mukai: *First Detection of an Optical Dust Trail along the Orbit of 22P/Kopff*, Astrophys. J. **572**, (1) L117 (2002)

[35] D. Janches, J.D. Mathews, D. David D.D. Meisel, and Q.H. Zhou: *Micrometeor Observation using the Arecibo 430 MHz Radar: I. Determination of the Ballistic Parameter from measured Doppler Velocity and Deceleration Results*, Icarus **145**, 2 (2000)

[36] E.K. Jessberger: *Meteors and Meteorites*. In: Landolt-Börnstein VI/2a, ed. by H.H. Voigt and K. Schaifers (Springer, Berlin, 1981) pp. 187–202.

[37] E.K. Jessberger: *On the elemental, isotopic and mineralogical ingredients of ROCKY cometary particulates*, The Origin and Composition of Cometary

Materials, ed. by K. Altwegg, P. Ehrenfreund, J. Geiss, and W. Huebner, Space Sci. Series of ISSI, (Kluwer, Dordrecht), Space Sci. Rev. **90**, 91 (1999)

[38] E.K. Jessberger, T. Stephan, D. Rost, P. Arndt, M. Maetz, F.J. Stadermann, D.E.Brownlee, J.P. Bradley, G. Kurat, In: ed. by E. Grün, B.A.S. Gustafson, S. Dermott, and H. Fechtig: *Interplanetary dust* (Springer, Berlin, New York, 2001) pp. 253-294.

[39] E.K. Jessberger, A. Christoforidis, and J. Kissel: *Aspects of the major element composition of Halley's dust*, Nature **332**, 691 (1988)

[40] H. Kimura and I. Mann: *Radiation pressure on porous micrometeoroids in interplanetary space.* In: Meteoroids ed. by W.J. Baggaley and V. Porubcan (Astronomical Institute, Slovak, Academy of Sciences, Bratislava, 1999) pp. 283–286

[41] H. Kimura, I. Mann, and A. Wehry: *Interstellar dust in the solar system,* Astrophys. Space Sci. **264**, 213 (1999]

[42] H. Kimura, I. Mann, and E.K. Jessberger: *Elemental Abundances and Mass Densities of Dust and Gas in the Local Interstellar Cloud,* Astrophys. J. 582,000 (2003a)

[43] H. Kimura, I. Mann, and E.K. Jessberger: *Composition, Structure and Size Distribution of Dust in the Local Interstellar Cloud,* Astrophys. J. 583,000 (2003b)

[44] J. Kissel, F.R. Krüger, J. Silén, and G. Haerendel: *The probable chemical nature of interstellar dust particles detected by CIDA on Stardust.* In: The Outer Heliosphere: The Next Frontiers, ed. by K. Scherer, H. Fichtner, H.J. Fahr, and E. Marsch (COSPAR Colloquia Series, Pergamon Press, Amsterdam 2001), pp. 351-354.

[45] J. Kissel, D.E. Brownlee, K. Buchler, B.C. Clark, M. Fechtig, E. Grün, K. Hornung, E.B. Igenbergs, E.K. Jessberger, F.R. Krueger, H. Kuczera, J.A.M. McDonnell, G.M. Morfill, J. Rahe, G.H. Schwehm, Z. Sekanina, N.G. Utterback, H.J. Völk, and H.A. Zook: *Composition of comet Halley dust particles from Giotto observations,* Nature, **321**, 336 (1986a)

[46] J. Kissel, R.Z. Sagdeev, J.L. Bertaux, V.N. Angarov, J. Audouze, J.E. Blamont, K. Buchler, E.N. Evlanov, H. Fechtig, M.N. Fomenkova, H. von Hoerner, N.A. Inogamov, V.N. Khromov, W. Knabe, F.R. Krueger, Y. Langevin, B. Leonasv, A.C. Levasseur-Regourd, G.G. Managadze, S.N. Podkolzin, V.D. Shapiro, S.R. Tabaldyev, and B.V. Zubkov: *Composition of comet Halley dust particles from VEGA observations,* Nature, **321**, 280 (1986b)

[47] H. Krüger, E. Grün, A. Heck, and S. Lammers: *Analysis of the sensor characteristics of the Galileo dust detector with collimated Jovian dust stream particles,* Planet. Space Sci. **47**, 1015 (1999)

[48] H. Krüger, E. Grün, M. Landgraf, S. Dermott, H. Fechtig, B.A. Gustafson, D.P. Hamilton, M.S. Hanner, M. Horányi, J. Kissel, B.A. Lindblad, D. Linkert, G. Linkert, I. Mann, J.A.M. McDonnell, G.E. Morfill, C. Polanskey, G. Schwehm, R. Srama, and H.A. Zook: *Four years of Ulysses dust data: 1996-1999,* Planet. Space Sci. **49**, (13) 1303 (2001)

[49] P. Lamy, S. Perruchot, J.-L. Reynaud, M.R. Leese, J.A.M. McDonnell, S.F. Green, E. Bussoletti, L. Colangeli, M. Fulle, A. Rotundi, F. Giovane, B. Gustafson, and J.-M. Perrin: *DFA-the dust flux analyzer for the Rosetta orbiter*, Adv. Space Res. **21**, (11) 1557 (1998)

[50] M. Landgraf: *Modeling the motion and distribution of interstellar dust inside the heliosphere*, J. Geophys. Res. **105**, 10303 (2000)

[51] M.R. Leese, J.A.M. McDonnell, S.F. Green, E. Bussoletti, B.C. Clark, L. Colangeli, J.F. Crifo, P. Eberhardt, F. Giovane, E. Grün, B. Gustafson, D.W. Hughes, D. Jackson, P. Lamy, Y. Langevin, I. Mann, S. McKenna-Lawlor, W.G. Tanner, P.R. Weissman, and J.C. Zarnecki: *Dust Flux Analyser experiment for the Rosetta mission*, Adv. Space Res. **17**, (12) 137 (1996)

[52] A.C. Levasseur-Regourd, D.W. Schuerman, R.H. Zerull, and R. H. Giese: *Cometary dust observations by optical in-situ methods*, Adv. Space Res. **1**, (8) 113 (1981)

[53] A.C. Levasseur-Regourd, N. McBride, E. Hadamcik, and M. Fulle: *Similarities between in-situ measurements of local dust light scattering and dust flux impact data within the coma of 1P/Halley*, Astron. Astrophys. **348**, 636 (1999)

[54] A.C. Levasseur-Regourd, J.L. Bertaux, R. Dumont, M. Festou, R.H. Giese, F. Giovane, P. Lamy, J.M. Le Blanc, A. Llebaria, and J.L. Weinberg: *Optical probing of comet Halley from the Giotto spacecraft*, Nature **321**, 341 (1986)

[55] A. Li and J. M. Greenberg: *A unified model of interstellar dust*, Astron. Astrophys. **323** 566 (1997)

[56] P.C. Liewer, R.A. Mewaldt, J.A. Ayon, and R.A. Wallace: *NASA's Interstellar Probe Mission*, Space Technology and Applications International Forum 2000, ed. by M.S. El-Genk (American Institute of Physics Press, New York) AIP Conference Proceedings **504**, 911 (2000)

[57] I. Mann: *Interstellar Grains in the Solar System, Requirements for an Analysis*, In: The Heliosphere in the Local Interstellar Medium, ed. by R. von Steiger, R. Lallement, and M. Lee, Space Sci. Series of ISSI (Kluwer, Dordrecht) Space Sci. Rev. **78**, 259 (1996)

[58] I. Mann: *Zodiacal Cloud Complexes*, Earth, Planets and Space **50**, (6,7) 465 (1998)

[59] I. Mann and E. Grün: *Dust Studies on a Solar Probe*, Adv. Space Res. **17**, (3) 99 (1995)

[60] I. Mann and H. Kimura: *Interstellar dust properties derived from mass density, mass distribution, and flux rates in the heliosphere*, J. Geophys. Res. **105**, 10317 (2000)

[61] I. Mann and H. Kimura: *Dust Properties in the Local Interstellar Medium*, Space Sci. Rev. **97**, (1/4) 389 (2001)

[62] I. Mann, H. Okamoto, T. Mukai, H. Kimura, and Y. Kitada: *Fractal aggregate analogues for near solar dust properties*, Astron. Astrophys. **291**, 1011 (1994)

[63] I. Mann, H. Kimura, E.K. Jessberger, M. Fehringer, and H. Svedhem: *Dust in the inner solar system*, Solar encounter, First Solar Orbiter Workshop 2001, ed. by B. Battrick and H. Sawaya-Lacoste (ESA Publications Division, Noordwijk, The Netherlands) ESA **SP-493**, 445 (2001)

[64] J.S. Mathis: *Dust models with tight abundance constraints*, Astrophys. J. **472**, 643 (1996)

[65] J.S. Mathis, W. Rumpl, and K.H. Nordsieck: *The size distribution of interstellar grains*, Astrophys. J. **217**, 425 (1977)

[66] M. Maurette, C. Olinger, M.C. Michel-Levy, G. Kurat, M. Pourchet, F. Brandstatter, and M. Bourot-Denise: *A collection of diverse micrometeorites recovered from 100 tonnes of Antarctic blue ice*, Nature **351**, 44 (1991)

[67] J.A.M. McDonnell: *Microparticle Studies by Space Instrumentation*. In: Cosmic Dust, ed. by J.A.M. McDonnell (Wiley-Interscience, Chichester, Sussex, England and New York, 1978) pp. 337-419.

[68] J.A.M. McDonnell, W.M. Alexander, W.M. Burton, E. Bussoletti, D.H. Clark, R.J.L. Grard, E. Grün, M.S. Hanner, Z. Sekanina, and D.W. Hughes: *Dust density and mass distribution near comet Halley from Giotto observations*, Nature, **321**, 338 (1986)

[69] J.A.M. McDonnell, G.C. Evans, S.T. Evans, W.M. Alexander, W.M. Burton, J.G. Firth, E. Bussoletti, R. Grard, M.S. Hanner, and Z. Sekanina: *The dust distribution within the inner coma of comet P/Halley–Encounter by Giotto's impact detectors*, Astron. Astrophys. **187**, 719 (1987)

[70] J.A.M. McDonnell, M.J. Burchell, S.F. Green, M. Leese, D. Wallis, J.C. Zarnecki, D.J. Catling, D.E. Brownlee, P. Tsou, L. Colangeli, E. Bussoletti, G. Drolshagen, C.R. Maag, and H. Yano: *APSIS - Aerogel Position–Sensitive Impact Sensor: Capabilities for in-situ Collection and Sample Return*, Adv. Space Res. **25**, (2) 315 (1999)

[71] N. Meyer-Vernet, M.G. Aubier, and B.M. Pedersen: *Voyager 2 at Uranus - Grain impacts in the ring plane*, Geophys. Res. Letters **13**, 617 (1986a)

[72] N. Meyer-Vernet, P. Couturier, S. Hoang, C. Perche, J.L. Steinberg, J. Fainberg, and C. Meetre: *Plasma diagnosis from thermal noise and limits on dust flux or mass in comet Giacobini-Zinner*, Science **232**, 370 (1986)

[73] S. Mukai, T. Mukai, R.H. Giese, K. Weiss, and R.H. Zerull: *Scattering of radiation by a large particle with a random rough surface*, Moon and the Planets, **26**, 197 (1982)

[74] T. Mukai, H. Ishimoto, T. Kozasa, J. Blum, and J.M. Greenberg: *Radiation pressure forces of fluffy porous grains*, Astron. Astrophys. **262**, 315 (1992)

[75] F.M. Neubauer, K.-H. Glassmeier, A.J. Coates, R. Goldstein, and M.H Acuna: *Hypervelocity dust particle impacts observed by the Giotto magnetometer and plasma experiments*, Geophys. Res. Letters **17**, 1809 (1990)

[76] A. Pellinen-Wannberg, A. Westman, G. Wannberg, and K. Kaila: *Meteor fluxes and visual magnitudes from EISCAT radar event rates: A comparison with cross section based magnitude estimates and optical data*, Annales Geophysicae **16**, 1475 (1998)

[77] W. Riedler, K. Torkar, F. Rüdenauer, M. Fehringer, R. Schmidt, H. Arends, R.J.L. Grard, E.K. Jessberger, R. Kassing, H.St.C. Alleyne, P. Ehrenfreund, A.C. Levasseur-Regourd, C. Koeberl, O. Havnes, W. Klöck, E. Zinner, and M. Rott: *The MIDAS experiment for the Rosetta mission*, Adv. Space Res. **21**, (11) 1547 (1998)

[78] J. Romstedt, R. Schmidt, and E.K. Jessberger: *Microscopy in Space - Past and Future. From MACRO via micro to nano. Laboratory Astrophysics and Space Research*. In: ed. by P. Ehrenfreund, K. Krafft, H. Kochan, and V. Pirronello, Astrophys. Space Sci. Lib. **236** (Kluwer Academic Publishers, Dordrecht, 1998) pp. 483-506.

[79] J. Romstedt, F. Rüdenauer, M. Fehringer, and A. Jäckel: *Simulation Experiments of a Dust-collecting Surface for a Cometary Environment*, Meteoritics & Planetary Science **34**, Supplement, A98, (1999)

[80] J.A. Simpson and A.J. Tuzzolino: *Polarized Polymer Films as Electronic Pulse Detectors of Cosmic Dust Particles*, Nuclear Instruments and Methods in Physics Research **A236**, 187 (1985)

[81] J.A. Simpson, D. Rabinowitz, A.J. Tuzzolino, L.V. Ksanfomaliti, and R.Z. Sagdeev: *The dust coma of comet P/Halley - Measurements on the Vega-1 and Vega-2 spacecraft*, Astron. Astrophys. **187**, (1-2) 742 (1987)

[82] J.A. Simpson, R.Z. Sagdeev, A.J. Tuzzolino, M.A. Perkins, L.V. Ksanfomality, D. Rabinowitz, G.A. Lentz, V.V. Afonin, J. Ero, E. Keppler, J. Kosorokov, E. Petrova, L. Szabo, and G. Umlauft: *Dust counter and mass analyser (DUCMA) measurements of comet Halley's coma from VEGA spacecraft*, Nature **321**, 278 (1986)

[83] R. Srama and E. Grün: *The dust sensor for CASSINI* Adv. Space Res. **20**, (8) 1467 (1997)

[84] T. Stephan: *TOF-SIMS in Cosmochemistry*, Planet. Space Sci. **49**, 859 (2001)

[85] T. Stephan: *TOF-SIMS analysis of heavy-nitrogen-carrying phases in interplanetary dust*, 33rd Lunar Planet. Sci. Conf. #1352

[86] D. Tsintikidis, D. Gurnett, L.J. Granroth, S.C. Allendorf, and W.S. Kurth: *A revised analysis of micron-sized particles detected near Saturn by the Voyager 2 plasma wave instrument*, J. Geophys. Res. **99**, (A2) 2261 (1994)

[87] B.T. Tsurutani, K. Leschly, S. Nikzad, E.R. Fossum, G. Murphy, T.L. Killeen, B.C.Kennedy, A.M. Title, D.L. Chenette, G. Musmann, F. Gliem, S.L. Moses, I. Mann, and A.J. Tuzzolino: *An Integrated Space Physics Instrument (ISPI) for Solar Probe*, Space Technology and Applications International Forum 1997, ed. by M.S. El-Genk (American Institute of Physics Press, New York) AIP Conference Proceedings **387**, 131 (1997)

[88] A.J. Tuzzolino, R.B. McKibben, J.A. Simpson, S. BenZvi, H.D. Voss, and H. Gursky: *The Space Dust (SPADUS) instrument aboard the Earth-orbiting ARGOS spacecraft: I-instrument description*, Planet. Space Sci. **49**, (7) 689 (2001a)

[89] A.J. Tuzzolino, R.B. McKibben, J.A. Simpson, S. BenZvi, H.D. Voss, and H. Gursky: *The Space Dust (SPADUS) instrument aboard the Earth-*

orbiting ARGOS spacecraft: II-results from the first 16 months of flight, Planet. Space Sci. **49**, (7) 705 (2001b)

[90] N.G. Utterback and J. Kissel: *Attogram Dust Cloud a Million Kilometers from Comet Halley,* Astron. J. **100**, 1315 (1990)

[91] K. Weiss-Wrana: *Optical properties of interplanetary dust - Comparison with light scattering by larger meteoritic and terrestrial grains,* Astron. Astrophys. **126**, (2) 240 (1983)

[92] A. Wehry and I. Mann: *Identification of β–meteoroids from measurements of the dust detector onboard the ULYSSES spacecraft,* Astron. Astrophys. **341**, 296 (1999)

[93] K.P. Wenzel, R.G. Marsden, D.E. Page, and E.J. Smith: *The ULYSSES Mission,* Astron. Astrophys. **92**, Suppl., 207 (1992)

[94] M. Wilck and I. Mann: *Radiation Pressure Forces on Typical Interplanetary Dust Grains,* Planet. Space Sci. **44**, 493 (1996)

[95] M. Witte, H. Rosenbauer, M. Banaszkiewicz, and H. Fahr: *The ULYSSES neutral gas experiment - Determination of the velocity and temperature of the interstellar neutral helium* Adv. Space Res. **13**, (6) 121 (1993)

[96] H. Zscheeg, J. Kissel, G. Natour, and E. Vollmer: *CoMA - an advanced space experiment for in situ analysis of cometary matter,* Astrophys. Space Sci. **195**, 447 (1992)

The Astromineralogy
of Interplanetary Dust Particles

John Bradley

Institute for Geophysics and Planetary Physics, Lawrence Livermore National Laboratory, Livermore, CA 94551, USA

Abstract. Some chondritic interplanetary dust particles (IDPs) collected in the stratosphere are from comets. Because comets accreted at heliocentric distances beyond the giant planets, presolar grains or "astrominerals" both with solar and non-solar isotopic compositions are expected to be even more abundant in cometary IDPs than in primitive meteorites. Non-solar D/H and $^{15}N/^{14}N$ isotopic enrichments in chondritic IDPs are associated with a carbonaceous carrier. These H and N enrichments are attributed to extreme mass fractionation during chemical reactions in cold (10–100 K), dense interstellar molecular clouds. Nano-diamonds appear to be systematically depleted or even absent in some IDPs suggesting that some meteoritic nano-diamonds may not be (presolar) astrominerals. Enstatite ($MgSiO_3$) and forsterite (Mg_2SiO_4) crystals in IDPs are physically and compositionally similar to enstatite and forsterite grains detected around young and old stars by the Infrared Space Observatory (ISO), and large non-solar oxygen isotopic compositions recently measured in an IDP forsterite establish that they are presolar circumstellar silicates. The compositions, mineralogy, and optical properties of GEMS are consistent with those of interstellar amorphous silicates. Submicrometer FeNi sulfide astrominerals like those found in IDPs may be responsible for a broad $\sim 23.5\mu$m feature observed around protostars and protoplanetary discs by ISO. The first returned samples of contemporary interstellar dust as well as dust from comet Wild-2 will be returned to Earth in 2006 by the STARDUST mission, providing a mother lode of astrominerals for future laboratory investigations.

1 Introduction

For the purposes of this chapter astrominerals are defined as those materials of presolar origin that are found in primitive meteoritic materials that existed prior to collapse of the solar nebula. These grains may include preserved circumstellar grains, grains formed in supernovae outflows, grains formed or modified within the interstellar medium, including grains that were present in the presolar molecular cloud. They may have survived either because they were resistant to asteroidal parent body alteration processes within the inner solar system or they were sequestered in small primitive bodies like comets at much larger heliocentric distance where post-accretional alteration processes were minimal or essentially non existent. Astrominerals can be found in chemically primitive meteorites (e.g. carbonaceous chondrites), polar micrometeorites, and interplanetary dust particles (IDPs).

The carbonaceous chondrites are believed to be derived from the inner main belt of the asteroids located at a heliocentric distance of 2.5–3 astronomical units (AU) [1]. Their mineralogy and petrography indicate that most have experienced significant parent body alteration and that they are composed almost exclusively of minerals formed within the solar system. However, the discovery of trace quantities of surviving presolar grains within the matrices of carbonaceous chondrites shows that at least some astrominerals survived [2, 3]. They are indisputably presolar circumstellar astrominerals because their isotopic compositions differ significantly from solar system compositions. Other presolar interstellar astrominerals may have normal (solar) isotopic compositions [4]. Astrominerals so far identified in meteorites include diamond, silicon carbide (SiC), titanium carbide (TiC), graphite, corundum, spinel, and silicon nitride [5]. Although none of these astrominerals are cosmically abundant, the presence of some of them (e.g. SiC and TiC) in circumstellar environments has been confirmed by astronomical observations [6, 7, 8].

All of the astrominerals so far identified in carbonaceous chondrites are highly refractory and their abundance decreases with increasing petrologic grade of meteorite [3, 5, 9]. Thus it appears that only the most robust astrominerals have survived incorporation into the parent bodies of meteorites and, to date, not a single silicate astromineral has been identified in any meteorite. (Strong resonances in astronomical infrared (IR) spectra at ~ 10 and $\sim 18\mu m$ corresponding to the Si-O stretch and Si-O-Si bending mode vibrations in silicates are observed both in absorption and emission along most lines-of-sight [10]). Until recently, failure to find what is arguably the most cosmically abundant type of astromineral ranked among the most perplexing problems confronting meteoriticists and planetary scientists. The most likely explanation for the apparent absence of presolar silicates in meteorites is that they are simply not robust enough to have survived parent body aqueous alteration and thermal metamorphism.

Astrominerals are expected to be more abundant in interplanetary dust particles (IDPs) because IDPs are more chemically and isotopically primitive than carbonaceous chondrites and some of them are from comets [11, 12, 13, 14]. Since comets are small bodies located at extreme heliocentric distances they have likely undergone less alteration than asteroids [15, 16, 17]. High D/H enrichments indicate that some IDPs contain organic matter from a presolar molecular cloud [12, 18]. The first hint of surviving silicate astrominerals in an IDP was provided by the observation of a significant ^{16}O enrichment in a CP IDP, the largest "whole rock" ^{16}O enrichment ever observed in any meteoritic material [19]. (Silicates are the major carrier of O in IDPs). Stunning confirmation of the presence of silicate astrominerals in IDPs was recently provided by the discovery of a submicrometer forsterite (Mg_2SiO_4) grain with O isotopic compositions clearly indicating that they are presolar astrominerals [20].

This chapter discusses the astromineralogy of the anhydrous "chondritic porous" (CP) subset of IDPs. Measured helium release temperatures indicate that many of these IDPs enter the atmosphere at speeds > 16 km sec^{-1} and are therefore probably of cometary origin [21]. IDPs entering the atmosphere at speeds < 16 km sec^{-1} are more likely of asteroidal origin, although some low-speed (< 16 km sec^{-1}) IDPs may also be cometary [22, 23]. Although all IDPs

Fig. 1. Secondary electron image of chondritic porous (CP) interplanetary dust particle (IDP) U220B43.

are pulse heated during atmospheric entry, the survival of solar flare tracks and low-temperature minerals indicates that many of them survive with surprisingly little thermal alteration [24, 25, 26]. Studies of IDPs are providing exciting new insight about grain formation, processing, and destruction in space, and comparison of their properties with those of dust in space is a key aspect of the new discipline of astromineralogy.

2   Specimen Preparation and Analytical Methods

Typical CP IDPs are 5–20 μm in diameter (Fig. 1). They are composed of heterogeneous aggregates of submicrometer-to-nanometer sized crystalline and amorphous grains (Figs. 2-4). Analytical studies of IDPs typically require the use of micromanipulators, embedding and thin-sectioning equipment (ultramicrotomes), and analytical instruments offering both high spatial resolution and high sensitivity. Ultramicrotomy is used for preparing electron and optically transparent thin sections of IDPs [27]. Epoxy is the embedding medium of choice but, as ever more challenging analyses are being attempted (e.g. detection of C, N, and organics in IDPs) other embedding media like sulfur are increasingly utilized [28]. Embedding small particles in sulfur is difficult because (molten) sulfur has a much higher surface tension than epoxy. However, it offers a huge advantage in that it can be completely eliminated (by sublimation) after the thin section has been transferred to a TEM grid. In the absence of epoxy, spectral

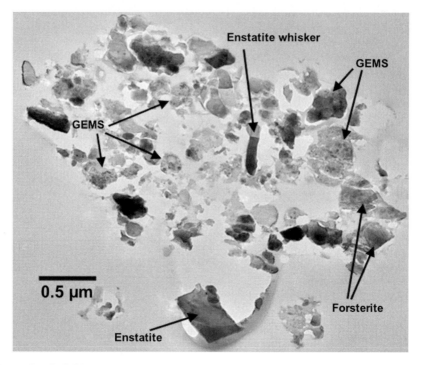

Fig. 2. Brightfield transmission electron micrograph of a thin section of cluster particle U219C3. The three major types of silicate grains found in CP IDPs are GEMS, enstatite, and forsterite.

backgrounds are reduced making it possible to begin to analyze smaller samples and investigate the nature of indigenous C, N, and organic species.

An in-situ acid "nano-etching" procedure has been developed where all minerals within an ultramicrotomed thin-section of an IDP except sulfides and carbonaceous minerals are removed [29]. This procedure greatly simplifies the search for acid-resistant astrominerals like nano-diamonds that are otherwise impossible to recognize among other more abundant nanometer-sized minerals (e.g. FeNi metal). The dissolution is carried by exposing the (gold) TEM grid to triple distilled concentrated hydrofluoric acid (HF). After a room temperature etch the grid is rinsed with water and dried. After etching, the slices are characterized in the TEM (Fig. 5).

The principal instruments used to analyze IDPs are the (analytical) transmission electron microscope (TEM), ion microprobe, and infrared micro spectrophotometer. A variety of other analytical methods are also applied to IDPs [30, 31, 32]. Emerging analytical techniques include scanning transmission x-ray microprobe (STYXM) and x-ray absorption near edge structure (XANES), both of which use a synchrotron radiation source [33, 34, 35]. The TEM offers both high image resolution (0.2–0.5 nm) and, in conjunction with focused nanoprobes, high compositional spatial resolution (1–5 nm) using energy-dispersive x-ray

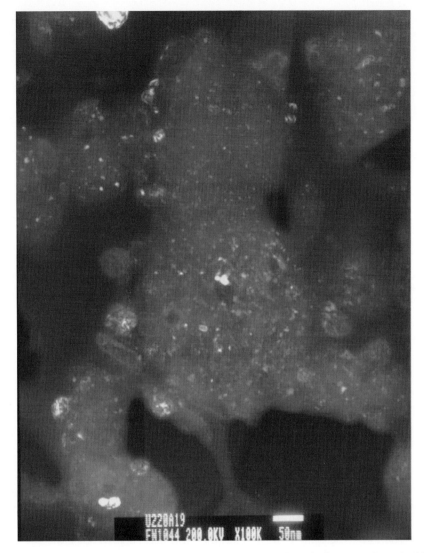

Fig. 3. Darkfield transmission electron micrograph of GEMS (glass with embedded metal and sulfides) in IDP U220A19. The bright inclusions are FeNi metal (see also Fig. 4) and FeNi sulfide crystals that are embedded in Mg-rich silicate glass (uniform grey matrix).

analysis (EDS) and electron energy-loss spectroscopy (EELS). State-of-the-art TEMs are now increasingly configured for compositional mapping of electron transparent thin sections with spatial resolution on a scale of 1-3 nm using EDS and EELS. In the future such capabilities may play a key role in the identification of astrominerals through compositional mapping of the ultrafine-grained matrices of IDPs at a whole new level of detail revealing nanoscale petrographic

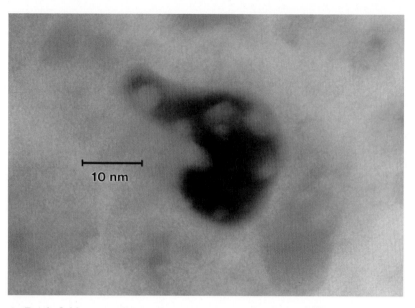

Fig. 4. Brightfield transmission electron micrograph of an FeNi metal grain within GEMS in IDP U220A19. The "polished" morphology and "swiss cheese" microstructure likely result from prolonged exposure to radiation sputtering.

information that has likely been overlooked using conventional brightfield and darkfield imaging techniques.

The isotopic compositions of IDPs are measured using secondary ion mass spectroscopy (SIMS). Most measurements have been made on whole particles pressed into gold foils [12, 36], as well as individual mineral grains (enstatite and GEMS) among fragments of an IDP that were crushed and dispersed on a gold substrate [37]. Recently the first in-situ measurements of submicrometer-sized grains within thin sections of an IDP mounted on a TEM grid were made using the new nanoSIMS [20]. These measurements represent a major breakthrough because they demonstrate that it is possible to first probe the compositions, mineralogy, and petrography of an IDP using TEM and then transfer the specimen to the nanoSIMS for complimentary high-spatial resolution isotope measurements. Already anomalies in H, N, and O have been observed in IDPs using the nanoSIMS in some cases at levels far above those previously reported. The discovery of preserved circumstellar silicates in IDPs underscores the vital role nanoSIMS will play in future IDP research [20].

The optical properties of IDPs have been measured in the infrared, visible, and ultraviolet spectral regions [38, 39, 40, 41]. Most have been IR transmission measurements in the 2–25 μm IR region using microscope spectrophotometers equipped with globar sources [38, 39]. More recently, the high-brightness synchrotron light source has proven to be ideal for spectral microanalysis of IDPs and even subcomponents of IDPs because it provides a $\sim 3\mu$m diameter IR beam spot > 100X brighter than laboratory IR spectrophotometers (equipped with

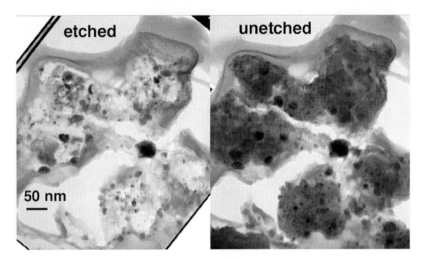

Fig. 5. A GEMS-rich thin section before and after in-situ acid etching. In addition to preserving carbonaceous phases (e.g. organics and nano-diamonds) FeNi sulfides (dark grains) are also preserved such that their nanoscale compositions and crystal structures can be measured without interference from other nanophase minerals (e.g. FeNi metal, see Figs. 3,4)). See [29].

glow-bar discharge sources) [42]. At present the synchrotron facility is equipped with a detector that covers the 2–25 μm wavelength range but optics extending the spectral range to 50 μm and longer wavelengths are planned. In principle, we will be able to directly compare IDPs with astronomical silicates detected by ISO over 2–100 μm wavelength range.

Reflectance spectra have been collected from IDPs over the visible 450 – 800 nm wavelength range [40]. In general chondritic IDPs are spectrally dark objects ($<$ 15% reflectivity). Hydrated IDPs that contain layer lattice silicates exhibit spectral characteristics similar to carbonaceous chondrites and main-belt C-type asteroids. Some hydrated IDPs are believed to be of asteroidal origin [26, 43, 44]. In contrast, most anhydrous CP IDPs dominated by enstatite, forsterite, and GEMS exhibit spectral characteristics similar to those of smaller, more primitive solar system objects (e.g. P and D asteroids). Carbon-rich IDPs are spectrally red with a redness comparable to the comet-like outer asteroid Pholus [40, 45].

3   Astrominerals in IDPs

3.1   Organic Matter

The bulk carbon content of chondritic IDPs is typically 2–5X higher than carbonaceous chondrites [46]. The carbon is predominantly amorphous and partially graphitic carbon exhibiting characteristic $\sim$ 3.4Å basal spacings that is common in carbonaceous chondrites is conspicuously rare in IDPs. The spectral redness of carbon-rich IDPs suggest that they contain organic carbon, which is confirmed

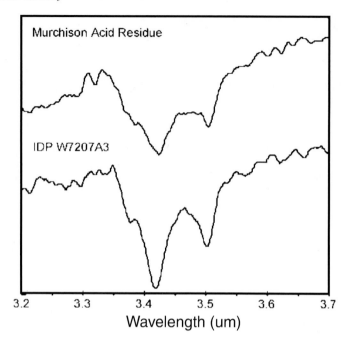

Fig. 6. Comparison of the $\sim 3.4\mu$m C-H stretch features from Murchison acid residue and a thin section of IDP W7207A3 following in-situ acid dissolution [29].

by IR spectra that exhibit a pronounced organic C-H stretch feature at $\sim 3.4\mu$m (Fig. 6) [29].

Some IDPs contain D/H and ^{15}N/^{14}N isotopic enrichments that are associated with an organic carrier [47]. The enrichments are attributed to isotope fractionation in cold (10–100 K) dense interstellar molecular clouds where extreme mass fractionation of H and N can occur during chemical reactions. Astronomical measurements of cold clouds show D/H ratios of $10^{-2} - 10^{-1}$ in the coldest clouds, two to three orders of magnitude higher than the solar value ($\sim 10^{-4}$). Much smaller D/H and ^{15}N/^{14}N enrichments observed in meteorites have long been attributed to the partial survival of presolar molecular cloud material that has been diluted and significantly altered by asteroidal parent body processes. However, in some IDPs the magnitudes of the enrichments are far larger and more common, in some cases with D/H values approaching those of molecular clouds, suggesting the complete preservation of molecular cloud material [12, 47].

3.2 Nano-diamonds

Nano-diamonds are the most abundant type of presolar grains in chondritic meteorites (by over two orders of magnitude) where their average abundance is $\sim$400 ppm [48]. Most nano-diamonds are between 1 and 10 nm in diameter with a lognormal size distribution and an average diameter of $\sim$3 nm [49]. Although the average ^{13}C/^{12}C composition of the nano-diamonds is normal (solar), they are widely considered to be presolar astrominerals because of their association of a noble gas carrier, specifically an anomalous Xe-HL component that is consistent with supernovae processes [48]. However, the abundance of Xe is so low that only one nano-diamond in a million contains a Xe atom. Therefore, while at least some nano-diamonds are likely presolar, it is an unclear whether most nano-diamonds in meteorites originated in the solar system or presolar environments [5]. In other words, it is not certain that all of the nano-diamonds in meteorites are astrominerals.

Since chondritic IDPs contain 2–5 times more carbon than carbonaceous chondrites [46] it is possible that they contain even higher abundances of nano-diamonds. Moreover, the abundance of nano-diamonds in IDPs may shed light on their origins for the following reason. If nano-diamonds are presolar astrominerals their abundance is expected to be higher in cometary IDPs because comets originate at extreme heliocentric distances. If nano-diamonds formed in the solar system they are expected to be more abundant in meteorites, asteroidal micrometeorites, and asteroidal IDPs because they originate within the inner solar system.

Nano-diamonds in meteorites are isolated from gram-mass samples by a multi-step chemical (acid digestion) process that has been described as "burning down the haystack to find the needle" [2]. Sub-nanogram mass thin sections of IDPs are acid etched in-situ using the simplified procedure described in the previous section [29]. Thin sections of fine-grained matrix from the Murchison (CM) and Orgueil (CI) meteorites, two "CM-like" polar micrometeorites, and nine chondritic IDPs were subjected to the in-situ procedure. Nano-diamonds were identified in the meteorite and micrometeorite thin sections and in only four of the nine IDPs (Fig. 7). One of the nano-diamond-containing IDPs (RB12A44-3) is mineralogically similar to CM meteorites and is likely an asteroidal particle [26], while the others are so-called "giant cluster particles". It has been suggested that cluster particles are from asteroids [50], although their mineralogical similarity to cometary IDPs means that they could be derived from comets, asteroids, or both. The other five non-cluster particles are typical of cometary IDPs although their atmospheric entry speeds have not been determined. In any case, the apparent absence or depletion of nano-diamonds in the five non-cluster IDPs appears to challenge the widely held view that nano-diamonds in meteorites are presolar astrominerals.

3.3   Silicates

Silicates have long been known to be an important constituent of dust in circumstellar environments, the interstellar medium, and solar system comets. The structure of the astronomical 10 and 20 μm Si-O stretch and Si-O-Si bending features indicate that in general these silicate astrominerals are almost exclusively amorphous (glassy) [10, 51]. This canonical picture of astronomical silicates changed radically when the Infrared Space Observatory (ISO) detected ubiquitous submicrometer crystalline forsterite (Mg_2SiO_4) and enstatite ($MgSiO_3$)

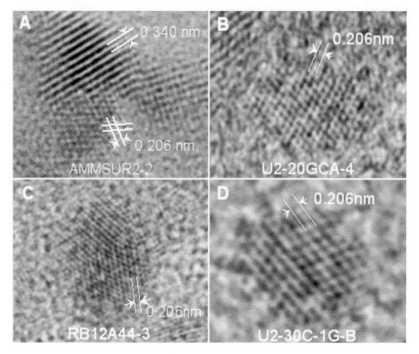

Fig. 7. Lattice-fringe images of nano-diamonds (0.206 nm spacings) in Antarctica micrometeorite AMMSRU2-2 and IDPs U2-20-GCA, RB12A44,and U2-30C-1G-B. See [89].

around young and old stars and in comet Hale-Bopp [52, 53, 54]. The short-wavelength and long-wavelength spectrometers on ISO provided coverage of the mid IR region where the ~ 10 and $\sim 20\mu$m "silicate" features are observed and, for the first time, the 20–100μm region where features due to crystalline silicate minerals occur. Another significant finding is the similarity between the silicate mineralogy of comet Hale-Bopp's dust and dust around the isolated Herbig Ae/Be star HD100546 [16, 17, 55, 56].

Silicate astrominerals were undoubtedly a major component of the dust from which the solar system was formed. However, as discussed above not a single (isotopically anomalous) presolar silicate grain has been unambiguously identified in a meteorite. ISO provides information useful in the search for crystalline silicate astrominerals in primitive meteoritic materials. The discovery that crystalline circumstellar silicates are Mg-rich predominantly submicrometer-sized grains greatly simplifies distinguishing presolar silicates from the huge "background" of solar system Mg-Fe silicates in primitive meteoritic materials [52]. There are rare Mg-rich forsterite (and enstatite) grains in chondritic meteorites, some of them exhibiting morphological, compositional, or crystallographic evidence of vapor phase growth, that appear to have survived since the earliest stages of solar system formation [57, 58]. Some "relict" forsterite grains were

among the earliest high-temperature solids in the solar nebula and they exhibit
^{16}O enrichments (to 50 parts per thousand) consistent with formation from or
mass exchange with an (^{16}O-enriched) reservoir of grains and/or gas [59]. However, most of these grains are much larger than those detected by ISO.

Whereas submicrometer forsterite and enstatite crystals (like those detected
by ISO) are exceedingly rare in meteorites and micrometeorites, they are the
most abundant crystalline silicates in the fluffy chondritic porous class IDPs
[60, 61]. Grain sizes range from 0.05–5 μm, although most grains are between
0.1 and 0.75 μm (100–750 nm) in diameter (Fig. 2). The crystalline silicates
detected by ISO presumably form by vapor phase growth in stellar outflows,
and there is crystallographic evidence that some enstatite grains in IDPs formed
by direct gas-to-grain condensation [62]. These crystals exhibit highly unusual
whisker and ultrathin platelet morphologies and internal crystallographic defects
(axial screw dislocations) that are typical of crystals grown from the vapor phase
(Fig. 2). There is also compositional evidence that forsterite and enstatite grains
in IDPs grew by direct gas-to-grain condensation from a nebula gas [63]. Low Fe,
Mn enriched (LIME) enstatite and forsterite grains in IDPs contain up to 5 wt.
% MnO, in contrast to the majority of pyroxenes and olivines in meteorites which
contain < 0.5 % MnO. It is likely that the high Mn contents reflect condensation
from a nebula gas. A first attempt to measure the Mg isotopic composition of a
single ∼ 5μm long enstatite whisker in a CP IDP failed to reveal a significant
isotope anomaly [37].

ISO provides little information about the nature of amorphous silicate astrominerals in primitive meteoritic materials. Rare amorphous silicates (glasses) in
primitive meteorites are believed to have formed in the solar system. In contrast,
IDPs contain abundant glassy silicates known as GEMS (glass with embedded
metal and sulfides) as well as other glassy grains [42, 64, 65]. They are 0.1–0.5 μm
glassy spheroids with nanometer-sized FeNi metal and Fe-rich sulfide inclusions.
They have aroused much interdisciplinary interest because their properties are
similar to the exotic properties that astronomers have inferred for interstellar
"amorphous silicates" [66, 67, 68]. If they are indeed interstellar silicates one of
the major building blocks of the solar system has been found since, with the
exception of carbon, most of the condensable atoms in the solar system were
likely carried within amorphous silicate astrominerals prior to the collapse of
the solar nebula.

Although the compositions and mineralogy of GEMS are consistent with
interstellar amorphous silicates, rigorously proving that they are presolar requires measurement of non-solar isotope abundances. An attempt to detect (Mg)
isotope anomalies in individual GEMS was unsuccessful [37], although a high-
precision bulk oxygen isotope measurement of a GEMS-rich IDP yielded the
highest "whole rock" ^{16}O enrichment yet measured in any chondritic object [19].
Since silicates are the major carrier of oxygen in IDPs, the bulk measurements
provided the first hint that GEMS-rich IDPs contain presolar silicates.

The petrography of GEMS links them to presolar astrominerals. Using electron energy- loss spectroscopy, it was shown that nitrogen in a ^{15}N-enriched IDP

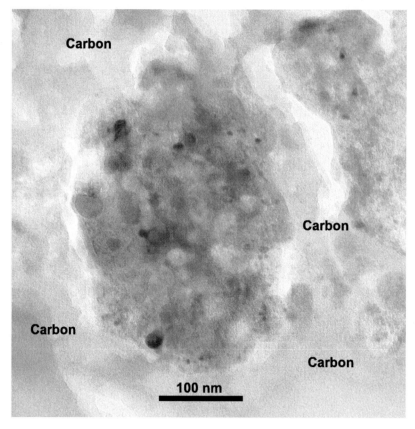

Fig. 8. Brightfield transmission electron micrograph of a GEMS embedded a amorphous carbon material in IDP L2011R11. The remnants of a deeply eroded relict grain with "swiss cheese" microstructure can be seen towards the center of the grain.

is localized within a carbonaceous phase that mantles and encapsulates GEMS (Fig. 8) [36]. It is possible that the (isotopically anomalous) mantles were deposited on the GEMS in a presolar molecular cloud environment. Similarly, a deuterium "hot spot", first identified within an IDP pressed into a gold ion microprobe mount, was subsequently thin-sectioned using ultramicrotomy and examined using TEM. The "hotspot" is dominated by poorly ordered organic carbonaceous material with embedded GEMS and Fe-rich sulfides [69].

The infrared spectral properties of GEMS provide further insight about their origin(s). GEMS-rich thin sections of IDPs, i.e. sections with GEMS plus submicrometer enstatite and forsterite crystals, exhibit a $\sim 10\mu m$ silicate feature similar to solar system comets and dust around some Herbig Ae/Be stars [42]. Some essentially pure GEMS exhibit an Si-O stretch feature with an absorption minimum at $\sim 9.7\mu m$, a bandwidth of $\sim 3.5\mu m$ (FWHM), and a pronounced excess or asymmetry on the long wavelength side of the feature. Interstellar silicates also exhibit an Si-O stretch feature with an absorption minimum at

$\sim 9.7\mu\text{m}$, a bandwidth of 2.5–3.5 μm (FWHM), and a pronounced excess or asymmetry on the long wavelength side of the feature. These GEMS are the first identified natural silicate to provide such a match, adding to the exotic set of compositional and mineralogical properties required to match those of interstellar amorphous silicates [42, 66, 67, 68]. However, preliminary nanoSIMS measurements of several GEMS show that their O isotope compositions are normal (solar). This may mean that GEMS are not presolar grains. Alternatively, they are indeed presolar grains but their isotopic compositions, like their chemical compositions and structures have been homogenized by prolonged exposure to irradiation and other erosional phenomena during their $\sim 10^8$ lifetimes in the interstellar medium. The isotopic compositions of relict forsterite or enstatite crystals that are found in some GEMS may clarify their origin(s) [60].

3.4   FeNi Sulfides

FeNi sulfide grains are the major carrier of sulfur in primitive chondritic meteorites and chondritic IDPs [70, 71, 72]. They were also identified as a component of comet Halley's dust [73, 74]. Although they are a crystallographically complex group of minerals [75, 76], sulfides were one of the first minerals identified in IDPs [77, 78, 79]. TEM studies have shown that they span an enormous size range $(10^{-1} - 10^4$ nm) [11, 71, 80]. Grain morphologies include irregular, hexagonal, rectangular,and rounded shapes, and a variety of different compositions and structures may coexist within a single IDP. Most sulfide grains in IDPs are low-Ni pyrrhotite [78]. Pentlandite, troilite, and sphalerite (ZnS) have also been reported [71, 72, 81].

Recently, an FeNi sulfide with a "spinel-like" cubic structure was identified in chondritic IDPs [80]. The composition of this sulfide is similar to those of pyrrhotite and pentlandite, and it is commonly closely associated with hexagonal pyrrhotite. Electron diffraction patterns and high-resolution lattice-fringe images indicate that both minerals are sometimes coherently intergrown on a unit cell scale but, when the spinel sulfide is mildly heated in the electron beam, it transforms into hexagonal pyrrhotite (Fig. 9). Therefore, hexagonal pyrrhotite, the dominant sulfide in IDPs, may be a secondary thermal alteration product of the spinel-like sulfide. All IDPs are pulse heated above 500° C during atmospheric entry [81].

The spinel-like sulfide shows that the sulfide mineralogy of CP IDPs is significantly different from that of the primitive chondritic meteorites. Therefore, it is possible that some or all of the sulfides in IDPs formed in a different kind of environment. A crystallographically similar sulfide was synthesized by low-temperature (< 200° C), low-pressure vapor phase growth [82]. There are other indications that the sulfide mineralogy of IDPs differs from that of other meteoritic sulfides. In a recent comparative study of sulfides within IDPs and meteorites showed selenium levels in IDPs are 60% higher than those in meteoritic sulfides, implying that "the IDP sulfides are likely to have formed in a different environment than the (Orgueil) meteoritic pyrrhotite" [83].

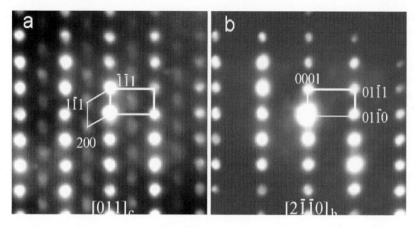

Fig. 9. Time-resolved electron micro-diffraction patterns taken from an FeNi sulfide grain in IDP W7020A-8D. In pattern (a), the reflections with strong intensity are indexed as simple hexagonal pyrrhotite (thick line box). The extra weak reflections (thin line box) are indexed as cubic spinel sulfide. After illumination in the electron beam for several tens seconds, the weak cubic spinel reflections disappear. See [80].

FeNi sulfides in IDPs may be highly relevant to question of astromineralogy. We have proposed that sulfides are responsible for a broad $\sim 23\mu$m feature recently detected around young and old stars by the Infrared Space Observatory (ISO) [84]. Laboratory spectra were obtained from pyrrhotite ([Fe,Ni]$_{1-x}$S) grains in IDPs, as well as pyrrhotite and troilite (FeS) mineral standards using a high-brightness synchrotron light source and they were compared with astronomical spectra from the Infrared Space Observatory (ISO). All of the sulfides produce a broad Fe-S stretch feature centered at $\sim 23.5\mu$m and a similar broad feature centered $\sim 23.5\mu$m is observed in the ISO spectra. Thus Fe sulfide grains may be an important, previously unrecognized component of circumstellar dust. Although S is not significantly depleted from the gas phase in the interstellar medium, it is highly depleted in cold, dense interstellar molecular clouds indicating that it is in solid grains [85, 86, 87].

4   Conclusions

The focus of this chapter has been the anhydrous "chondritic porous" (CP) subset of IDPs, some or all of which are likely of cometary origin [21]. Despite the technical difficulties inherent in analyzing tiny nanogram-mass IDPs a large database of compositional, mineralogical, and isotopic data have been acquired. These data establish that IDPs differ fundamentally from even the most primitive meteorites and micrometeorites, leading to the conclusion that "there is a very real possibility that some CP IDPs are well-preserved aggregates of (presolar) circumstellar and interstellar materials " [88]. This conclusion was strengthened by the detection of a significant bulk ^{16}O enrichment in an CP IDP [19], followed by the detection using the new nanoSIMS instrument of multiple silicate grains in several IDPs that exhibit large O non-solar isotopic anomalies [20].

Although isotope measurements will undoubtedly play a dominant role in identifying new presolar astrominerals in IDPs, the limitations of using isotopic composition to recognize presolar astrominerals need to be acknowledged. A non-solar isotopic composition is not always proof of a non-solar origin. For example, the large D/H and $^{15}N/^{14}N$ excesses observed in some IDPs, widely assumed to indicate the preservation of carbonanceous astrominerals [12, 18, 47], could equally have been produced by similar low temperature chemical reactions in the outer regions of the solar nebula. It has been widely accepted that nano-diamonds are presolar even though their $^{13}C/^{12}C$ isotopic compositions are normal (solar) [2, 5, 9]. The apparent depletion or absence of nano-diamonds in some IDPs suggests that many of them may have formed within the solar system [89]. Although some presolar grains clearly have non-solar isotopic compositions there is evidence that typical interstellar grains have compositions similar to those of solar system materials [4]. R. M. Walker best articulated the dilemma over the significance of isotopic composition: .... "pre-existing interstellar grains might not, on the average, be very different from solar-system material. Although there is strong evidence that some circumstellar grains with distinctive isotopic signatures have survived intact in meteorites, theoretical calculations indicate that most grains are quickly destroyed in the interstellar medium. Thus the grains found in a protostellar dust cloud may themselves consist of interstellar grains whose compositions have been homogenized in the interstellar medium to give compositions similar to solar system values" [4].

Interplanetary dust research may be on the verge of an exciting new era. Techniques for sample micromanipulation have been highly refined, optical and electron-transparent thin-sections are now routinely produced, the compositions, mineralogy and petrography of individual subcomponents of IDPs are determined with sub-nanometer scale resolution, and the isotopic compositions of grains $\sim$ 100nm diameter can now be measured. Synchrotron light sources are being used to measure the optical properties of (thin sections of) IDPs and they are directly compared with those of dust in cometary, interstellar, and circumstellar environments. The STARDUST spacecraft recently collected contemporary interstellar dust. This dust, together with dust grains from comet Wild-2, will be returned to earth in 2006. The returned samples will provide ground truth testing of our collective hypotheses (and prejudices) concerning the nature of astrominerals.

Acknowledgements

This research is supported by NASA grants NAG5-7450 and NAG5-9797.

References

[1] G. W. Wetherill, C. R. Chapman: "Asteroids and meteorites". In: Meteorites and the Early Solar System, ed. by J.F. Kerridge, M.S. Matthews

(University of Arizona Press, Tucson, 1988), pp 35-70.

[2] T.J. Bernatowicz, R. M. Walker: Physics Today, 50, 26 (1997).

[3] U. Ott: this volume.

[4] J. P. Bradley, S. Sandford, R. M. Walker: "Interplanetary Dust Particles," In: Meteorites and the Early Solar System, eds. J. Kerridge & M.S. Mathews (Univ. Arizona Press, Tucson, 1988) pp. 861-895.

[5] P. Hoppe, E. Zinner: J. Geophys. Res. 105, 10,371 (2000).

[6] A. C. Andersen, J. Jäger, H. Mutsche, A. Braatz, D. Clément, Th. Henning, U. G. Jorgensen, U. Ott: Astron. Astrophys. 343, 933, (1999).

[7] H. Mutschke, A. C. Anderson, D. Clément, Th. Henning: Astron. Astrophys. 345, 187, (1999).

[8] G. von Helden, A.G.G.M. Tielens, D. van Heijnsbergen, et al: Science 288, 313, (2000).

[9] E. Anders, E. Zinner: Meteoritics 28, 490 (1993).

[10] J. S. Mathis: Rep. Prog. Phys. 56, 605 (1993).

[11] J. P. Bradley: Geochim. Cosmochim. Acta 52, 889 (1988).

[12] S. Messenger: Nature 404, 968 (2000).

[13] L. S. Schramm, D. E. Brownlee, M. M. Wheelock: Meteoritics 24, 99 (1989).

[14] P. Arndt, J. Bohsung, M. Maetz, E. K. Jessberger: Met. Planet. Sci. 31, 817 (1996).

[15] Greenberg, J. M. (1982) "What are comets made of?". In: Comets, ed. by L. L. Wilkening (Univ. Arizona Press, Tucson, 1982) pp 131-162.

[16] D. H. Wooden, D. E. Harker, C. E. Woodward, H. M. Butner, C. Koike, F. C. Witteborn, C. W. McMurty: Astrophys. J. 517, 1034 (1999).

[17] M. Hanner: this volume.

[18] S. J. Clemett, C. R. Maechling, R. N. Zare, P. D. Swan, R. M. Walker: Science 262, 721 (1993).

[19] C. Engrand, K. D. McKeegan, L. A. Leshin, J. P. Bradley, D. E. Brownlee: Lunar Planet. Sci. XXX, 1690-1691 (1999).

[20] S. Messenger, L. P. Keller, R. M. Walker: Lunar Planet. Sci. XXXIII, 1887 (2002).

[21] D. E. Brownlee, D. J. Joswiak, D. J. Schlutter, R. O. Pepin, J. P. Bradley, S. G. Love: Lunar Planet Sci. XXVI, 183 (1995).

[22] G. J. Flynn: "Sources of 10 m interplanetary dust: possible importance of the Kuiper belt" . In: Physics, Chemistry, and Dynamics of Interplanetary Dust, ed. by Bo. S. Gustavson, M. S. Hanner (ASP Conference Series, Vol. 104, 1996) pp 274-282.

[23] J. C. Liou, H. A. Zook: Lunar Planet. Sci. XXVII, 763(1996).

[24] J. P. Bradley, D. E. Brownlee, P. Fraundorf: Science 226, 1432 (1984).

[25] S. A. Sandford, J. P. Bradley: Icarus 82, 146(1989).

[26] J. P. Bradley, D. E. Brownlee: Science 251, 549 (1992).

[27] J. P. Bradley, D. E. Brownlee: Science 231, 1542 (1986).

[28] J. P., Bradley, L. P. Keller, K. L. Thomas, T.B. VanderWood, D. E. Brownlee: Lunar Planet. Sci. XXIV, 173 (1993).

[29] D. E. Brownlee, D. J. Joswiak, J. P. Bradley, J. C. Gezo, H. G. M. Hill: Lunar Planet. Sci. XXXI, 1921 (2000).

[30] M. E. Zolensky, T. L. Wilson, F. J. M. Rietmeijer, G. J. Flynn (eds): Analysis of Interplanetary Dust. AIP Conf. Proc. 310. (Am. Inst. Phys. New York,1994) p357.

[31] M. E. Zolensky, C. Pieters, B. Clark, J. J. Papike: Met. Planet. Sci. 35, 9 (2000).

[32] E. Grün, B. Å. S. Gustavson, S. F. Dermott, H. Fechtig (eds.): Interplanetary Dust (Springer, New York, 2000) p804.

[33] S. R. Sutton, G. J. Flynn, M. Rivers, P. Eng, M. Newville, G. Shea-McCarthy, A. Lanzirotti: Met. Planet. Sci. 34, A113 (1999).

[34] G. J. Flynn, L. P. Keller, C. Jacobsen, S. Wirick: Lunar Planet Sci. XXI, LPI Houston CD-ROM, Abstract #1904 (2000).

[35] G.J. Flynn, L. P. Keller, C. Jacobsen, S. Wirick: Met. Planet. Sci: 34, A36 (1999).

[36] L. P. Keller, S. Messenger, J. P. Bradley: J. Geophys. Res: 165, 10,397 (2000).

[37] J. P. Bradley, T. Ireland: "The search for interstellar components in interplanetary dust particles", In: Physics, Chemistry, and Dynamics of Interplanetary Dust, eds. Bo Å. S. Gustavson, M. S. Hanner (ASP Conference Series, Vol. 104, 1996), pp 274-282.

[38] S. A. Sandford, R. M. Walker: Astrophys.. J. 291, 838 (1982).

[39] J. P. Bradley, H. Humecki, M. S. Germani: Astrophys. J. 394, 643 (1992).

[40] J. P. Bradley, L. P. Keller, D. E. Brownlee, K. L. Thomas: Met. Planet. Sci,. 31, 394 (1996).

[41] J. C. Gezo, J. P. Bradley, D. E. Brownlee, K. Kaleida, L. P. Keller: Lunar planet. Sci. XXX1, 1816 (2000).

[42] J. P. Bradley, L. P. Keller, T. P. Snow, M. S. Hanner, G. J. Flynn, J.C. Gezo, S. J. Clemett, D. E. Brownlee, J. E. Bowey: Science 285, 1716 (1999).

[43] F. J.. M. Rietmeijer: Met. Planet. Sci. 31, 278 (1996).

[44] L. P. Keller, K. L. Thomas, D. S. McKay: Geochim. Cosmochim. Acta 56, 1409 (1992).

[45] R. P. Binzel: Icarus 99, 238 (1993).

[46] K. L. Thomas, G. E. Blanford, L. P. Keller, W. Klöck, D. S. McKay: Geochim. Cosmochim. Acta 57, 1551 (1993).

[47] S. Messenger, R. M. Walker: "Evidence for molecular cloud material in meteorites and interplanetary dust". In: Astrophysical implications of the laboratory study of presolar materials (American Institute of Physics, Woodbury, N.Y.1997) pp 545-564.

[48] E. Anders, E. Zinner: Meteoritics 28, 490 (1993).

[49] T. L. Daulton, D. D. Eisenhour, T. J. Bernatowicz, R. S. Lewis, P. R. Buseck: Geochim. Cosmochim. Acta 60, 4853 (1996).

[50] K. L. Thomas, G. E. Blanford, S. J. Clemett, D. S. McKay, S. Messenger, R. N. Zare: Geochim. Cosmochim. Acta 59, 2797 (1995)

[51] D. C. B. Whittet: Dust in the Galactic Environment (Inst. Phys., New York, 1992).

[52] L. Hendecourt, C. Joblin, A. Jones (eds.): Solid Interstellar Matter: The ISO Revolution. (Les Houches No 11, EDP Sciences, Les Ullis, 1999) p 315.

[53] L. B. F. M. Waters, F. J. Molster, T. de Jong, et al.: Astron. Astrophys. 315, L361 (1996).

[54] F. J. Molster: Crystalline silicates in circumstellar dust shells. (PhD thesis, Univ. Amsterdam, 2000).

[55] M. S. Hanner, R. D. Gehrz, D. E. Harker, T. L. Hayward, D. K. Lynch, C. C. Mason, R. W. Russell, D. M. Williams, D. H. Wooden, C. E. Woodward: Earth, Moon & Planets 79, 247 (1998).

[56] J. Crovisier, K. Leech, D. Bockelee-Morvan, T. Y. Brooke, M. S. Hanner, B. Altieri, H. U. Keller, E. Lellouch: Science 275, 1904 (1997).

[57] I. M. Steele: Geochim. Cosmochim. Acta 50,1379 (1986).

[58] I. M. Steele: Am. Mineral. 71, 966 (1986).

[59] R. H. Jones, J. M. Saxton, I. C. Lyon, G. Turner: Lunar Planet. Sci. XXIX, 1795 (1988).

[60] J. P. Bradley: "Mg-rich olivine and pyroxene grains in primitive meteoritic materials: comparison with crystalline silicate data from ISO". In: Solid Interstellar Matter: The ISO Revolution, eds. L. d'Hendecourt, C. Joblin & A. Jones. (Les Houches No 11, EDP Sciences, Les Ullis, 1999) pp 297-315.

[61] M. E. Zolensky, R. Barrett (1994) "Olivine and pyroxene compositions of chondritic interplanetary dust particles". In: Analysis of interplanetary Dust, eds. M. E. Zolensky, T. L. Wilson, F. J. M. Rietmeijer, G. J. Flynn (AIP Conf. Proc. 310 Am. Inst. Phys. New York) pp 105-114.

[62] J. P. Bradley, D. E. Brownlee, D. R. Veblen: Nature 301, 473 (1983).

[63] W. Klöck, K. L. Thomas, D. S. McKay, H. Palme: Nature 339, 126 (1989).

[64] J. P. Bradley: Science 265, 925 (1994).

[65] Rietmeijer, F. J. M: "Interplanetary dust particles". In: Planetary Materials, ed. by J. J. Papike (Min. Soc. Am. Washington DC) pp (2)1- (2)95.

[66] G. J. Flynn: Nature 371, 287 (1994).

[67] A. A. Goodman, D. C. B. Whittet: Astrophys. J. 455, L181 (1995).

[68] P. G. Martin: Astrophys. J. 445, L63 (1995).

[69] L. P. Keller, S. Messenger, G. J. Flynn, S. Wirick, C. Jacobsen Lunar Planet. Sci. XXXIII (in press).

[70] P. R. Buseck, X. Hua: Annu. Rev. Earth Planet. Sci. 21, 255 (1993)

[71] M. E. Zolensky, K. L. Thomas: Geochim. Cosmochim. Acta 59, 4707(1995).

[72] K. Tomeoka, P. R. Buseck: Earth. Planet. Sci. Lett. 69, 243- (1984).

[73] E. K. Jessberger, A. Christoforidis, J. Kissel: Nature 332, 691 (1988)

[74] M. E. Lawler, D. E. Brownlee, S. Temple, M. M. Wheelock: Icarus 80, 225 (1989).

[75] M. Pósfai, T. G. Sharp, A. Kontny: Am. Min. 85, 1406 (2000).

[76] P. Ribbe: Sulfide mineralogy. Volume 1, Reviews in Mineralogy Series. (Mineralogical Society of America , Washington DC, 1974).

[77] D. E. Brownlee: "Interplanetary dust: Possible implications for comets and presolar interstellar grains". In: Protostars and Planets, ed. T. Gehrels (Univ. Arizona Press, Tucson. 1978) pp. 134-150.

[78] P. Fraundorf: Geochim. Cosmochim. Acta 45, 915 (1981).

[79] G. J. Flynn, P. Fraundorf, J. Shirck, R. M. Walker: Proc. 9th Lunar Planet. Sci. Conf., 1187(1978)

[80] Z. R. Dai, J. P. Bradley: Geochim. Cosmochim. Acta 65, 3601 (2001).
[81] S. G. Love, D. E. Brownlee: Icarus 89, 26 (1991).
[82] H. Nakazawa, T. Osaka, K. Sakaguchi: Nature 242, 13 (1973).
[83] G. J. Flynn: Met. Planet. Sci. 35, A54 (2000).
[84] L. P. Keller, J. P. Bradley, F. J. Molster, L. B. F. M. Waters, J. Bouwman, D. E. Brownlee, G. J. Flynn, Th. Henning, H. Mutsche: Nature (submitted).
[85] B. D. Savage, K. R. Sembach: Ann. Rev. Astron. Astrophys. 34, 279 (1996).
[86] T. P. Snow, A. N. Witt: Interstellar depletions updated: Where all the atoms went. Astrophys. J. 468, L65 (1996).
[87] U. J. Sofia, J. A. Cardelli, B. D. Savage: Astrophys. J. 430, 650 (1994).
[88] J. P. Bradley: "Interstellar dust - evidence from interplanetary dust particles". In: Formation of Solids in Space, eds. J. M. Greenberg and A. Li (Kluwer Academic Press, Netherlands, 1997), pp 485-503.
[89] Z. R. Dai, J. P. Bradley, D. J. Joswiak, D. E. Brownlee, M. J. Genge: Lunar Planet. Sci. XXXIII (in press).

The Most Primitive Material in Meteorites

Ulrich Ott

Max-Planck-Institut für Chemie, Postfach 3060, D-55020 Mainz, Germany

Abstract. The most primitive matter known to occur in meteorites are circumstellar condensates. They were found as result of the search for the carrier phases of isotopically anomalous noble gases and consist of thermally and chemically highly resistant materials. Diamond was the first to be detected and may occur with an abundance as high as 1 per mil, but it is not clear what fraction of the observed nanometer-sized diamonds is truly presolar. Others (graphite, silicon carbide, refractory oxides and silicon nitride) occur on the sub-ppm to ~10 ppm level. Isotope abundance anomalies are the key feature based on which the presolar nature of a given grain is ascertained. They allow to draw conclusions about nucleosynthesis and mixing in stars as well as galactic chemical evolution, while at the same time it is possible in many cases to establish likely stellar sources for the grains. The major source of SiC and oxide grains are AGB stars, while some (~1%) of the SiC and all of the silicon nitride grains appear to be linked to supernovae, as are probably the presolar nanodiamonds. Graphite grains come from a number of different stellar sources. Astronomical detection around specific sources, as in the case of SiC, would be helpful if it could be achieved also for other grains. Deriving an age for the grains and establishing a detailed history between time of formation in stellar outflows and arrival in what was to become the solar system remains a task to be accomplished.

There have been limits to our ability to identify presolar grains in meteorites. The approach taken so far has mostly relied on chemical resistance. But it is likely that there are also acid-soluble presolar grains, e.g. of silicate composition, as indicated by isotope anomalies found during stepwise dissolution of primitive meteorites and the recent identification of presolar silicates in interplanetary dust particles. The latter has been achieved by a new generation of analytical equipment that surely will be applied to meteorites in the near future.

1 The Most Primitive Material in Meteorites

What is primitive? In a sense, all the material making up the major group of meteorites - the chondrites (~ 85 % of all falls) - is primitive. All their constituents, including thermally and/or aqueously metamorphosed ones, formed within the first few tens of millions of years after formation of the solar system (e.g., [1, 2]). They thus constitute old material formed 4.5 to 4.6 Ga ago and essentially unaltered since then - they are primitive. In fact, chondritic meteorites, and in particular the subgroup of the carbonaceous chondrites represent the most pristine material that is available for detailed laboratory study in large amounts. Others, less abundantly available materials include interplanetary dust picked up by aircraft in the stratosphere, micrometeorites collected in deep sea sediments/polar ice (cf. contribution [3]), plus hopefully in the future material collected in space.

Yet one of the more astonishing recent developments in the study of meteorites has been the realization that they contain small amounts of even more primitive material - presolar dust which predates the solar system. These grains the first of which (diamond and silicon carbide) were detected in 1987, formed during the final stages of stellar evolution from material ejected by stars in strong stellar winds or during explosions. Subsequently they survived passage through the interstellar medium and became part of the parcel of gas and dust from which the solar system formed. The grains then survived what violent events there may have been in the early stages of solar system formation and were incorporated into meteorite parent bodies where they spent the following $\sim$ 4.5 Ga. Collisions that liberated the meteorites-to-be from their parent bodies and sent them towards Earth made it possible that we can have those presolar grains in the laboratory now, where we can investigate them with the full scrutiny of modern analytical techniques (Fig. 1). The implications of the laboratory analysis of presolar grains have been the subject of an excellent recent Conference Book [4] and the reader can find many more details in this book than can be discussed here. Also useful may be the additional study of two recent review articles [5, 6] dealing in detail with the isotopic structures of presolar dust grains from meteorites and the inferred stellar sources.

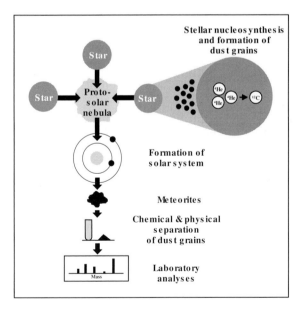

Fig. 1. The path of a circumstellar grain from stellar source to terrestrial laboratory. Sketch courtesy of P. Hoppe, modified from [36].

2   Overview: Identification and Isolation

2.1   Identification

Recognizing presolar grains as such is not a trivial task. Many distinguishing features, even those characteristic for presolar origin, are not unique enough to clearly rule out a more commonplace origin. A case in point is that of the GEMS (Glass with Embedded Metal and Sulfides) found in interplanetary dust particles [7, 124]. Many features, such as amorphous structure, size and non-stochiometry, point to an interstellar origin, but nevertheless there exists general agreement that for confirmation isotopic analysis of their constituent elements is required. This task, because of their submicron size, has been impossible with existing analytical equipment, but may be achieved with a new generation of ion microprobes coming into operation [8].

The situation reflects the general agreement that the only reliable indicator of a non-solar-system origin is the isotopic composition of a grain. Elements heavier than and including carbon are produced in stars, and grains forming from matter expelled by a specific star will carry isotopic signatures diagnostic for the processes that occurred in this star. A grain of "circumstellar" origin will thus show - relative to solar system isotopic abundances - clear isotope abundance anomalies based upon which it can be recognized. Inter- rather than circumstellar grains, present at the time and place of birth of the solar system, on the other hand, are likely to be isotopically similar to solar system matter. Using the strict isotopic criterion, we may not be able to recognize such material as presolar. Therefore, all the grain types and the features to be discussed below refer to grains of circumstellar origin only, and not to presolar/interstellar grains in general.

It is important to also keep in mind that isotope abundance anomalies have been found in meteorites that have *not* been linked to presolar grains. One type, most conspicuous in the elements in the mass range of iron (Ca, Ti, Cr, Fe, Ni, Zn) is found in calcium-aluminum-rich inclusions (CAIs) in primitive meteorites [9]. These inclusions are macroscopic objects (mm- to cm-size), may be the first solids that formed within the solar system (e.g, [1]) and anomalies are likely to be due to incomplete homogenization of solar system precursor materials. Anomalies are much smaller than comparable anomalies in true circumstellar grains (e.g., [10]). Other anomalies - in elements that came into the solar system primarily as gas - are also conserved in elements such as hydrogen and nitrogen [11, 12].

2.2   Isolation

A paradigm in the 1960s stated that the planets and other objects in the solar system had formed from "a well-mixed primordial nebula of chemically and isotopically uniform composition" [13] because the assumption was that the nebula had gone through a phase when it was so hot that no solids could survive. However, already then, key evidence to the contrary was evolving: isotopic abundance

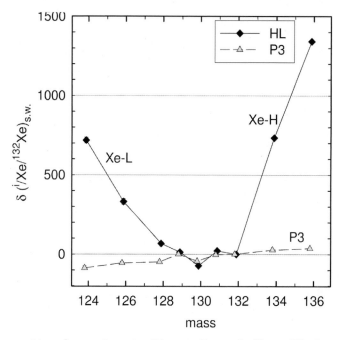

Fig. 2. Composition of xenon in meteoritic nanodiamonds. Xenon-HL shows overabundances in light (Xe-L) and heavy (Xe-H) isotopes and indicates a connection to supernovae. The presence of Xe-P3, an isotopically almost normal component of Xe, indicates complexity (cf. also Chap. 3.5). Shown are deviations of isotope ratios $^{i}Xe/^{132}Xe$ from solar xenon (in per mil) as inferred from solar wind implanted in lunar material [73].

anomalies were found in the noble gas elements xenon (Fig. 2) and neon that ultimately were traced to the presence of presolar grains. The anomalous xenon first seen by [14] - Xenon-HL in today's nomenclature (e.g., [15]) - is characterized by approximately two-fold enhancement in the abundances of the heaviest (hence H) and lightest (hence L) isotopes; the anomalous neon (Neon-E) first recognized by [16] consists virtually of only one isotope, ^{22}Ne. As we know now, Ne-E occurs in two varieties Ne-E(L) and Ne-E(H), where the former is carried by grains of presolar graphite, the latter by grains of presolar silicon carbide (SiC). Xenon-HL, to the best of our knowledge, is carried by grains of presolar nanometer-sized diamonds.

It was the search for the carrier phases of noble gases that led, in 1987, to the identification of the first two types of presolar grains, diamond and silicon carbide [17, 18]. Key to the isolation was the realization that they are resistant to acids [19, 20]. Following up on this observation, a kind of "standard recipe" has developed, which is the one described by [21] for their K series separates. In short, the most important steps for isolating the most abundant of the presolar phases (Table 1) involve -a) dissolving the bulk of the meteorite (silicates and metal) by treatment with hydrofluoric and hydrochloric acid, -b) treatment by, e.g. chromate, to oxidize the more reactive part of the mostly carbonaceous acid residue, -c) colloidal extraction of grains of nanodiamonds, -d) density separation of the non-colloidal part to recover graphite in the light fraction, and e) treatments of the diamond as well as the heavy part of the non-colloidal fraction with perchloric acid. The latter fraction is, in addition, treated with hot sulfuric acid which dissolves spinels. After this it consists mostly of silicon carbide.

Because the procedure is lengthy and time-consuming, attempts have been made to find shortcuts. For example, accepting loss of graphite, [22] have developed a more aggressive microwave technique which allows isolation of diamonds within one week. But unexpectedly not only graphite, but also silicon carbide was found to be lost in the procedure. Still another modification of the standard procedure where hydrofluoric acid is replaced by other fluorine compounds has been recently developed by [23].

It is important to keep in mind, though, that even application of the "standard" procedure can lead to significant losses. For example, [24] noted that silicon carbide in sample KJ isolated from the Murchison meteorite by [21] using the standard procedure seemed to be more coarse-grained than SiC in other primitive meteorites. This led [24] to consider the possibility of size-sorting in the solar nebula, but it is almost certain that during preparation of KJ fine-grained SiC was lost (maybe as much as three quarters of all SiC), as shown by analysis of less-processed Murchison acid-resistant residues [25, 26]. Along the same lines, during TEM observations fine-grained SiC was found to be present in non-negligible amounts in diamond samples [27].

Ideally, of course, the grains could be extracted with only physical means, avoiding chemical treatments altogether. In fact, [28] succeeded in locating SiC grains in unprocessed meteoritic material, and later [29] in extracting grains purely physically. But up to now at least their methods have been shown to work for SiC only and only above a certain size limit which still leaves most grains of interest undetected.

2.3   Overview

An overview on the most abundant of the presolar grain types identified so far is given in Table 1. Among the properties listed are approximate "typical" abundances for the most primitive meteorite types, size, and also characteristic isotopic abundance patterns and inferred stellar sources (see also Chap. 3). The most abundant grains are those that have been found because they carry isotopically anomalous noble gases - diamond, silicon carbide and graphite (Fig. 3). By far most abundant are the nanometer-sized diamonds; their abundance surpasses that of all others combined by two orders of magnitude. This is nominal, however, only, because in contrast to the typically μm-sized other grains, the nanodiamonds cannot be analyzed individually and so it is not really known if all or which fraction of them is presolar (cf. Chap. 3.5).

Also included in Table 1 are aluminum oxide (Al_2O_3) and silicon nitride (Si_3N_4). These grains are not known to carry any noble gases, but were discovered due to the fact that they too are resistant to acids and ended up in the same residue samples as the silicon carbide grains. Their presolar nature was

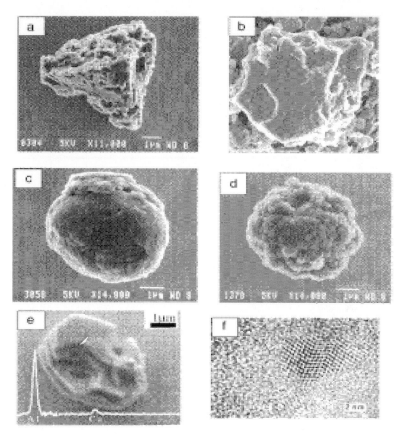

Fig. 3. Images of several types of presolar grains in meteorites: silicon carbide (a, b), graphite (c, d), an oxide (hibonite) [81] (e) by SEM, diamond (f) by TEM. The SiC grain shown in (a) has been extracted by chemical means, the one shown in (b) purely physically. The graphite grains shown represent different morphologies, "onion" (c) and "cauliflower" (d), cf. Chap. 5. Pictures were provided by S. Amari (a, c, d), T.J. Bernatowicz (b), G.R. Huss (e) and F. Banhart (f).

discovered when grains of these "contaminants" were individually analyzed with the ion microprobe [30, 31, 32]. Few grains were also found of presolar spinel, hibonite and rutile ($MgAl_2O_4$, $CaAl_{12}O_{19}$, TiO_2). In contrast to the carbonaceous grains, presolar oxide grains are swamped by isotopically normal oxide grains of solar system origin (cf. [33]). Most were detected only after development of the ion imaging technique for the ion microprobe that allows to identify few candidates among a large number of grains for subsequent detailed analysis. Finally, tiny grains of heavier carbides (TiC, ZrC) were found within mostly graphite [34] but also SiC grains [35] of presolar origin and therefore must be presolar in origin as well.

Table 1. Major types of identified presolar dust grains and some of their properties

mineral	"typical" abundance [ppm][a]	"typical" size [μm][b]	isotopic signatures[c]	stellar source[c]	contribution of stellar source[c]
diamond	~ 1400	~ 0.0026	Xe-HL	supernovae	?
silicon carbide	~ 14	~ 0.3 - 20	enhanced ^{13}C, ^{14}N, ^{22}Ne, s-process elements	AGB stars (1-3 $M_\odot$, different metallicities)	>90%
			low $^{12}C/^{13}C$, often enhanced ^{15}N	J-type C stars (?)	< 5%
			enhanced ^{12}C, ^{15}N, ^{28}Si; extinct ^{26}Al, ^{44}Ti	supernovae	1%
			low $^{12}C/^{13}C$, low $^{14}N/^{15}N$	novae	0.1%
graphite	~ 10	~ 0.8 - 20	enhanced ^{12}C, ^{15}N, ^{28}Si; extinct ^{26}Al, ^{41}Ca, ^{44}Ti	supernovae, Wolf-Rayet stars (?)	> 80%
			Kr-S	AGB stars	< 10%
			low $^{12}C/^{13}C$	J-type C stars (?)	< 10%
			low $^{12}C/^{13}C$; Ne-E(L)	novae	2%
corundum	~ 0.01	~ 0.5 - 5	enhanced ^{17}O, moderatly depleted ^{18}O	RGB and AGB stars (1-9 $M_\odot$, different metallicities)	> 70%
			enhanced ^{17}O, strongly depleted ^{18}O	AGB stars(<1.5$M_\odot$, cool bottom processing)	20%
			enhanced ^{16}O	supernovae	1%
silicon nitride	≥ 0.002	~ 1	enhanced ^{12}C, ^{15}N, ^{28}Si; extinct ^{26}Al	supernovae	100%

[a]Listed abundances are from the compilation [118] and refer to the CI chondrite Orgueil (exception: silicon nitride, where the number is for CM Murchison). Generally the abundances scale with the matrix fraction of the meteorites and decrease with increasing metamorphism of the host meteorites [118]

[b]Size ranges are from [5, 6]

[c]Isotopic signatures as well as stellar and relative contribution are infered and listed in [6].
Enhancements and depletions are expressed relative to the solar system composition. See also discussion in Chap. 3.

3   Isotopic sTructures and Stellar Sources

Study of isotopic structures has been the central theme in the investigation of presolar grains in meteorites. Reasons for this are: a) isotopic analyses are central as proof of presolar origin; and b) they have the potential to answer - in combination with stellar nucleosynthesis theory - the basic question of the stellar sources of the grains. Accordingly the field so far has been dominated by researchers working in the field of isotopic analyses.

Several excellent reviews have appeared dealing with the subject of isotopic structures and inferred stellar sources of presolar grains in meteorites (recently, e.g., [5, 6]). The topic cannot be addressed in the same detail here and the reader is referred to these reviews for more information. An exception is the case of the nanodiamonds, where the situation is less straightforward and which is only sparingly discussed in existing reviews. They are discussed in some detail towards the end of this section (Sect. 3.5). The other grains are discussed in the order in which they are listed in Table 1.

3.1   Silicon Carbide

Silicon carbide has been studied in most detail, a fact that has been helped by its abundance, grain size and high content of trace elements. Isotopic analyses were performed on single grains > 0.5 μm by secondary ion mass spectrometry (SIMS) for Si and C as well as abundant (N and Al with up to several weight-%) and other minor elements up to Ti [5, 6, 36]. Trace elements were isotopically measured by SIMS [37] and thermal ionization mass spectrometry (TIMS; see [36] for a summary) on aggregates of grains and in single grains by laser-ablation resonance ionization mass spectrometry (RIMS; e.g., [38, 39]). Heavy element abundance patterns for single grains have also been measured by SIMS [40]. Noble gases typically have been determined on grain aggregates, but larger grains have also been measured individually for He and Ne [41]. Remarkably, each grain of SiC analyzed individually has turned out to be of presolar origin, and the variations of isotopic compositions are enormous (Fig. 4; Table 2).

The wealth of data has allowed a classification into different categories based on C, N, and Si isotopic compositions. Most abundant ($\sim$ 90%) are the 'main-stream' grains, other types have been named A, B, X, Y, Z (Table 2). For most a rather clear understanding of their stellar origins exists, and for all but the X grains - of likely supernova origin - an origin from Red Giant carbon stars is highly probable [5, 6, 36]. Arguments are a) carbon stars provide the necessary chemical environment because the abundance of carbon exceeds that of oxygen; b) AGB stars are the main contributors of carbonaceous dust to the interstellar medium [42, 43]; c) AGB stars show in their atmospheres the 11.2 μm emission feature of SiC grains ([44, 45]; Chap. 6); d) the distribution of $^{12}C/^{13}C$ ratios in singly analyzed SiC grains is similar to that observed for carbon star atmospheres [46]; e) nucleosynthesis in the He shell produces s-process nuclides as well as Ne dominated by ^{22}Ne, i.e. the observed Ne-E(H) which led to the isolation of presolar SiC [47]. Signatures of the individual subtypes are discussed below.

Table 2. Distribution of single SiC grains among subgroups and diagnostic isotopic signatures of the light to intermediate-mass elements (C to Si). From [36], with updates according to [5, 6, 51, 56]

group	abundance^a (%)	$^{12}C/^{13}C$	$^{14}N/^{15}N$	$^{26}Al/^{27}Al^b$	Si
mainstream	87-94	10-100	50-20000	10^{-5}-10^{-2}	slope 1.34 line
A	2	2-3.5	40-600	10^{-4}-9×10^{-3}	slope 1.34 line
B	2-5	3.5-10	70-12000	10^{-4}-5×10^{-3}	slope 1.34 line
X	1	18-7000	13-180	up to 0.6	excess in ^{28}Si (up to 5×)
Y	1-2	140-260	400-5000	10^{-4}-4×10^{-3}	slope 0.35 line
Z	0-3	8-180	1100-19000	up to 10^{-3}	^{30}Si-rich side of slope 1.34 line

^{a} ranges refer to abundances observed in different size separates.
^{b} approximate ranges.

Mainstream Grains. Mainstream grains range in $^{12}C/^{13}C$ from ~10 to ~100, and from ~50 to ~20,000 in $^{14}N/^{15}N$ (Table 2). Most show the input from hydrogen burning via the CNO cycle, with $^{12}C/^{13}C$ between 40 and 80 (below the solar system value of ~89), and $^{14}N/^{15}N$ between 500 and 5000, higher than solar. They also contain s-process elements, with a signature similar to that of the "main component" that contributes to the solar system inventory of s-process nuclides [48, 49]. This points to an origin in low-mass (1-3 $M_\odot$) carbon stars in the thermally pulsing asymptotic giant branch (TP-AGB) phase. During this phase, the s-process operates in the He burning shell of the star and the products are brought to the surface by convective mixing in thermal pulses. At the same time Ne-E(H) is produced (Chap. 2.2. and above) which is found concentrated in a few % of SiC grains [41].

Silicon isotopes typically show enrichments of the heavy isotopes ^{29}Si and ^{30}Si up to ~ +200 per mil compared to normal silicon, i.e. $\delta^{29,30}Si/^{28}Si$ up to ~ +200 per mil (Fig. 5). In a plot $\delta^{29}Si/^{28}Si$ vs. $\delta^{30}Si/^{28}Si$ data points follow an approximately linear relationship with slope 1.34 [36]. The trend is not due to nucleosynthesis in the parent star, but rather the result of galactic chemical evolution and the fact that the grains derive from a large number of individual stars (e.g., [50]).

Also compatible - although details remain to be addressed - are results for the Mg/Al system as well as calcium and titanium [5, 36]. Most grains show detectable excesses of ^{26}Mg, the decay product of ^{26}Al($T_{1/2} = 0.7$ Ma). Inferred ratios for $^{26}Al/^{27}Al$ at the time of grain formation range from ~ 10^{-5} to ~ 10^{-2}. Calcium isotopes were mostly measured on aggregates of grains. Observed are enhancements at ^{42}Ca and ^{43}Ca as expected from s-process contributions, the larger enhancements at ^{44}Ca are likely to be caused by the presence of X grains from supernovae (see below) which carry radiogenic ^{44}Ca from decay of ^{44}Ti ($T_{1/2} = 60$ a). In titanium typically lighter and heavier isotopes are enriched relative to the most abundant isotope, ^{48}Ti. As the Si isotopes, the Ti isotopes

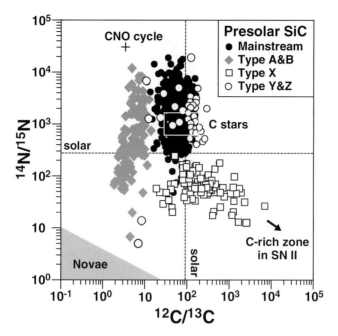

Fig. 4. Carbon and nitrogen isotopic compositions measured in single presolar SiC grains. Solar isotopic compositions are indicated by lines. Most analyses plot in the mainstream field (cf. text and Tables 1, 2), displaced from normal in a direction indicating the influence of hydrogen burning via the CNO cycle. Also indicated are trends to be expected from novae and supernovae contributions. Graph courtesy of P. Hoppe [119].

are influenced by galactic chemical evolution, and variations in Ti and Si isotopes correlate with each other [36].

A and B Grains. A and B grains are distinguished by their very low $^{12}C/^{13}C$ ratios ≤ 10 (Table 2). The dividing line is the equilibrium value for CNO cycle H burning, with $^{12}C/^{13}C \leq 3.5$ for grains of type A. The range in $^{14}N/^{15}N$ is from ~ 40 to ~ 10000, and inferred $^{26}Al/^{27}Al$ at the time of grain formation - as for the mainstream grains - ranges up to 10^{-2} [51]. Also Si and Ti isotopic ratios behave rather similarly as in mainstream grains. The rare A and B grains that have been analyzed for s-process-specific elements suggest that about 2/3 contain overabundances of s-process nuclides, while $\sim 1/3$ do not. The observed features point to an origin from specific types of carbon stars, probably J stars for the s-process rich grains, and possibly "born-again" AGB stars for the s-process poor grains [51].

X Grains. X grains (about 1% of all individually analyzed SiC grains) are clearly distinct from all other types and almost certainly derive from supernovae (Table 2). Ratios $^{12}C/^{13}C$ are typically higher and $^{14}N/^{15}N$ ratios lower than

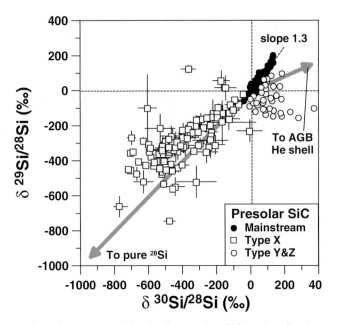

Fig. 5. Si isotopic ratios measured in single presolar SiC grains. Ratios are shown as δ-values, i.e. deviations from normal values in per mil. In this representation most grains plot within 200 per mil of normal, close to a line with slope 1.34 (Table 2, see text). X grains of likely supernova origin plot in the lower left showing large relative enhancements of ^{28}Si. Graph courtesy of P. Hoppe [119].

solar (Fig. 4), which is the signature of He burning. More telling even are the Si isotopes that show enormous relative enhancements of ^{28}Si (Fig. 5). These features as well as the huge excesses of ^{26}Mg and ^{44}Ca from decay of ^{26}Al and ^{44}Ti, with inferred ^{26}Al/^{27}Al and ^{44}Ti/^{48}Ti ratios at the time of grain formation ranging up to ∼0.6 and ∼0.7, resp., are diagnostic for an origin as condensates from supernova ejecta [5, 6]. Ti-44, in particular can only be produced in supernovae [6, 52, 53]. To provide suitable conditions for formation of SiC and inclusion of ^{44}Ti in supernovae of type II requires complex mixing among supernova ejecta material and it remains to be seen whether supernovae type II or Ia provide a more suitable environment [6, 54].

An interesting pattern of Mo isotopes has been found in single X grains by RIMS ([39; Chap. 4). In contrast to the mainstream grains that show the expected s- process signature, molybdenum in SiC grains of type X shows a new pattern that seems best explained as the result of a neutron burst [55], a process intermediate between the traditional s- and r-processes of nucleosynthesis. This process which had originally been invoked in order to explain xenon in presolar diamonds (see Chaps. 3.5 and 4) also is thought to occur during the explosion of a supernova.

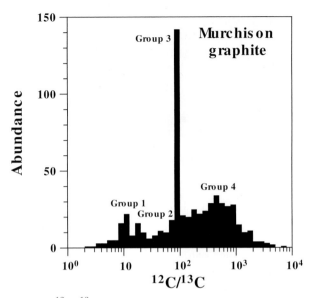

Fig. 6. Distribution of $^{12}C/^{13}C$ ratios measured in single graphite grains (courtesy P. Hoppe, [6]). Based on this ratio four populations can be distinguished.

Y and Z Grains. Y and Z grains, like A and B grains, share many properties of the mainstream grains and must have a similar origin. Characteristic is the difference in Si isotopes ([36]; Table 2): both fall to the ^{30}Si-rich side of the mainstream grains. Y grains have $\delta^{30}Si/^{28}Si > 0$ and $^{12}C/^{13}C > 100$. Z grains, on the other hand, have always lower than solar $^{29}Si/^{28}Si$ and $^{12}C/^{13}C$, and their deviations from the mainstream grains are larger than for the Y grains. Y grains probably come from low-mass lower-than-solar metallicity AGB stars that experienced strong He shell dredge up [6, 56]. As for the Z grains, for most the likely source are low-mass, low-metallicity AGB stars that experienced strong cool bottom processing (Chap. 4) during the red giant phase [6, 57], although some fraction with low $^{12}C/^{13}C$ and $^{14}N/^{15}N$ may come from novae [6].

3.2   Graphite

The case of graphite is more complex than that of silicon carbide. Overall $^{12}C/^{13}C$ spans the same wide range of ~ 2 to ~ 7000 as in SiC, the range in $^{14}N/^{15}N$ is however from ~ 30 to 700 only, being higher than solar in most grains [6, 58]. The primary parameter based on which the grains can be assigned to one out of four different groups is the carbon isotopic ratio (Fig. 6; [58]), and remarkably there seems to be a correlation with density and morphology: a) group 3 grains which are characterized by normal $^{12}C/^{13}C$ are of "non-round compact" or "spherulitic aggregate" type; they presumably are condensates from the molecular cloud from which the solar system formed [59]; and b) among circumstellar grains the relative proportion assigned to each group varies with density and morphology.

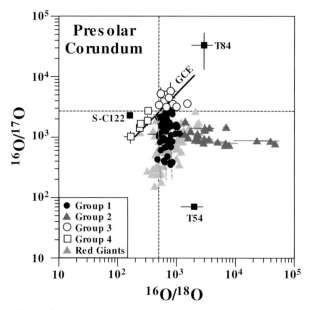

Fig. 7. Oxygen isotopic ratios measured in single presolar corundum grains. Based on oxygen isotopes all but three grains can be assigned to one of four groups [63]. Also shown are measured values for the atmospheres of red giant stars taken from the literature. Figure courtesy P. Hoppe.

Group 1 grains are characterized by $^{12}C/^{13}C$ in the range 2-20, similar to type A and B SiC grains, while group 2 grains with $^{12}C/^{13}C$ between 20 and 80 resemble mainstream SiC grains. Group 4 grains are most abundant ($\sim 50\%$ of all singly analyzed graphite grains) and isotopically light, with $^{12}C/^{13}C > 100$, up to 7000. With these isotopic features, plus observed excesses of ^{18}O, it is likely that the majority of graphite grains derives from massive stars [6]. Considering also the high inferred $^{26}Al/^{27}Al$ ratios (up to 0.15) at the time of grain formation, the fact that most grains show deficits in ^{29}Si and ^{30}Si (similar to type X SiC grains), as well as the large excesses in ^{44}Ca from ^{44}Ti decay (again, as SiC X grains) and ^{41}K (from ^{41}Ca, $T_{1/2} = 0.1$ Ma), type II supernovae seem the most likely sources [5, 6, 60].

However, other sources must have contributed as well. Heavy noble gases measured in grain aggregates show the presence of material that experienced the s- process [61], so some fraction must derive from AGB stars. And the presence of Ne- E(L) (cf. Chap. 2.2) in a small fraction of graphite grains [41, 62] is indicative of contributions from novae.

3.3   Corundum

Isotopic information on corundum grains is much more limited, both because of its lower abundance and lower content of trace elements. In addition most grains analyzed in detail were selected after ion imaging indicated unusual $^{18}O/^{16}O$, so there may be a bias in the inferred distribution among different recognized populations (see below).

Oxygen is the isotopically most diagnostic element, with $^{16}O/^{17}O$ ranging [5, 6, 63] from ~70 to ~30,000 (0.025 to 11x solar) and $^{16}O/^{18}O$ from 150 to 50,000 (0.3 to 100x solar), and based on oxygen four populations have been recognized (Fig. 7). Al and Mg have also been analyzed in many grains and inferred $^{26}Al/^{27}Al$ ratios at the time of grain formation show distinct distributions in the different populations, with the highest values (up to ~0.02) observed in type 1 and 2 grains. Most group 1 and 2 grains have lower than solar $^{16}O/^{17}O$ and higher than solar $^{16}O/^{18}O$ ratios. The composition of the former is similar to ratios observed in the atmospheres of red giant and AGB stars (1 to 9 $M_\odot$) which makes these the most likely stellar sources [6]. Group 3 grains may be related to group 1 grains and may originate from red giant stars of low mass [6]. Deficiencies of ^{18}O in grains of group 2 are larger than in group 1 grains; their parent stars may have been low-mass stars that experienced cool bottom processing during the AGB phase (5, 6, 64; Chap. 4). Oxygen isotopic signatures of type 4 grains are also compatible with a low-mass AGB star origin [5, 6].

Here, as in many other cases, a complex interplay of variable stellar mass and metallicity, galactic chemical evolution and "new" nucleosynthesis (e.g., cool bottom processing; Chap. 4) is required in order to explain the observations. In addition, three grains analyzed do not fall in any of the four groups and cannot be explained by conventional red giant star models. Two of these [65, 66] may come from supernovae, in one case consistent with Ti isotopic data obtained on the same grain [66]. This is a surprisingly small fraction [6].

3.4   Silicon Nitride

Silicon nitride grains have isotopic signatures similar to those of the silicon carbide X grains [32]. Nitrogen is characterized by low $^{14}N/^{15}N$ (range 18 to 100) and silicon by relative enhancements of ^{28}Si, with $\delta^{29}Si/^{28}Si$ ranging down to 200 and $\delta^{30}Si/^{28}Si$ down to -350 per mil [5, 6]. Coupled with the high level of radiogenic ^{26}Mg (up to inferred $^{26}Al/^{27}Al$ at grain formation of 0.2) these features clearly indicate formation from supernova ejecta, as for the SiC X grains [5, 6].

3.5   The Problems with the Diamonds

The case of the most abundant of the minerals listed in Table 1, the diamonds, is not straightforward (cf. Chap. 2.3). The major problem is that we do not really know how many of them are presolar. A strong argument that a major part is not, comes from the isotopic ratio for the structural element, $^{12}C/^{13}C$. Consisting of on the average about 1000 atoms only, the diamonds cannot be analyzed individually, and "bulk analyses" of many grains ("aggregates") yields $^{12}C/^{13}C$ ~92, within the range of isotopic ratios observed in solar system matter, even on Earth. While stellar scenarios can be constructed that result in such a ratio [67], given the large variations observed in stellar atmospheres and other presolar grain types, such a close agreement with "normal" within % would seem a remarkable coincidence. Nitrogen is the most abundant trace element in the diamonds, with an abundance close to 1% [68]. The isotopic ratio $^{14}N/^{15}N$ is $\sim$ 400, i.e. $\delta^{15}N$ is $\sim$ -330 per mil relative to the standard, air, value of 3.67×10^{-3}. Is this isotopic ratio anomalous and might it indicate that at least a % or so of the diamonds are presolar? We do not really know, because there are uncertainties about the true $^{14}N/^{15}N$ characteristic for the solar system. While recent results for solar wind nitrogen obtained by spacecraft have given results compatible within error with the air value [69], $^{14}N/^{15}N$ in the solar wind may have changed over the course of the 4.6 Ga history of the solar system [70]. Also in conflict with the spacecraft results - if both are taken at face value - are new measurements by ion microprobe of surface-correlated, presumably solar wind- implanted, nitrogen in lunar grains that suggest higher $^{14}N/^{15}N$ [71]. This result seems corroborated by a recent re-evaluation of nitrogen in the Jovian atmosphere [72] which suggests protosolar nitrogen to be similarly "light" with $^{14}N/^{15}N$ possibly as high as $\sim$400. If correct, this suggests that nitrogen in macroscopic samples of meteoritic nanodiamonds, like carbon, may be normal after all.

Another complication arises from the noble gases themselves that constituted the evidence for the diamonds' presolar nature in the first place. Diamonds not just contain isotopically anomalous traces like Xe-HL, but also others which are essentially normal in their isotopic composition (Fig. 1). The situation is best studied for xenon [15, 73]: diamonds contain the anomalous xenon (Xe-HL), an approximately normal xenon component (Xe-P3) and a third one that differs from both (Xe-P6). The P6 component has been observed only mixed in variable degrees with the HL component and its endmember (pure) composition is not known. It may be isotopically anomalous or, similar to P3, isotopically approximately normal. Taken the isotopic argument for identification of presolar grains (Chap. 2.1) to the extreme, it can be argued that - because Xe-HL occurs in an abundance that on the average one Xe atom occurs per about a million diamonds grains - we can confidently assume only that one out of a million diamonds is of presolar nature. But, of course, Xenon-HL is accompanied by other noble gases. Neon and helium are especially high in abundance, so high that diamonds in fact dominate the helium and neon inventory of many primitive meteorites - the major part of what has been called "planetary" helium and neon. Helium occurs in an abundance that a He atom is contained in about every hundredth of the diamond grains. Following the same logic as for Xe, this may indicate that the fraction of presolar diamond grains actually is on the order of % or more, similar to what one might conclude from nitrogen, if indeed nitrogen is isotopically anomalous. Interestingly, the abundance level of truly presolar diamond grains inferred from this exercise is quite similar to that for SiC and graphite, but the conclusion is anything but firm.

Information on isotopic anomalies in trace elements other than the noble gases that may help shed light on the issue is sparse. The only convincing case is that of tellurium where [74] identified a Te-H component (but not Te-L) consisting of the two most neutron-rich isotopes, ^{128}Te and ^{130}Te, only. This has been recently confirmed, although the effect was smaller (probably because of dilution with other matter), by [75], who also observed a barely resolved effect in palladium, with an overabundance of ^{110}Pd of (9±6) per mil. Similarly small and poorly understood effects had been seen in barium and strontium before [76].

Overall, those isotopic anomalies in nanodiamonds that show a clear pattern (Xe, Te) are characterized by overabundances of the most neutron-rich isotopes. This suggests an origin from a kind of rapid neutron capture process, although not the unadulterated standard r-process of nucleosynthesis (Chap. 4). They point to an origin from type II supernovae as the most likely source (e.g., [77-79, 55]), where also the simultaneous excesses in the most proton-rich isotopes of Xe (Xe-L) may come from [80, 79], although the situation is less clear. In spite of all the efforts, presently all that these isotopic effects in trace elements indicate is that some fraction of the diamonds is presolar, but we just do not know how large this fraction is.

3.6   Other Grains

In addition to the types of grains discussed above only few other grains have been identified as being of presolar origin. These include few grains of spinel, $MgAl_2O_4$, and hibonite, $CaAl_{12}O_{19}$ [66, 81], possibly a titanium oxide (TiO_2) grain [65] and carbides of Ti, Mo and Zr and Fe-Ni inside grains of graphite ([34]; cf. Chap. 5) but also SiC grains [35]. The oxides, like most corundum grains, appear to come from AGB stars [5, 6]. As for the carbides, TiC grains in particular are likely to be the host of the often very large enhancements of ^{44}Ca from decay of ^{44}Ti. They, like the other enclosed carbides, have been too small for individual analysis but there is hope that this will change with a new generation of ion microprobes coming into operation now [8].

4   Nucleosynthesis Inferred

While general isotopic features observed in presolar grains in many cases quite clearly point to specific stellar sources, details often point to deficiencies in our understanding of stellar evolution and nucleosynthesis. This naturally concerns mostly the AGB and supernova sources that supply most of the known grains. Generally additional information found within the isotopic structures falls into one (or several) of the following categories: (a) quality of nuclear data; (b) (old) nuclear processes under (new) non-standard conditions; (c) mixing processes. In addition, with an ever increasing number of analyzed grains from many stars, it has become obvious that in some cases the initial composition of the parent stars (and, in turn, galactic chemical evolution) is still reflected in the grains. A case in point is the array of mainstream SiC grains along the slope 1.34 line in Fig. 5 [82, 50]. Also apparent are isotopic differences in the inherited compositions of parent stars in oxygen [83, 6].

4.1   Mixing in Supernova Ejecta (or New Chemistry?)

Mixing in stars larger than previously expected may be required in several cases. While this includes red giants and AGB stars, the case based on supernova grains (SiC type X, graphite, silicon nitride) is clearest [5, 6]. In type II supernovae, the radionuclide ^{44}Ti, which is the most diagnostic feature, and the observed excesses of ^{28}Si are produced in the innermost zones. The bulk of the grains, however, should have formed from material derived from the outer zones because this is where the abundance of carbon exceeds that of oxygen so that carbonaceous grains can form. So, if the grains derive from SN type II, outer and inner zones must have mixed and achieved a higher abundance of carbon than of oxygen in spite of the fact that they are separated by a large intermediate region consisting largely of oxygen [5, 6, 84]. If derived from type Ia instead problems in achieving all required conditions simultaneously are quite similar [54,85]. An alternative to extreme mixing may be to invoke a kind of "new chemistry" that in the strong radiation field of a supernova inhibits formation of the CO molecule and allows formation of carbonaceous grains even in an oxygen-rich environment [86].

4.2   Red Giant Light Element Nucleosynthesis

Strong indications for "new kinds" of "old" nucleosynthesis are provided by the oxygen data for the corundum grains. In order to explain them, [87] and [64] had to invoke two processes of hydrogen burning under special conditions. A) "Hot Bottom Burning " (HBB; [87]) occurs in AGB stars with masses greater than 4 $M_\odot$ where the convecting envelope is deep enough during interpulse periods to reach the star's hydrogen burning shell. B) Red giants of lower mass, on the other hand, experience - following the first dredge-up - "extra mixing" that transports material from the cool bottom of the envelope down to the hydrogen shell where its composition can be modified and from where it can be mixed back into the envelope. This process has been called "Cool Bottom Processing" (CBP; [64]).

4.3   Nuclear Cross Sections

For the light elements the situation is complicated by many contributing factors, so the question of nuclear cross sections has rarely been addressed based on presolar grain data. The only serious case is that of the ^{18}O(p,α)^{15}N reaction where [88] have suggested that the rate may be a factor of $\sim$1,000 higher than commonly assumed, in order to solve the problem of the SiC A and B grains with ^{12}C/^{13}C and ^{14}N/^{15}N both low. The case is more straightforward in the case of the heavy nuclides produced by the s-process, where cross sections can be measured and checked down to precisions on the percent level. Over a large mass range for neighboring s-process nuclides the approximate relationship

$$N \times \sigma = const. \tag{1}$$

holds, where N is the abundance of a nuclide manufactured by the s-process and σ its Maxwellian-averaged neutron capture cross section at the relevant

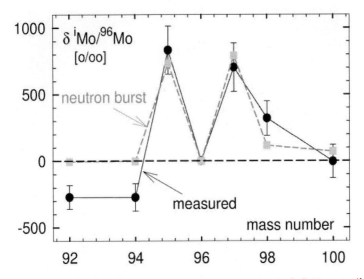

Fig. 8. Molybdenum isotopic composition measured in a single SiC-X grain ([39]; data slightly revised according to A. Davis, pers. commun). Shown are the deviations of isotopic ratios $^iMo/^{96}Mo$ (normalized to the s-process only nuclide ^{96}Mo) – from isotopically normal Mo. The composition is largely in agreement with expectations for a mixture of 83% normal Mo and 17% of Mo as produced in a neutron burst [55].

temperature (e.g., [89, 90]; see also [91]). Discrepancies between this relation where it should hold and observations in presolar SiC have been largely erased whenever the neutron capture cross sections have been re-measured ([36, 5]; but see [92]).

4.4   Old (and New?) Neutron Capture Processes

Results for the composition of the s-process component in presolar silicon carbide grains have allowed to refine the constraints on physical conditions during the process. Basis for this are nuclide ratios that are affected by branchings in the s-process (i.e. where nuclides are encountered at which there is competition between neutron capture and β-decay). Effective values for neutron density, temperature and electron density can be inferred from the results in the framework of the "classical" model of the s-process (parameters held constant, e.g., [36, 89]), but the results of course put also constraints on more realistic "stellar models" [91]. In contrast to the clear signature of the s-process that is seen in the SiC grains, no clear signature of the rapid neutron capture process (r-process) has been seen in presolar grains so far. R-process like signatures are seen in the diamonds (xenon, tellurium) and in SiC grains of type X (molybdenum), but differ in detail (Fig. 8; Chap. 3). The SiC-X Mo data seem well explained by the action of a neutron burst, some kind of neutron capture process intermediate between the canonical s- and r- neutron capture processes [55]. The process does, however, considerably worse in explaining diamond Xenon-HL [79], although it was

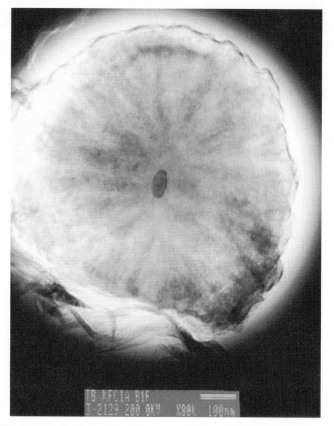

Fig. 9. TEM micrograph of an ultrathin section of an onion-type circumstellar graphite grain. Included in the graphite grain is a TiC subgrain. Figure courtesy of T.J. Bernatowicz.

developed specifically for this purpose [78, 93]. Therefore, for the diamonds an alternative "rapid separation" scenario has also been considered [79] which combines the average r-process with separation of stable r-process products from radioactive precursors on a time scale of hours. While the predictions of the model fit the observed isotopic pattern [79] and possibly also the tellurium pattern [75], it faces the problem of how to achieve the required separation. The question of heavy trace elements in diamond and their implications remains an open questions at this time.

5   Mineralogy and Morphology

Probably the single most important fact regarding the mineralogy of the presently identified presolar grains in meteorites is that they are made of minerals - diamond, graphite, silicon carbide, refractory oxides - that are both thermally and chemically highly resistant. They are among the earliest solids formed upon cooling [94] from material expelled from stars, and their high chemical resistance (under the right oxygen fugacity) may have helped their survival in the interstellar medium and the early solar system. Crucial, of course, was that several are carrying noble gases that could be used as "beacons" in the search for them (cf. Chap. 2). Finding presolar grains that are acid-soluble and do not carry noble gases, if they exist, may require new techniques for isolation [8, 95].

Another simple observation is that generally different grain types have different primary sources: red giants for SiC and corundum grains, supernovae (probably) for diamonds and for silicon nitride, and no single source dominating contributions to graphite in a similar manner. But generally little detailed investigations have been performed that would allow to connect isotopes with other properties of the grains. Graphite is the exception in that such information exists (Chap. 3.2): a) only round grains have a non-solar carbon isotope ratio and clearly constitute circumstellar material; b) grains can be distinguished based on surface features as "onion" or "cauliflower" type (Fig. 3), a distinction reflected also in their trace element contents and interior structure [34] - the "onions" contain cores of graphene or small refractory (Ti, Zr, Mo) carbide grains (Fig. 9); and c) graphite grains of different density show different relative contributions from the various stellar sources [6, 58]. As for silicon carbide, the successful extraction without chemical means of a number of grains has allowed to compare features and it is obvious that the surfaces are etched during chemical isolation [29]. However, there is no evidence that the grains are altered otherwise during this process. Early investigations [96] seemed to indicate a prevalence among the SiC grains of the β polymorph, but more recent work [27] has revealed the presence also of α-SiC. This finding has partly eliminated another earlier apparent paradox, namely that meteoritic SiC seemed to be of the β type, while spectroscopic observations were interpreted as indicating α-SiC in the atmospheres of carbon stars [45]. More of this apparent puzzle has gone away also with the realization that spectroscopic features of fine-grained SiC may more strongly depend on grain size and shape than on the type of polymorph [97, 122, 123]. Diamonds have been studied in detail in the TEM [98], but their size (average 2.6 nm) does not allow any individual isotopic analysis and the only clear trend associated with morphology is the size dependence of noble gas concentration [99]. Given the rare occurrence of diagnostic trace elements within the diamonds we cannot even be sure that any of the grains imaged by TEM [98, 100] has been one of circumstellar origin (cf. Chap. 3.5). Little is known about the morphology of oxides and silicon nitride. Most oxide and Si_3N_4 trains were found by ion imaging in the ion probe and only few have been imaged by SEM, so there is little morphological information to correlate with features diagnostic of their stellar origin.

5.1 Grain Sizes

Grain size is, on the average, ~ 2.6 nm for the diamonds [98], while the other grains in Table 1 have sizes in the μm range (from fraction of μm up to ~ 20 μm). Size distributions have been determined for SiC and graphite by [21] on

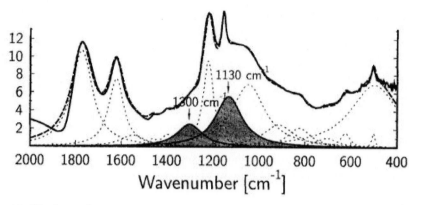

Fig. 10. IR absorption spectra measured on a sample of diamond extracted from the Murchison meteorite. The absorption expected due to nitrogen as single atoms in the diamonds (gray area) - probably the only signatures independent from chemical history and current environment - may be present, but are masked by the presence of much larger peaks due to attached surface groups. Figure provided by A. Braatz [107].

grains from the Murchison meteorite. For both graphite and SiC the results can be explained as a sum of log-normal distributions. For SiC, an alternative is a power-law distribution (truncated at the upper and lower end) of the type

$$N = C \times [a \; exp(-b)] \tag{2}$$

in the size range a from 0.7 to 3.2 μm and $b = 5.7$ for the cumulative distribution. It is not clear how representative this result is. Earlier work gave some hints that SiC in the Murchison meteorite from which the samples of [21] were derived might be unusually coarse-grained compared to other meteorites [24, 6], but, more realistically, fine-grained SiC grains seem to have been lost from the SiC sample of [21], presumably during chemical isolation [25, 26]. Taken at face value, nevertheless, the exponent in the distribution of 5.7 is notably different from that in the size distribution for interstellar grains derived by [101] from astronomical observations. It has to be kept in mind, though, that the latter may apply primarily to other types of grains and definitely was derived for a different size range (0.025 to 0.25 μm). It is similar, however, to the size distribution of particles expected in a dust-driven wind [120]. That large sizes are observed in the meteorites, however, is compatible with the sizes of grains that have been observed to enter the solar system today [102].

6   Detection in Space

The detectability in space of those minerals that are present as presolar grains in meteorites is of special interest. Where stellar sources have been inferred from isotopic systematics, it will be a welcome confirmation. And, where no clear indication exists based on isotopes, detection in space may yield the decisive piece of evidence.

Again, the situation is clearest in the case of silicon carbide. As pointed out earlier (Chap. 3), the identification of the ~11.2 μm absorption line of SiC in carbon star atmospheres even before their identification in meteorites [44] has been used as major argument for a carbon star origin of most SiC grains. As for the apparent discrepancy (cf. Chap. 5) of α-SiC in C star atmospheres [45] vs. β-SiC in meteorites [96] this has largely disappeared with the recognition that α-SiC is present in meteorites as well [27] and that grain size may have a larger influence on details of the absorption feature than the type of polymorph [97]. The diamonds are only weakly – via rare trace elements – linked to a supernova source. As a consequence, much effort has been spent trying to establish what the absorption features are of the meteoritic nanodiamonds, a fact helped by their relatively high abundance (Table 1). However, the many studies [103-107] that have been performed have hardly been conclusive so far. The major problem is that pure diamonds do not show any absorption in the optical or infrared, so generally features that have been seen are either due to contaminants or to surface groups that – because of the small size – are abundantly attached to the diamonds (cf. [108]). Consequently, features that might be observed in space will depend on the chemical environment (and then might not be specific to diamonds), and what is observed on the meteoritic diamonds in the laboratory depends on the chemical extraction procedure. The latter is abundantly demonstrated by changes observed between various stages of the extraction [107] and the fact that different investigations found different sets of absorption features.

Under these circumstances the only specific features will be those that are caused by the presence of trace elements that are indigenous (rather than surface-correlated) to the diamonds, such as nitrogen [107]. These, in turn, depend on the configuration of nitrogen, which could be determined via electron paramagnetic resonance (EPR) to occur in the meteoritic nanodiamonds as single atoms. The results suggest that nanodiamonds of the type found in meteorites should show major nitrogen-related absorption features at 1130 and 1300 and 1344 cm^{-1} (8.85, 7.69 and 7.44 μm; Fig. 10) as well as in the UV at 270 nm [107]. No such features seem to have been reported in the astronomical literature [107]. More promising may be a search in emission, and indeed emission features observed in the dusty envelopes of some stars have been interpreted as indicative for the presence of small diamonds [109]. However again, as the absorption features seen in [108], these features are caused by terminal groups at the diamonds' surfaces.

7   Age and History

7.1   Ages

Determination of ages for presolar grains is not straightforward. A major reason is what just characterizes the grains: anomalies in isotope abundances. As a consequence, in the application of long-lived chronometers it is often not clear in what ratio parent and daughter nuclides would have been present at the time of grain formation (the starting composition that then would have changed as a function of time due to radioactive decay). Thorough understanding of nucleosynthesis and mixing in the parent star and/or galactic chemical evolution (which may define the ratios of the nuclides when the parent star formed) would be necessary to overcome this difficulty. An exception may be those "simple" cases in which stellar nucleosynthesis totally resets the composition, such as an r-process in a supernova which totally resets the distribution of the heavy nuclides. So in principle grains like the SiC-X grains might be datable via determination of, e.g. the uranium isotopic ratio, by mass spectrometry or via fission tracks [110]. But not only will this be a difficult experiment, there is the problem that the X grains (and also the diamonds, for that matter), in lighter elements at least, do not show the signature of a simple unadulterated r-process but such from a very specific process, which, in addition is mixed with some fraction of isotopically approximately normal matter (Chap. 4). So interpretation of any obtained results in terms of an age will not be a trivial task. Since uranium is produced in the r-process only, the same approach applied to mainstream SiC grains would effectively date the time the parent star was born.

Another approach relies on determining a presolar cosmic ray exposure age that – added to the 4.6 Ga age of the solar system – would correspond to the total (absolute) age [111]. Among the known presolar grains this approach may be limited to the SiC mainstream grains, both because they occur in high enough abundance and contain the right target elements to produce measurable amounts of cosmogenic nuclides (mostly rare noble gas nuclides that can be sensitively determined). Using this approach [111, 112] arrived at presolar cosmic ray exposure ages ranging from $\sim$10 to $\sim$100 Ma, increasing with grain size. These results are flawed, however, because recoil losses of the cosmogenic product ^{21}Ne were seriously underestimated [113], and the apparent ^{21}Ne excesses that were attributed to cosmic ray production may predominantly have a nucleosynthetic origin instead. Losses of xenon isotopes produced by cosmic rays on the abundant trace Ba from $\sim$ μm- sized SiC grains are significantly smaller than those of neon isotopes so that cosmogenic xenon offers a better hope for determining an age [114], but other problems remain: exact determination of the cosmogenic contribution, and relevant production rate by cosmic rays.

7.2   History

There are two things that currently can be said with certainty about the history of the presolar grains in meteorites. One is that many must have formed rather rapidly. This is documented by the signature of extinct radionuclides (Chap. 3) such as ^{44}Ti ($T_{1/2} = 60$ a) or ^{22}Na ($T_{1/2} = 2.6$ a). The other is that they experienced the same type of metamorphism as the other constituents of their host meteorites. This is shown by the correlation of their abundance with petrologic subtype, with no presolar grains present in types more thermally metamorphosed than type 3.8 chondrites [115]. The sensitivity to thermal metamorphism increases in the order graphite, silicon carbide, diamond [115] and a fine resolution is possible based on the abundance of various noble gas components in the diamonds [116]. Apart from that, those that survived did so largely intact. But

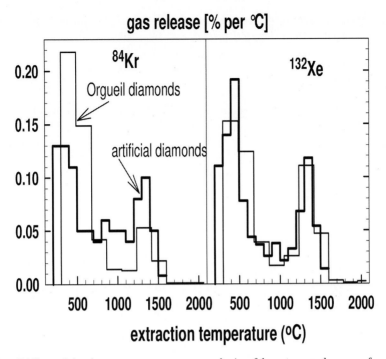

Fig. 11. Differential release upon vacuum pyrolysis of krypton and xenon from diamonds extracted from the Orgueil meteorite [15] compared with the relase of gases implanted into artificial terrestrial nanodiamonds [117]. The agreement can be used to construct a history for the interstellar diamonds [117].

the time it took from the stellar sources to the solar system (cf. Chap. 7.1) and the conditions they survived there are poorly defined.

Only in one case, that of the diamonds, a very specific scenario has been suggested [117]. This scenario is based on the occurrence of multiple components of noble gases ([15]; Fig. 1]), the evidence that introduction of trace gases was by ion implantation (e.g., [17, 99]) and results from laboratory simulation experiments (Fig. 11; [117]). In this scenario, after their formation the diamonds were irradiated in a circumstellar environment with trace elements having a supernova isotopic signature, and subsequently heated (up to $\sim$700 °C) which led to loss of part of the implanted atoms. The same diamonds were irradiated again at some later time with trace gases of about normal isotopic composition; or at some other place and time a different population of diamonds was irradiated with this "normal" type material and became mixed later with the ones carrying the "anomalous" matter. After that the diamonds found in the most primitive meteorites did not experience elevated temperatures any more for any significant length of time (e.g., not more than $\sim$ 10,000 years at a temperature higher than 100°C or equivalent combinations [92]).The scenario involves a number of assumptions and therefore has to be regarded with some caution, but currently is the only case where a detailed history has been suggested.

8   Summary

The most primitive material occurring in meteorites are circumstellar condensates several of which occur on the sub- to some 10 ppm level in the most primitive ones: graphite, silicon carbide, silicon nitride and refractory oxides, primarily corundum. Diamond was the first to be detected and may occur in much higher abundance ($\sim$ per mil), but it is not clear what fraction of the observed nanometer-sized diamonds is truly presolar.

The presently known presolar grain types were found as the result of the search for the carrier phases of isotopically anomalous noble gases: diamond, graphite and silicon carbide are these carriers, and the other known presolar minerals were found because they followed silicon carbide in the isolation procedure. A most remarkable feature - though maybe not surprising given the history of the detection - is that all consist of thermally and chemically highly resistant materials.

Based on isotopic features it is possible in many cases to establish likely stellar sources, and the same features also allow to draw conclusions about nucleosynthesis and mixing in stars as well as galactic chemical evolution. The major source of SiC and oxide grains are AGB stars, while some ($\sim$ 1 %) of the SiC and all of the silicon nitride grains appear to be linked to supernovae, as are probably the presolar nanodiamonds. Graphite grains come from a number of different stellar sources and the distribution varies with morphology. Astronomical detection around specific sources, as in the case of SiC, would be helpful if it could be achieved also for other grains. Deriving an age for the grains and establishing a detailed history between time of formation in stellar outflows and arrival in what was to become the solar system remains a task to be accomplished.

Isotope abundance anomalies are the key feature based on which the presolar nature of a given grain is ascertained. Based on this criterion, all of the SiC grains contained in primitive meteorites are presolar, but only a small fraction of oxide grains. Applying the isotope anomaly criterion has also its limits, as is shown by the case of the nanodiamonds which consist on the average of $\sim$1000 atoms of carbon only. Requiring that a given grain (which cannot be analyzed singly anyway) that is presolar must carry an atom of anomalous xenon - as sometimes is done - misses a point: in nature there is no such mineral that in bulk contains as much as $\sim$ 1% of xenon by weight. In addition xenon would be accompanied by at least some 100 atoms of other elements of the same neutron capture origin, so that such a grain would no longer be anything resembling a diamond any more. While this shows our limit in ascertaining a presolar origin for a given grain, there have also been limits to our ability to isolate presolar grains. So far we have mostly relied on their chemical resistance. But it is likely that there are also acid-soluble presolar grains, e.g. of silicate composition, as indicated by isotope anomalies found during stepwise dissolution of primitive meteorites (e.g., [95]) and their recent discovery in interplanetary dust particles [121]. The hope exists that such grains - which are probably also very fine-grained - can be found by a new generation of analytical equipment [8].

Acknowledgments

Thanks go to a large number of colleagues in the field for many discussions on topics relating to presolar grains in meteorites. Special thanks go to Peter Hoppe, who also contributed Figs. 1, 4-7, and to Sachiko Amari who also supplied Figs. 3a , 3c and 3d. Figure 3b and 9 are from T.J. Bernatowicz, Fig. 3e from G.R. Huss, Fig. 3f from F. Banhart and Fig. 10 from A. Braatz.

References

[1] G.W. Lugmair, A. Shukolyukov, Meteoritics Planet Sci. **36**, 1017 (2001)
[2] J.A. Whitby, R. Burgess, G. Turner, J. Gilmour, J. Bridges, Science **288**, 1819 (2000)
[3] I. Mann, E.K. Jessberger: this volume
[4] T.J. Bernatowicz, E. Zinner (eds.): Astrophysical Implications of the Laboratory Study of Presolar Materials (American Institute of Physics, Woodbury, N.Y. 1997)
[5] E. Zinner, Annu. Rev. Earth Planet. Sci. **26**, 147 (1998)
[6] P. Hoppe, E. Zinner, Journ. Geophys. Res. **105**, 10371 (2000)
[7] J.P. Bradley, Science **265**, 925 (1994)
[8] E. Zinner, Meteoritics Planet. Sci. 33, 549 (1998)
[9] T. Lee: 'Implications of Isotopic Anomalies for Nucleosynthesis'. In: Meteorites and the Early Solar System, ed. by J.F. Kerridge, M.S. Matthews (University of Arizona Press, Tucson, 1988), pp 1063-1089
[10] U. Ott, Nature **364**, 25 (1993)
[11] F. Robert, S. Epstein, Geochim. Cosmochim. Acta **46**, 81 (1982)
[12] F. Robert, Science **293**, 1056 (2001)
[13] H.E. Suess, Annu. Rev. Astron. Astrophys. 3, **217** (1965)
[14] J.H. Reynolds, G. Turner, Journ. Geophys. Res. **69**, 3263 (1964)
[15] G.R. Huss, R.S. Lewis, Meteoritics **29**, 791 (1994)
[16] D.C. Black, R.O. Pepin, Earth Planet. Sci. Lett. **6**, 395 (1969)
[17] R.S. Lewis, M. Tang, J.F. Wacker, E. Anders, E. Steel, Nature **326**, 160 (1987)
[18] T. Bernatowicz, G. Fraundorf, M. Tang, E. Anders, B. Wopenka, E. Zinner, P. Fraundorf, Nature **330**, 728 (1987)
[19] R.S. Lewis, B. Srinivasan, E. Anders, Science **190**, 1251 (1975)
[20] B. Srinivasan, E. Anders, Science **201**, 51 (1978)
[21] S. Amari, R.S. Lewis, E. Anders, Geochim. Cosmochim. Acta **58**, 459 (1994)
[22] S. Merchel, U. Ott, Meteoritics Planet. Sci. **34**, A81 (1999)
[23] L.R. Nittler, C.M.O'D. Alexander, F. Tera, Meteoritics Planet. Sci. **36**, A 149 (2001)

[24] S.S. Russell, U. Ott, C.M.O'D. Alexander, E.K. Zinner, J.W. Arden, C.T. Pillinger, Meteoritics Planet. Sci. **32**, 719 (1997)

[25] U. Ott, S. Merchel, Lunar Planet. Sci. XXXI, #1356 (CD-ROM) (2000)

[26] G.R. Huss, A.P. Meshik, C.M. Hohenberg, Lunar Planet. Sci. XXXII, #1685 (CD-ROM) (2001)

[27] T.L. Daulton, R.S. Lewis, S. Amari, Meteoritics Planet. Sci. **33**, A37 (1998)

[28] C.M.O'D. Alexander, P. Swan, R.M. Walker, Nature **348**, 715 (1990)

[29] R.J. Macke, T. Bernatowicz, P. Swan, R.M. Walker, E. Zinner, Lunar Planet. Sci. XXX, #1435 (CD-ROM) (1999)

[30] I.D. Hutcheon, G.R. Huss, A.J. Fahey, G.J. Wasserburg, Astrophys. Journ. **425**, L97 (1994)

[31] L.R. Nittler, C.M.O'D. Alexander, X. Gao, R.M. Walker, E.K. Zinner, Nature **370**, 443 (1994)

[32] L.R. Nittler, P. Hoppe, C.M.O'D. Alexander, S. Amari, P. Eberhardt, X. Gao, R.S. Lewis, R. Strebel, R.M. Walker, E. Zinner, Astrophys. Journ. **453**, L25 (1995)

[33] A. Virag, E. Zinner, S. Amari, E. Anders, Geochim. Cosmochim. Acta **55**, 2045 (1991)

[34] T.J. Bernatowicz, R. Cowsik, P.C. Gibbons, K. Lodders. B. Fegley Jr., S. Amari, R.S. Lewis, Astrophys. Journ. **472**, 760 (1996)

[35] T.J. Bernatowicz, S. Amari, R.S. Lewis, Lunar Planet. Sci. XXIII, 91 (1992)

[36] P. Hoppe, U. Ott, 'Mainstream silicon carbide grains from meteorites'. In: Astrophysical Implications of the Laboratory Study of Presolar Materials, ed. by T.J. Bernatowicz, E. Zinner (American Institute of Physics, Woodbury, N.Y. 1997), pp. 27-58

[37] E. Zinner, S. Amari, R.S. Lewis, Astrophys. Journ. **382**, L47 (1991)

[38] G.K. Nicolussi, A.M. Davis, M.J. Pellin, R.S. Lewis, R.N. Clayton, S. Amari, Science **277**, 1281 (1997)

[39] M.J. Pellin, A.M. Davis, R.S. Lewis, S. Amari, R.N. Clayton, Lunar Planet. Sci. XXX, #1969 (CD-ROM) (1999)

[40] S. Amari, P. Hoppe, E. Zinner, R.S. Lewis, Meteoritics **30**, 679 (1995)

[41] R.H. Nichols Jr., C.M. Hohenberg, P. Hoppe, S. Amari, R.S. Lewis, Lunar Planet. Sci. XXIII, 989 (1992)

[42] D.C.B. Whittet, Dust in the Galactic Environment (Inst. Phys., New York, 1992), 295 pp

[43] Th. Henning, F. Salama, Science **282**, 2204 (1998)

[44] I.R. Little-Marenin, Astrophys. Journ. **307**, L15 (1986)

[45] A.K. Speck, M.J. Barlow, C.J. Skinner, Monthly Not. R. Astron. Soc. **234**, 79 (1997)

[46] C.M.O'D. Alexander, Geochim. Cosmochim. Acta **57**, 2869 (1993)

[47] R. Gallino, M. Busso, G. Picchio, C.M. Raiteri, Nature **348**, 298 (1990)

[48] U. Ott, F. Begemann, Astrophys. Journ. **353**, L57 (1990)

[49] R. Gallino, C.M. Raiteri, M. Busso, Astrophys. Journ. **410**, 400 (1993)

[50] F.X. Timmes, D.D. Clayton, Astrophys. Journ. **472**, 723 (1996)

[51] S. Amari, L.R. Nittler, E. Zinner, K. Lodders, R.S. Lewis, Astrophys. Journ. **559**, 463 (2001)

[52] S.E. Woosley, T.A. Weaver, Astrophys. Journ. Suppl. **101**, 181 (1995)

[53] F.X. Timmes, S.E. Woosley, D.H. Hartmann, R.D. Hoffman, Astrophys. Journ. **464**, 332 (1996)

[54] D.D. Clayton, W.D. Arnett, J. Kane, B.S. Meyer, Astrophys. Journ. **486**, 824 (1997)

[55] B.S. Meyer, D.D. Clayton, L.-S. The, Astrophys. Journ. **540**, L49 (2000)

[56] S. Amari, L.R. Nittler, E. Zinner, R. Gallino, M. Lugaro, R.S. Lewis, Astrophys. Journ. **546**, 248 (2001)

[57] P. Hoppe, P. Annen, R. Strebel, P. Eberhardt, R. Gallino, M. Lugaro, S. Amari, R.S. Lewis, Astrophys. Journ. **487**, L101 (1997)

[58] P. Hoppe, S. Amari, E. Zinner, R.S. Lewis, Geochim. Cosmochim. Acta **59**, 4029 (1995)

[59] E. Zinner, S. Amari, B. Wopenka, R.S. Lewis, Meteoritics **30**, 209 (1995)

[60] S. Amari, E. Zinner, R.S. Lewis, Astrophys. Journ. **470**, L101 (1995)

[61] S. Amari, R.S. Lewis, E. Anders, Geochim. Cosmochim. Acta **59**, 1411 (1995)

[62] R.H. Nichols Jr., K. Kehm, R. Brazzle, S. Amari, C.M. Hohenberg, R.S. Lewis, Meteoritics **29**, 510 (1994)

[63] L.R. Nittler, C.M.O'D. Alexander, X. Gao, R.M. Walker, E. Zinner, Astrophys. Journ. **483**, 475 (1997)

[64] G.J. Wasserburg, A.I. Boothroyd, I.-J. Sackmann, Astrophys. Journ. **447**, L37 (1995)

[65] L.R. Nittler, C.M.O'D. Alexander, J. Wang, X. Gao, Nature **393**, 222 (1998)

[66] B.-G. Choi, G.R. Huss, G.J. Wasserburg, R. Gallino, Science **282**, 1284 (1998)

[67] D.D. Clayton, B.S. Meyer, C.I. Sanderson, S.S. Russell, C.T. Pillinger, Astrophys. Journ. **447**, 894 (1995)

[68] S.S. Russell, J.W. Arden, C.T. Pillinger, Meteoritics Planet. Sci. **31**, 343 (1996)

[69] R. Kallenbach, J. Geiss, F.M. Ipavich, G. Gloeckler, P. Bochsler, F. Gliem, S. Hefti, M. Hilchenbach, D. Hovestadt, Astrophys. Journ. **507**, L185 (1998)

[70] J.F. Kerridge, Rev. Geophys. **31**, 423 (1993)

[71] K. Hashizume, M. Chaussidon, B. Marty, F. Robert, Science **290**, 1142 (2000)

[72] T. Owen, P.R. Mahaffy, H.B. Niemann, S. Atreya, M. Wong, Astrophys. Journ. **553**, L77 (2001)

[73] U. Ott: 'Noble gases in meteorites - trapped components'. In: Noble Gases and Cosmochemistry (Reviews in Mineralogy and Geochemistry), ed. by D. Porcelli, C.J. Ballentine, R. Wieler, in press (2002)

[74] S. Richter, U. Ott, F. Begemann, Nature **391**, 261 (1998)

[75] R. Maas, R.D. Loss, K.J.R. Rosman, J.R. DeLaeter, R.S. Lewis, G.R. Huss, G.W. Lugmair, Meteoritics Planet. Sci. **36**, 849 (2001)

[76] R.S. Lewis, G.R. Huss, G. Lugmair, Lunar Planet. Sci. XXII, 807 (1991)

[77] D. Heymann, M. Dziczkaniec, Proc. Lunar Sci. Conf. 10th, 1943 (1979)

[78] D.D. Clayton, Astrophys. Journ. **340**, 613 (1989)

[79] U. Ott, Astrophys. Journ. **463**, 344 (1996)

[80] D. Heymann, M. Dziczkaniec, Meteoritics **15**, 15 (1980)

[81] B.-G. Choi, G.J. Wasserburg, G.R. Huss, Astrophys. Journ. **522**, L133 (1999)

[82] R. Gallino, C.M. Raiteri, M. Busso, F. Matteucci, Astrophys. Journ. **430**, 858 (1994)

[83] G.R. Huss, A.J. Fahey, R. Gallino, G.J. Wasserburg, Astrophys. Journ. **430**, L81 (1994)

[84] C. Travaglio, R. Gallino, S. Amari, E. Zinner, S. Woosley, R.S. Lewis, Astrophys. Journ. **510**, 325 (1999)

[85] S. Amari, E. Zinner, D.D. Clayton, B.S. Meyer, Meteoritics Planet. Sci. **33**, A10 (1998)

[86] D.D. Clayton, W. Liu, A. Dalgarno, Science **283**, 1290 (1999)

[87] A.I. Boothroyd, I.-J. Sackmann, G.J. Wasserburg, Astrophys. Journ. **442**, L21 (1995)

[88] G.R. Huss, I.D. Hutcheon, G.J. Wasserburg, Geochim. Cosmochim. Acta **61**, 5117 (1997)

[89] F. Käppeler, H. Beer, K. Wisshak, Rep. Prog. Phys. **52**, 945 (1989)

[90] F. Käppeler, R. Gallino, M. Busso, G. Picchio, C.M. Raiteri, Astrophys. Journ. **354**, 630 (1990)

[91] C. Arlandini, F. Käppeler, K. Wisshak, R. Gallino, M. Lugaro, M. Busso, O. Straniero, Astrophys. Journ. **525**, 886 (1999)

[92] U. Ott, Proc. Indian Aca. Sci. (Earth Planet. Sci.) **107**, 379 (1998)

[93] W.M. Howard, B.S. Meyer, D.D. Clayton, Meteoritics **27**, 404 (1992)

[94] K. Lodders, B. Fegley Jr., Meteoritics **30**, 661 (1995)

[95] F.A. Podosek, U. Ott, J.C. Brannon, C.R. Neal, T.J. Bernatowicz, P. Swan, S.E. Mahan, Meteoritics Planet. Sci. **32**, 617 (1997)

[96] A. Virag, B. Wopenka, S. Amari, E. Zinner, E. Anders, R.S. Lewis, Geochim. Cosmochim. Acta **56**, 1715 (1992)

[97] A.C. Andersen, C. Jäger, H. Mutschke, A. Braatz, D. Clément, Th. Henning, U.G. Jørgensen, U. Ott, Astron. Astrophys. **343**, 933 (1999)

[98] T.L. Daulton, D.D. Eisenhour, T.J. Bernatowicz, R.S. Lewis, P. Buseck, Geochim. Cosmochim. Acta **60**, 4853 (1996)

[99] A.B. Verchovsky, A.V. Fisenko, L.F. Semjonova, I.P. Wright, M.R. Lee, C.T. Pillinger: Science **281**, 1165 (1998)

[100] F. Banhart, Y. Lyutovich, A. Braatz, C. Jäger, T. Henning, J. Dorschner, U. Ott, Meteoritics Planet. Sci. **33**, A12 (1998)

[101] J.S. Mathis, W. Rumpl, K.H. Nordsieck, Astrophys. Journ. **217**, 425 (1977)

[102] P.C. Frisch et al., Astrophys. Journ. **525**, 492 (1999)

[103] R.S. Lewis, E. Anders, B.T. Draine, Nature **339**, 117 (1989)

[104] L. Colangeli, V. Mennella, J. Stephens, E. Bussoletti, Astron. Astrophys. **248**,583 (1994)

[105] H. Hill, L. D'Hendecourt, C. Perron, A. Jones, Meteoritics Planet. Sci. **32**, 713 (1997)

[106] A.C. Andersen, U.G. Jørgensen, F.M. Nicolaisen, P.G. Sørensen, K. Glejbjøl, Astron. Astrophys. **330**, 1080 (1998)

[107] A. Braatz, U. Ott, Th. Henning, C. Jäger, G. Jeschke, Meteoritics Planet. Sci. **35**, 75 (2000)

[108] L. Allamandola, S. Sandford, A. Tielens, T. Herbst, Science **260**, 64 (1993)

[109] O. Guillois, G. Ledoux, C. Reynaud, Astrophys. Journ. **521**, L33 (1999)

[110] R.M. Walker, comm. to U.B. Marvin, Meteoritics Planet. Sci. **36**, A275 (2001)

[111] M. Tang, E. Anders, Astrophys. Journ. **335**, L31 (1988)

[112] R.S. Lewis, S. Amari, E. Anders, Geochim. Cosmochim. Acta **58**, 471 (1994)

[113] U. Ott, F. Begemann, Meteoritics Planet. Sci. **35**, 53 (2000)

[114] U. Ott, M. Altmaier, U. Herpers, J. Kuhnhenn, S. Merchel, R. Michel, R.K. Mohapatra, Meteoritics Planet. Sci. **36**, A155 (2001)

[115] G.R. Huss, R.S. Lewis, Geochim. Cosmochim. Acta **59**, 115 (1995)

[116] G.R. Huss, R.S. Lewis, Meteoritics **29**, 811 (1994)

[117] A.P. Koscheev, M.D. Gromov, R.K. Mohapatra, U. Ott, Nature **412**, 615 (2001)

[118] G.R. Huss: 'The survival of presolar grains in solar system bodies'. In: Astrophysical Implications of the Laboratory Study of Presolar Materials, ed. by T.J. Bernatowicz, E. Zinner (American Institute of Physics, Woodbury, N.Y. 1997), pp. 721-748

[119] P. Hoppe, Nuclear Physics **A688**, 94 c (2001)

[120] C. Dominik, H.-P. Gail, E. Sedlmayr, Astron. Astrophys. **223**, 227 (1989)

[121] S. Messenger, L.P. Keller, R.M. Walker, Lunar Planet. Sci. XXXIII, #1887 (CD-ROM) (2002)

[122] H. Mutschke, A.C. Andersen, D. Clément, Th. Henning, G. Peiter, Astron. Astrophys. **345**, 187 (1999)

[123] Th. Henning, H. Mutschke, Spectrochimica Acta Part A **57**, 815 (2001)

[124] J. Bradley, this volume

Laboratory Astrophysics
of Cosmic Dust Analogues

Thomas Henning

Astrophysical Institute and University Observatory (AIU),
Schillergässchen 3, D-07745 Jena, Germany
and
Max Planck Institute for Astronomy,
Königstuhl 17, D-69117 Heidelberg, Germany

Abstract. In this chapter, the main techniques for producing and characterizing cosmic dust analogues in the laboratory will be discussed. It will be shown how optical data of astronomically relevant materials can be measured and how such data can be applied to interpret astronomical spectra. The identification of minerals in space from infrared spectroscopy will be summarized.

1 Introduction

The identification of distinct dust components in astronomical spectra requires the knowledge of basic optical data, which have to be provided by laboratory measurements. In contrast to bulk material, solids in space form a system of mostly submicron-sized isolated particles. Absorption and scattering of light by such particles depend on their shape and size distribution as well as on the material of which they are composed. Surface modes in small particles cause interesting absorption features [5]. Spectra of small metallic particles can show features where none are present in the bulk material. Non-conducting grains show often several features with the wavelength positions and widths depending on both the dielectric function and the shape of the particles. Only spherical particles are characterized by a single resonance, whereas a distribution of shapes leads to broad absorption bands with smaller peak absorption. A condition for the presence of surface modes is a negative value of the real part of the dielectric function. A variety of experimental methods exist to determine the complex dielectric function. In this contribution, we will discuss how such particles can be produced and characterized.

The dust particles not only absorb light, but also thermally re-radiate energy at infrared and millimetre wavelengths. Therefore, spectral energy distributions of dusty astronomical objects also depend on the temperature of the particles, which is determined by the balance of absorption and re-emission of radiation. Grains of different size and chemical composition will have distinct temperatures. The treatment of radiation transfer through a dusty medium requires the solution of a transfer equation, which complicates the analysis of astronomical spectra [19].

2   What Are Cosmic Dust Analogue Materials?

First constraints on the nature of cosmic grains come from an analysis of abundance patterns, namely the depletion of elements in the interstellar medium with respect to a given reference abundance, and the chemical composition of stellar outflows from evolved stars. Presently, there is a discussion if the solar abundances can be taken as the reference values or if the photospheric abundances of recently formed B stars better represent the interstellar medium [53] [29]. The latter values are about 2/3 of the solar abundances and would impose strong constraints on dust models. The potential dust-forming elements can be divided in three major groups [29] [11]:

1. Primary dust-forming elements (abundance $\geq$ 100 ppm): carbon, oxygen, and nitrogen
2. Abundant dust-forming elements (abundance $\sim$ 30 ppm): magnesium, silicon, sulphur, iron
3. Minor dust-forming elements (abundance $\sim$ 3 ppm): aluminium, calcium, sodium, nickel

The abundance values are given relative to 10^6 atoms of hydrogen (ppm). With the exception of sulphur, all these elements are heavily depleted in the cool atomic hydrogen clouds ('diffuse clouds') of the galactic disk [47]. Therefore, the presence of silicon and iron particles, silicates and various metal oxides, carbonaceous grains with various hybridization states, and carbides can be expected in space. The abundant element nitrogen is usually bound to N_2 molecules, but may form nitrides in nitrogen-rich circumstellar environments.

Cosmic dust grains are part of a lifecycle of interstellar matter [12]. New ones form in the outflows of evolved stars. Depending on the chemical composition of these outflows, especially the C/O ratio, different types of grains are produced from the gas phase (cf. contribution by Gail). The small transition group of S-type stars is characterized by a C/O ratio very close to 1. As a consequence of nucleosynthesis and dredge-up processes during AGB evolution, 'oxygen-rich' stars with masses smaller than 4 $M_\odot$ will evolve into carbon stars and the molecular composition of the outflows will change. Other factors influencing the chemical state of the outflows, apart from the C/O ratio, can be the formation of molecules by grain surface reactions, the dissociation of some of the CO and N_2 molecules by chromospheric UV radiation and the influence of shock chemistry.

In oxygen-rich envelopes, the most abundant elements available for grain formation are O, Fe, Si, Mg, Al, and Ca. Without considering the kinetics of grain formation, we therefore expect the presence of silicates and other oxides (silica, corundum, spinel), iron and silicon grains. Silicates of olivine and pyroxene structure are of special interest. Olivines are solid solutions of forsterite (Mg_2SiO_4) and fayalite (Fe_2SiO_4), whereas pyroxenes are solid solutions of ferrosilite $(FeSiO_3)$ and enstatite $(MgSiO_3)$. The carbon-rich outflows lead to the production of carbonaceous grains as well as SiC and TiC particles. Si_3N_4, FeSi, and FeS/MgS grains are among other possible circumstellar grain components.

'Stardust' particles are modified during their journey through the interstellar medium by erosion in interstellar shock fronts, UV and ion irradiation, and surface reactions with atomic hydrogen, leading most probably to amorphous materials. In dense and UV-shielded regions such as the cold molecular cloud cores and the interiors of protoplanetary accretion disks, molecular ices form on the grain surfaces and particles grow by mutual collisions. In this paper, I will only deal with refractory materials and not with molecular ices or larger molecules such as carbon chains and polycyclic aromatic hydrocarbons (PAHs).

Laboratory dust analogues should reflect the expected chemical composition of the particles as well as their structural state. Silicates and other oxides, carbonaceous grains, carbides, nitrides, and sulphides are the major material classes which have to be considered. The investigation of 'cosmic dust analogues' does not necessarily aim at exactly producing particles with the properties of cosmic dust. It also includes the investigation of bulk matter or thin films prepared for the derivation of the dielectric material functions or the investigation of the annealing behaviour of different materials. Here one should note that some properties of small particles cannot be studied with bulk material. In addition, certain systems may not exist as bulk material at all. This is especially true for carbonaceous particles (e.g. carbon onions and nanotubes) and silicon nanoparticles.

3   Material Production and Characterization

The experimental investigation of cosmic dust analogues is generally based on three major steps: (i) synthesis of the material, (ii) chemical and structural characterization of the samples, and (iii) measurement of the optical properties over a wavelength range as broad as possible. The latter step is far from trivial because it requires different light sources and spectroscopic facilities.

Fig. 1. Jena laser pyrolysis system for the production of small particles

Production methods for refractory solids, applied in laboratory astrophysics, include: laser ablation and thermal evaporation and subsequent condensation in an inert (Ar, He) or reducing (H_2) atmosphere, infrared laser-induced and microwave-induced gas pyrolysis, arc discharges under various atmospheric conditions, burning of hydrocarbons, plasma deposition techniques, and sol-gel reactions in the case of silicatic materials [22] [11]. Fig. 1 shows the Jena laser pyrolysis system which has been used for the production of carbide, nitride, and silicon nanoparticles. Here, one has to keep in mind that "good" cosmic dust analogues may considerably deviate in stoichiometry and crystallinity from materials which are easily available for laboratory analysis. Therefore, dedicated preparative work is often required for producing cosmic dust analogues.

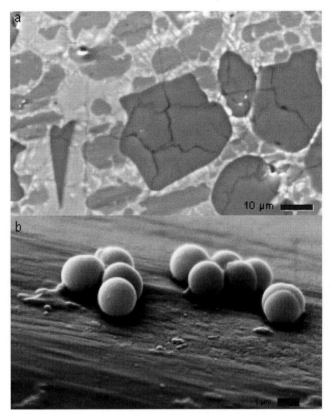

Fig. 2. a) SEM image of an inhomogeneous pyroxene (see text for a discussion). b) Spherical SiO_2 particles on a mineral surface. The average diameter of the spheres is 1 μm.

To understand the relation between structural and optical properties of dust grains, they have to be thoroughly characterized. Depending on the material, different methods have been used. Table 1 summarizes the different structural characterization techniques for silicates and their pros and cons. For carbonaceous grains, we refer the reader to the review by Henning & Schnaiter [24]. As examples for the characterization of silicon-based materials, Fig. 2a shows an SEM (scanning electron microsocopy) image of an inhomogeneous silicate of average pyroxene stoichiometry ($Mg_{0.4}Fe_{0.6}SiO_3$) and Fig. 2b some SiO_2 particles. The pyroxene has been produced by melting SiO_2, $MgCO_3$, and $FeC_2O_4 \cdot 6H_2O$ in the right stoichiometric ratio and subsequent slow cooling of the melt [28]. The sample is characterized by phase separation which can be traced by the backscattering electron image. The dark gray phase is a SiO_2 component, the gray phase is made of a Mg-rich pyroxene and the light phase is an Fe-rich pyroxene. The chemical composition of the different phases could be determined by Energy Dispersive X-ray (EDX) analysis. Any determination of optical data of such a sample has to take into account that the measurement is an average over different phases with different optical properties. SEM images can also be used to characterize the morphology of smoke or powder particles. Fig. 2b shows an SEM image of SiO_2 spheres on a surface, containing information on the size, shape, and agglomeration state of grains. These parameters have to be known for the interpretation of spectroscopic measurements.

Table 1. Characterization methods for silicate dust analogues

Method	Sensitivity	Pros	Cons
Wet-chemical analysis	ratio of polyvalent ions	very accurate	time-consuming, average over sample
Energy dispersive X-ray analysis	elemental composition	check of stoichiometry and purity	no elements lighter than carbon
(HR)TEM	global ordering	direct insight into the meso- and macroscopic solid-state structure	'expensive' search for representative structures
X-ray diffraction	local ordering	detailed coordination and bonding study, analysis of crystal phases	large amounts of material required
X-ray absorption spectroscopy	local- and medium-sized ordering	distance of next neighbour and coordination study	
IR spectr.	local and global ordering	very sensitive to small structural differences	difficult to interpret for mixtures

4   Measurement of Optical Properties

The description of the interaction of radiation with small particles requires the knowledge of the frequency-dependent extinction, absorption, and scattering cross sections C_{ext}, C_{abs}, and C_{sca}. These quantities depend on the material properties and the size, shape and agglomeration state of the particles. The fundamental material properties can be described either by the optical constants $m = n + ik$ or the dielectric function $\epsilon = \epsilon_1 + i\ \epsilon_2$, where $\epsilon_1 = n^2 - k^2$ and $\epsilon_2 = 2\ nk$ are the relations for non-magnetic materials. The optical constants are directly related to the phase velocity and attenuation of plane waves in the material. There are two different approaches for the determination of the optical properties of particles (see Fig. 3):

1. Calculation of cross sections from optical constants. The optical constants are usually determined from transmission measurements on thin films or reflection measurements on bulk materials (strong bands) or transmission measurements on bulk samples (small k). Reflection spectroscopy works with polished surfaces, using a variety of angles of incidence. A special case of reflection spectroscopy is ellipsometry which is characterized by a measurement of the change of polarization state. Transmission spectroscopy on thin films can be combined with reflection spectroscopy.
   The limitation of this approach is the fact that the bulk material may not have the structure of the nano- and microparticles. In addition, the calculation of cross sections from optical constants is not yet possible for complicated grain shapes or agglomerates in regions of strong resonances. In [22] and [17], a short summary of the procedure for calculating cross sections from optical constants is given.
2. Direct measurement of the cross sections on particle samples. This is certainly the way that should be preferred, but also the more complicated approach. One has to produce particles of the desired composition and structure in a narrow size and shape range and has to characterize them by spectroscopy. This is often difficult because conventional infrared spectroscopy (i) uses matrices in which the particles are embedded and (ii) is not able to prevent the agglomeration of primary particles to larger structures. Possibilities to reduce these effects are measurements of isolated particles in argon or neon matrices or direct spectroscopy by intense infrared radiation coming from synchrotron or free-electron laser facilities.
   Furthermore, transmission spectroscopy on a system of small isolated particles can be used to determine optical constants if the size and shape of particles are known.

The frequency dependence of the optical constants (dispersion relation) is determined by resonances of the electronic system, of the ionic lattice and, at very low frequencies, by relaxation of permanent dipoles. Here, we should note that many crystals, including the astronomically interesting crystalline silicates, show an anisotropy in their optical constants. In these cases, measurements with polarized light along the different axes of the crystal have to be carried out. These

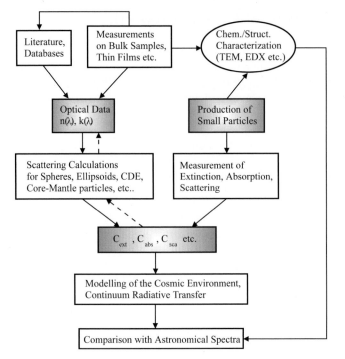

Fig. 3. Different approaches for determining the optical properties of cosmic dust. After [22]

measurements require a careful orientation of the crystal and an alignment of the polarizers. An example of such data is given in Fig. 4. In addition, the optical constants of cosmic dust analogues may depend on the temperature, leading to a change of the width and wavelength position of sharp bands of crystalline material or a change of the far-infrared continuum absorption [1] [21] [36] [6] [10].

It should be stressed that optical constants are macroscopic quantities. They loose their meaning for small clusters and molecules. In the transition region from solids to molecules, the introduction of size-dependent optical constants may be a reasonable way to describe the behaviour of the systems. This approach has been used for the description of the confinement of charge carriers to the limited volume of small metallic particles [33].

Compilations of frequency-dependent optical constants of solids can be found in the various editions of the 'Handbook of Optical Constants' [42] [43] [44]. An electronic database [26], especially dedicated to cosmic dust analogues, can be accessed via the web page http://www.astro.uni-jena.de.

It has been already mentioned that the absorption and scattering cross sections (or the mass absorption coefficients) not only depend on the optical constants, but also on the shape and size of the particles. This is demonstrated in Fig. 5 where the mass absorption coefficent of forsterite is shown for spheres and two different distributions of ellipsoids. In addition, powder spectra of the

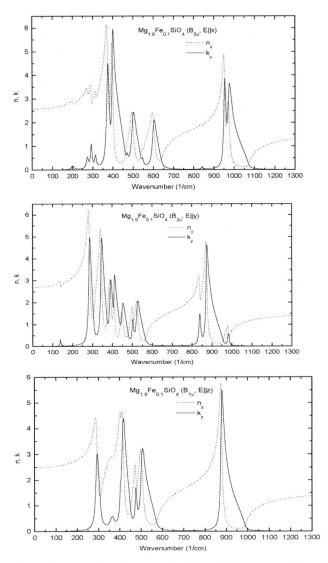

Fig. 4. Complex index of refraction $n + ik$ of an olivine crystal for three different crystallographic orientations. After [16]

small particles and observationally-determined band positions are plotted. It can be seen that the width, shape and position of strong 'crystalline' bands depend on the shape of the particles. This effect is much less important for the weaker far-infrared bands, which are practically not influenced by the particle shape. A comparison between mass absorption coefficients and astronomical spectra of optically thin envelopes around evolved stars indicates that both sphere-like

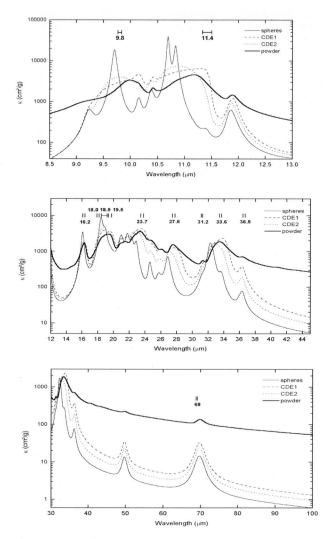

Fig. 5. Mass absorption coefficients of forsterite spheres and two different distributions of ellipsoids (CDE 1 and CDE 2) in vacuum and the Rayleigh limit. In addition, the results of transmission measurements on small particles (diameter < 1 μm) dispersed in a KBr/PE pellet are shown. Peak positions assigned to forsterite on the basis of ISO spectra from F. Molster [38] have been indicated. After [16].

particles and ellipsoids elongated along the z-axis are necessary to obtain a convincing band assignment [16].

5   Interpretation of Astronomical Data

The infrared spectral region has been often considered to be the 'fingerprint' region for the identification of the chemical composition of cosmic dust particles. Apart from the numerous diffuse interstellar bands (DIBs) at optical wavelengths [27], probably caused by carbon-containing molecules, and the extended red emission (ERE), recently related to silicon nanoparticles [59] [35], the UV/optical wavelength range contains only one strong resonance at 217 nm. This 'UV bump' is widely attributed to an electronic transition in carbonaceous grains [24]. An interesting additional fact is that observations show that the UV bump is located between 240 and 250 nm in the spectra of R Coronae Borealis stars - a special class of hydrogen-poor post-AGB stars which eject large amounts of carbon dust ar random times [14, 18].

Infrared spectra of solid particles reflect the fundamental vibrations of molecular bonds present in the material as well as the lattice modes of the solid. The broad solid-state bands are easily distinguishable from the rotational-vibrational bands of gas-phase molecules. A limitation of infrared spectroscopy is the fact that the identification of a specific carrier is difficult if only one band is observed. In addition, complex materials which contain similar molecular species may produce very similar spectra. This is especially true for carbonaceous grains.

Table 5 summarizes identifications of refractory materials based on infrared bands together with some key references which lead to more complete reference lists. We should stress again that the identification based on a single band has to be taken with caution. In the case of the 21 μm feature, SiS_2 is probably not the carrier of the band because it shows a second feature which is not present in spectroscopic data from the *Infrared Space Observatory ISO*.

Apart from the components discussed in Table 5, silicates are a widespread component of cosmic dust [11] [20]. Spectroscopic evidence for the presence of amorphous silicates comes from broad bands at 10 and 18 μm which are attributed to Si–O stretching and Si–O–Si bending modes. The depletion of oxygen and silicon is roughly in agreement with these atoms being bound in SiO_4 tetrahedra [54]. The *Infrared Space Observatory ISO* found clear evidence for the presence of crystalline silicates. Sharp and distinct features typical of Fe-poor olivines (i.e. Mg_2SiO_4) and pyroxenes (i.e. $MgSiO_3$) could be observed for wavelengths beyond 20 μm (cf. contribution by Molster & Waters).

Cosmic dust analogues investigated so far include a wide variety of carbonaceous grains, silicates, metal oxides, carbides, sulphides, and carbonates. We will discuss some of the relevant experimental investigations in the following sections, but will concentrate on infrared studies of carbon- and silicon-based materials. For more complete reviews and extensive compilations of references, the reader is referred to the papers by Colangeli et al. [11] and Henning [20] for silicates, Henning & Mutschke [23] for carbides, and Henning & Schnaiter [24] and Pendleton & Allamandola [45] for carbonaceous solids.

Table 2. Refractory Dust Components in Space

Wavelength	Component	Sources	References
3.4 μm, 6.85, 7.25 μm	Aliphatic hydrocarbons	C-rich CSE, diffuse ISM[b]	Schnaiter et al. 1999, Mennella et al. 2002 Pendleton & Allamandola 2002
5.5 μm	Metal carbonyls	Galactic centre	Tielens et al. 1996
10, 18 μm	Silicates[a]		Draine & Lee 1984, Ossenkopf et al. 1992, Henning 2002
11.3 μm	SiC	C-rich CSE[c]	Speck et al. 1997, 1999 Mutschke et al. 1999
13 μm	α-Al$_2$O$_3$ (CDE)	O-rich CSE	Sloan et al. 1996, Koike et al. 1995, Begemann et al. 1997
and 17, 32 μm	MgAl$_2$O$_4$		Posch et al. 1999, Fabian et al. 2001
21 (20.1) μm	Carbonaceous material SiS$_2$	C-rich CSE	Kwok et al. 1995, Volk et al. 1999 Henning et al. 1996, Begemann et al. 1996
	TiC nanocrystals		von Helden et al. 2000
23 μm	FeO (CDE) FeS	HAEBE stars[d]	van den Ancker et al. 2000 Keller et al. 2001
30 μm	MgS	C-rich CSE	Omont et al. 1995 Begemann et al. 1994 Szczerba et al. 1999
92.6 μm	CaCO$_3$	planetary nebulae	Kemper et al. 2002
		young stellar objects	Ceccarelli et al. 2002

a) Only main features listed; enstatites and forsterites securely identified;
b) ISM - interstellar medium;
c) CSE - circumstellar envelopes,
d) HAEBE stars- Herbig Ae/Be stars

5.1   Carbon-Based Solids

The observation of an absorption band at 3.4 μm along different lines of sight through the galactic diffuse interstellar medium and in some other galaxies provides direct evidence for the presence of hydrogenated carbonaceous grains with aliphatic character. The profile of this band shows substructure at about 3.38, 3.42 and 3.48 μm produced by C–H stretching vibrations in the methyl (–CH$_3$) and methylene (–CH$_2$–) groups. The average –CH$_2$–/–CH$_3$ ratio is about 2.5. The corresponding C–H deformation modes at 6.85 μm and 7.25 μm could be detected by observations with the Kuiper Airborne Observatory (only at 6.85 μm [55]) and ISO [9]. The structure of the 'interstellar' 3.4 μm band is quite similar to the band found in the C-rich protoplanetary nebula CRL 618 [8]. This type of band is completely absent in dense molecular clouds.

The identification of a specific material, producing the features, is not simple because the C–H modes only reflect the local bonding environment and are present in a wide variety of carbonceous materials. Recently, Schnaiter et al. [48] produced carbonaceous nanoparticles which reproduced both the 3.4 μm profile and the UV absorption band at 217 nm (see Fig. 6). It would be interesting to find out if these two features are indeed produced by the same material.

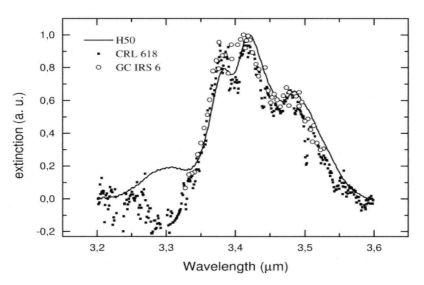

Fig. 6. Comparison of the 3.4 μm feature of the diffuse interstellar medium (dots) with measurements on hydrogenated carbon nanoparticles (solid line). After [48].

A very comprehensive comparison between astronomical spectra in the 2.5-10 μm range and various laboratory dust analogues has been performed by Pendleton & Allamandola [45]. Their general conclusion is that materials produced by plasma processes reproduce the astronomical data better than ices processed by the irradiation with energetic ions or UV light. The material is characterized by a hydrocarbon, with carbon distributed between aliphatic and aromatic structures and only minor contributions from oxygen and nitrogen. The investigation of UV irradiation of carbonaceous materials and the interaction with atomic hydrogen, expected to occur in the diffuse interstellar medium, led to the proposal that the material is formed in the diffuse medium with an equilibrium value for its degree of hydrogenation [37].

Apart from hydrocarbon particles, SiC (11.3 μm feature) and TiC grains (20.1 μm feature) have been identified as a dust component of carbon-rich circumstellar envelopes. The identifcation of a specific SiC material as the carrier of the 11.3 μm feature is complicated by the lack of an intrinsic astronomical band shape. This is difficult to obtain because molecular bands and other carbonaceous grains contribute to the measured fluxes. Based on a comparison of ground-based spectra with some laboratory data, it has been concluded that the observed band can be fitted best by β-SiC. However, a detailed laboratory study [39], based on bulk and thin-film optical data, demonstrated that the dielectric functions of the various SiC polytypes should not cause observable differences in the phonon band profile. In addition, this study showed that effects of particle shape and agglomeration state, as well as conductivity introduced by dopants like nitrogen, produce the differences seen in laboratory spectra and are very probably influencing astronomical band profiles. Another feature at about 20.1 μm (21 μm band), observed in the spectra of post-AGB stars, has recently attracted new attention because high-quality ISO data became available and allowed the derivation of an intrinsic band profile [56]. This profile is very smooth and does not show any substructure due to molecular bands. Titanium carbide clusters or nanocrystals [57] have been proposed as a very good candidate for the explanation of the 20.1 μm feature.

5.2   Silicates

Spectroscopic evidence for silicates has been found in such different environments as Seyfert galaxies, the galactic centre, disks and envelopes around young stellar objects, circumstellar regions around evolved stars, and also in comets and interplanetary dust grains. The broad features, usually observed at 10 and 18 μm, are carried by amorphous silicates. The features are observed in absorption, self-absorption, and emission depending on the optical depth of the astronomical object. First evidence for the presence of crystalline silicates in the envelopes of very young massive stars, comets, and β Pictoris-type objects came from the observations of a feature at 11.2 μm (see [28] for references). This feature is typical of crystalline olivine. ISO observations revealed the presence of a wealth of 'crystalline' silicate features in the wavelength range between 20 and 70 μm due to metal-oxyen vibrations (cf. chapter by Molster & Waters). These features are of great diagnostic value and allow a detailed characterization of the mineralogy of crystalline silicates. From the positions of the features, we have strong evidence that Mg-rich olivines and pyroxenes are the major crystalline silicates in space. Variations in band position and bandwidth are probably related to different dust temperatures and Mg/Fe ratios, the degree of crystallization, and size/shape effects among the particles [20]. All peaks of the olivine series (different Mg/Fe ratios) shift to longer wavlengths with growing iron content [28]. The same trend is visible in the great majority of the bands of the pyroxene series. The mass percentage of FeO is closely related to the wavenumber shift. A significant fraction of the features observed in the ISO spectra could not yet be identified, which shows the importance of additional investigations of relevant minerals.

6   Conclusions

Laboratory astrophysics, especially the investigation of cosmic dust analogues, has developed from a small field to an integral part of modern astrophysics. The high-quality spectra obtained with ISO could not have been analyzed without measured optical data of astronomically important materials. Combining detailed analytical characterization with spectroscopic measurements led to a much more comprehensive understanding of the properties of solids in space. New missions such as SIRTF, Herschel, and NGST will provide higher sensitivity to extend dust spectroscopy to fainter objects such as disks around young stellar objects of lower mass, exoplanet and brown dwarf atmospheres, and external galaxies.

Acknowledgements

I thank my collaborators from the Jena Laboratory Astrophysics Group for many discussion about cosmic dust analogues and Drs. J. Dorschner, J. Gürtler, and H. Mutschke for critically reading of the paper.

References

[1] N.I. Agladze, A.J. Sievers, S.A. Jones, J.M. Burlitch, S.V.W. Beckwith, Astrophys. J. **462**, 1026 (1996)

[2] B. Begemann, J. Dorschner, Th. Henning, H. Mutschke, E. Thamm, Astrophys. J. **423**, L71 (1994)

[3] B. Begemann, J. Dorschner, Th. Henning, H. Mutschke, Astrophys. J. **464**, L195 (1996)

[4] B. Begemann, J. Dorschner, Th. Henning et al., Astrophys. J. **476**, 199 (1997)

[5] C.G. Bohren, D.R. Huffman: *Absorption and Scattering of Light by Small Particles.* (John Wiley & Sons, New York 1983)

[6] J.E. Bowey, C. Lee, C. Tucker et al., Monthly Not. R.A.S. **325**, 886 (2001)

[7] C. Ceccarelli, E. Caux, A.G.G.M. Tielens, F. Kemper, L.B.F.M. Waters, Astro. Astrophys. **395**, L29 (2002)

[8] J.E. Chiar, Y.J. Pendleton, T.G. Geballe, A.G.G.M. Tielens, Astrophys. J. **507**, 281 (1998)

[9] J.E. Chiar, A.G.G.M. Tielens, D.C.B. Whittet et al., Astrophys. J. **537**, 749 (2000)

[10] H. Chihara, C. Koike, A. Tsuchiyama, Publ. Astron. Soc. Jap. **53**, 243 (2001)

[11] L. Colangeli, Th. Henning, J.R. Brucato et al.: Astron. Astrophys. Rev., in press (2002)

[12] J. Dorschner, Th. Henning: Astron. Astrophys. Rev. **6**, 271 (1995)

[13] B.T. Draine, H.M. Lee: Astrophys. J. **285**, 89 (1984)

[14] J.S. Drilling, J.H. Hecht, G.C. Clayton, J.A. Mattei, A.U. Landolt, B.A. Whitney, Astrophys. J. **476**, 865 (1997)

[15] D. Fabian, Th. Posch, H. Mutschke, F. Kerschbaum, J. Dorschner, Astron. Astrophys. **373**, 1125 (2001)

[16] D. Fabian, Th. Henning, C. Jäger, H. Mutschke, J. Dorschner, O. Wehrhan, Astron. Astrophys. **378**, 228 (2001)

[17] B.A.S. Gustafson, J.M. Greenberg, L. Kolokolova, Yun-lin Xu, R. Stognienko: 'Interactions with Electromagnetic Radiation'. In: *Interplanetary Dust*, ed. by E. Grün, B.A.S. Gustafson, S.F. Dermott, H. Fechtig (Springer, Berlin 2001), pp. 509–567.

[18] J.H. Hecht, A.V. Holm, B. Donn, C.-C. Wu, Astrophys. J. **280**, 228 (1984)

[19] Th. Henning: 'Frontiers of Radiative Transfer'. In: *The Formation of Binary Stars, IAU Symp. 200, Potsdam, April 10-15, 2000*, ed. by H. Zinnecker, R.D. Mathieu (ASP, San Francisco 2001) pp. 567–572.

[20] Th. Henning: 'Cosmic Silicates - A Review'. In: *Solid State Astrochemistry*, ed. by V. Pironello, J. Krelowski (Kluwer, Dordrecht 2002), in press.

[21] Th. Henning, H. Mutschke, Astron. Astrophys. **327**, 743 (1997)

[22] Th. Henning, H. Mutschke: 'Optical Properties of Cosmic Dust Analogs'. In: *Thermal Emission Spectroscopy and Analysis of Dust, Disks, and Regoliths, Houston, April 28-30, 1999*, ed. by M.L. Sitko, A.L. Sprague, D.K. Lynch (ASP Conf. Ser. 196, San Francisco, 2000) pp. 253–271.

[23] Th. Henning, H. Mutschke, Spectrochimica Acta A, **57**, 815 (2001)

[24] Th. Henning, M. Schnaiter: 'Carbon - From Space to the Laboratory'. In: *Laboratory Astrophysics and Space Research*, ed. by P. Ehrenfreund, H. Kochan, C. Krafft, V. Pironello (Kluwer, Dordrecht, 1998), pp.249–278.

[25] Th. Henning, S.J. Chan, R. Assendorp, Astron. Astrophys. **312**, 511 (1996)

[26] Th. Henning, V.B. Ilin, N.A. Krivova, B. Michel, N.V. Voshchinnikov, Astron. Astrophys. Suppl. **136**, 405 (1999)

[27] G.H. Herbig, Ann. Rev. Astron. Astrophys. **33**, 19 (1995)

[28] C. Jäger, F.J. Molster, J. Dorschner et al., Astron. Astrophys. **339**, 904 (1998)

[29] A.P. Jones, JGR **105**, 10257 (2000)

[30] L.P. Keller, J.P. Bradley, J. Bouwman et al., LPI **31**, 1860 (2000)

[31] F. Kemper, C. Jäger, L.B.F.M. Waters et al., Nature, **415**, 295 (2002)

[32] C. Koike, C. Kaito, T. Yamamoto, H. Shibai, S. Kimura, H. Suto, Icarus **114**, 203 (1995)

[33] U. Kreibig, M. Vollmer: *Optical Properties of Metal Clusters*, (Springer, Berlin 1995)

[34] S. Kwok, B.J. Hrivnak, T.R. Geballe, Astrophys. J. **454**, 394 (1995)

[35] G. Ledoux, O. Guillois, F. Huisken, B. Kohn, D. Porterat, C. Reynaud, Astron. Astrophys **377**, 707 (2001)

[36] V. Mennella, J.R. Brucato, L. Colangeli, P. Palumbo, A. Rotundi, E. Bussoletti, Astrophys. J. **496**, 1058 (1998)

[37] V. Mennella, J.R. Brucato, L. Colangeli, P. Palumbo, Astrophys. J. **569**, 531 (2002)

[38] F.J. Molster, PhD Thesis, Univ. of Amsterdam. 2000.

[39] H. Mutschke, A.C. Andersen, D. Clément, Th. Henning, G. Peiter, Astron. Astrophys. **345**, 187 (1999)

[40] A. Omont, S.H. Moseley, P. Cox et al.: Astrophys. J. **454**, 819 (1995)

[41] V. Ossenkopf, Th. Henning, J.S. Mathis: Astron. Astrophys. **261**, 567 (1992)

[42] E.D. Palik (ed.): *Handbook of Optical Constants*, (Academic Press, Orlando 1985)

[43] E.D. Palik (ed.): *Handbook of Optical Constants II*, (Academic Press, San Diego 1991)

[44] E.D. Palik (ed.): *Handbook of Optical Constants III*, (Academic Press, San Diego 1998)

[45] Y.J. Pendleton, L.J. Allamandola, Astrophys. J. Suppl., **138**, 75 (2002)

[46] Th. Posch, F. Kerschbaum, H. Mutschke, D. Fabian, J. Dorschner, J. Hron, Astron. Astrophys. **352**, 609 (1999)

[47] B.D. Savage, K.R. Sembach, Annu. Rev. Astron. Astrophys. **34**, 279 (1996)

[48] M. Schnaiter, Th. Henning, H. Mutschke, B. Kohn, M. Ehbrecht, F. Huisken, Astrophys. J. **519**, 687 (1999)

[49] A.K. Speck, M.J. Barlow, C.J. Skinner, Monthly Not. R. A.S. **288**, 431 (1997)

[50] A.K. Speck, A.M. Hofmeister, M.J. Barlow, Astrophys. J. **513**, L87 (1999)

[51] G.C. Sloan, P.D. Levan, I.R. Little-Marenin, Astrophys. J. **463**, 310 (1996)

[52] R. Szczerba, Th. Henning, K. Volk, P. Cox, Astron. Astrophys. **345**, L39 (1999)

[53] T.P. Snow, A.N. Witt, Science **270**, 1455 (1995)

[54] T.P. Snow, A.N. Witt, Astrophys. J. **468**, L65 (1996)

[55] A.G.G.M. Tielens, D.H. Wooden, L.J. Allamandola, J. Bregman, F.C. Witteborn, Astrophys. J. **461**, 210 (1996)

[56] K. Volk, S. Kwok, B.J. Hrivnak, Astrophys. J. **516**, 99 (1999)

[57] G. von Helden, A.G.G.M. Tielens, D. van Heijnsbergen et al., Science **288**, 313 (2000)

[58] M.E. van den Ancker, J. Bouwman, P.R. Wesselius, L.B.F.M. Waters, S.M. Dougherty, E.F. van Dishoeck, Astron. Astrophys. **358**, 1035 (2000)

[59] A.N. Witt, K.D. Gordon, D. Furton, Astrophys. J. **501**, L111 (1998)

Druck: Strauss Offsetdruck, Mörlenbach
Verarbeitung: Schäffer, Grünstadt